Y0-BXH-066

INHIBITORS TO COAGULATION FACTORS

ADVANCES IN EXPERIMENTAL MEDICINE AND BIOLOGY

Recent Volumes in this Series

Volume 378
DENDRITIC CELLS IN FUNDAMENTAL AND CLINICAL IMMUNOLOGY, Volume 2
Edited by Jacques Banchereau and Daniel Schmitt

Volume 379
SUBTILISIN ENZYMES: Practical Protein Engineering
Edited by Richard Bott and Christian Betzel

Volume 380
CORONA- AND RELATED VIRUSES: Current Concepts in Molecular Biology and Pathogenesis
Edited by Pierre J. Talbot and Gary A. Levy

Volume 381
CONTROL OF THE CARDIOVASCULAR AND RESPIRATORY SYSTEMS IN HEALTH AND DISEASE
Edited by C. Tissa Kappagoda and Marc P. Kaufman

Volume 382
MOLECULAR AND SUBCELLULAR CARDIOLOGY: Effects of Structure and Function
Edited by Samuel Sideman and Rafael Beyar

Volume 383
IMMUNOBIOLOGY OF PROTEINS AND PEPTIDES VIII: Manipulation or Modulation of the Immune Response
Edited by M. Zouhair Atassi and Garvin S. Bixler, Jr.

Volume 384
FATIGUE: Neural and Muscular Mechanisms
Edited by Simon C. Gandevia, Roger M. Enoka, Alan J. McComas, Douglas G. Stuart, and Christine K. Thomas

Volume 385
MUSCLE, MATRIX, AND BLADDER FUNCTION
Edited by Stephen A. Zderic

Volume 386
INHIBITORS TO COAGULATION FACTORS
Edited by Louis M. Aledort, Leon W. Hoyer, Jeanne M. Lusher, Howard M. Reisner, and Gilbert C. White II

A Continuation Order Plan is available for this series. A continuation order will bring delivery of each new volume immediately upon publication. Volumes are billed only upon actual shipment. For further information please contact the publisher.

INHIBITORS TO COAGULATION FACTORS

Edited by

Louis M. Aledort
Mount Sinai Medical School
New York, New York

Leon W. Hoyer
American Red Cross Biomedical Services
Rockville, Maryland

Jeanne M. Lusher
Wayne State University School of Medicine
and The Children's Hospital of Michigan
Detroit, Michigan

Howard M. Reisner
University of North Carolina at Chapel Hill
Chapel Hill, North Carolina

and

Gilbert C. White II
University of North Carolina at Chapel Hill
Chapel Hill, North Carolina

PLENUM PRESS • NEW YORK AND LONDON

Library of Congress Cataloging-in-Publication Data

On file

Proceedings of the Second International Symposium on Inhibitors to Coagulation Factors, held November 3–5, 1993, in Chapel Hill, North Carolina

ISBN 0-306-45196-4

A Division of Plenum Publishing Corporation
233 Spring Street, New York, N. Y. 10013

10 9 8 7 6 5 4 3 2 1

Printed in the United States of America

Contributors

Louis M. Aledort, Mt. Sinai Medical Center, New York, NY

Jeff Alexander, Cytel Corporation, San Diego, CA

M. Algiman, Centre des Hémophiles, Hôpital Cochin, Paris, France

Stylianos E. Antonarakis, Division of Medical Genetics, University of Geneva Medical School, and Cantonal Hospital, Geneva, Switzerland; and Center for Medical Genetics, The Johns Hopkins University School of Medicine, Baltimore, MD

William R. Bell, Johns Hopkins University School of Medicine, Division of Hematology, Baltimore, MD

R. Bellocco, Instituto Superiore di Sanita, Rome, Italy

David C. Benjamin, Department of Microbiology and the Beirne B. Carter Center for Immunology Research, University of Virginia Health Sciences Center, Charlottesville, VA

H. H. Brackmann, Institut Für Exp Hämatologie Und Transfusionmedizin - Der Universität Bonn, Bonn, Germany

Ernest Briët, Hemostasis and Thrombosis Center, Department of Hematology, University Hospital, Leiden, The Netherlands

J. Donald Capra, Department of Microbiology, UT Southwestern Medical Center, Dallas, TX

Adella Clark, Department of Pathology, School of Medicine, University of North Carolina at Chapel Hill, Chapel Hill, NC

Suzanne Courter, Baxter/Hyland, Glendale, CA

Philippe de Moerloose, Hemostasis Unit, Cantonal Hospital, Geneva, Switzerland

G. Dietrich, Unité INSERM 28, Hôpital Broussais, Paris, France

Pamela Esmon, Miles Inc., Berkeley, CA

Jorn F. Falch, Novo Nordisk A/S, Biopharmaceuticals Division, Gentofte, Denmark

A. B. Federici, Angelo Bianchi Bonomi Hemophilia and Thrombosis Center, Institute of Internal Medicine and IRCCS Maggiore Hospital, University of Milano, Milan, Italy

Matthew Felch, Holland Laboratory, American Red Cross, Rockville, MD

Alessandra Franco, Cytel Corporation, San Diego, CA

Christian Freiburghaus, Excorim AB, Lund, Sweden

A. Ghirardini, Instituto Superiore di Sanita, Rome, Italy

Jane Gitschier, Howard Hughes Medical Institute and University of California, San Francisco, CA

Steven Glazer, Novo Nordisk A/S, Biopharmaceuticals Division, Gentofte, Denmark

Edward Gomperts, Baxter/Hyland, Glendale, CA

Christopher Goodnow, Stanford University Medical Center, Stanford, CA

Howard M. Grey, Cytel Corporation, San Diego, CA

Charles Hay, Liverpool University Department of Haematology, Royal Liverpool University Hospital, Liverpool, England

Ulla Hedner, Novo Nordisk A/S, Biopharmaceuticals Division, Gentofte, Denmark

Stephen M. Hedrick, Department of Biology and the Cancer Center, University of California, San Diego, La Jolla, CA

Katherine A. High, Children's Hospital of Philadelphia, Philadelphia, PA

M. W. Hilgartner, Division of Pediatric Hematology, New York Hospital, New York, NY

Leon W. Hoyer, Holland Laboratory, American Red Cross, Rockville, MD

Deborah Hurst, Miles Inc., Berkeley, CA

Pierre Hutter, Division of Medical Genetics, University of Geneva Medical School and Cantonal Hospital, Geneva, Switzerland

Glenn Y. Ishioka, Cytel Corporation, San Diego, CA

Srinivas S. Kaveri, Service d'Immunologie and INSERM U28, Hôpital Broussais, Paris, France

Mark A. Kay, Markey Molecular Medicine Center, Division of Medical Genetics, University of Washington, Seattle, WA

M. D. Kazatchkine, Service d'Immunologie and INSERM U28, Hôpital Broussais, Paris, France

Haig H. Kazazian, Center for Medical Genetics, The Johns Hopkins University School of Medicine, Baltimore, MD

Craig Kessler, George Washington University School of Medicine, Washington, DC

Linda Levin, Department of Pathology, School of Medicine, and Department of Endodontics, School of Dentistry, University of North Carolina at Chapel Hill, Chapel Hill, NC

Pete Lollar, Division of Hematology-Oncology, Department of Medicine, Emory University, Atlanta, GA

Jay N. Lozier, University of North Carolina School of Medicine, Department of Medicine, Division of Hematology, Chapel Hill, NC

Jeanne M. Lusher, Children's Hospital of Michigan, Wayne State University School of Medicine, Detroit, MI

M. J. Manco-Johnson, University of Colorado Health Science Center, Denver, CO

P. M. Mannucci, Angelo Bianchi Bonomi Hemophilia and Thrombosis Center, Institute of Internal Medicine and IRCCS Maggiore Hospital, University of Milano, Milan, Italy

G. Mariani, University of Rome, Ematologia, Rome, Italy

Alison Moliterno, Johns Hopkins University School of Medicine, Division of Hematology, Baltimore, MD

Michael A. Morris, University of Geneva Medical School and Cantonal Hospital, Geneva, Switzerland

Yale Nemerson, Mt. Sinai School of Medicine of the City University of New York, New York, NY

Inga Marie Nilsson, Department for Coagulation Disorders, University of Lund, Malmö General Hospital, Malmö, Sweden

Urs E. Nydegger, Center for Blood Transfusion, Inselspital, Bern, Switzerland

Dawne M. Page, Department of Biology and the Cancer Center, University of California, San Diego, La Jolla, CA

Virginia Pascual, Department of Microbiology, UT Southwestern Medical Center, Dallas, TX

Richard Prescott, Holland Laboratory, American Red Cross, Rockville, MD

Howard M. Reisner, Department of Pathology, School of Medicine, University of North Carolina at Chapel Hill, Chapel Hill, NC

Harold R. Roberts, Center for Thrombosis and Hemostasis, School of Medicine, University of North Carolina at Chapel Hill, NC

M. A. Robinson, Laboratory of Immunogenetics, National Institute of Allergy and Infectious Diseases, National Institutes of Health, Rockville, MD

F. Rossi, Unité INSERM 28, Hôpital Broussais, Paris, France

Jörg Ruppert, Cytel Corporation, San Diego, CA

Dorothea Scandella, Holland Laboratory, American Red Cross, Rockville, MD

David W. Scott, University of Rochester Cancer Center and Department of Microbiology and Immunology, School of Medicine and Dentistry, Rochester, NY

Alessandro Sette, Cytel Corporation, San Diego, CA

Y. Sultan, Centre des Hémophiles, Hôpital Cochin, Paris, France

A. R. Thompson, Hemophilia Program, Puget Sound Blood Center, Seattle, WA

J. Tusell, Hospital de Traumatologia y Rehabilitacio, Barcelona, Spain

Gilbert C. White II, Department of Medicine and Pharmacology, University of North Carolina at Chapel Hill, Chapel Hill, NC

Preface

"For the blood is the life...." (Deut. 12:23)

"...because the blood, in its value as life, makes atonement" (Lev. 17:11)

Hemophilia is a rare disease, severe hemophilia rarer still, yet the written history of hemophilia extends back over a millennium and a half. In the ancient Middle East, blood and life were coupled. Blood was the primary substance necessary for life, given to God in sacrifice and forbidden as a food to mortals by Levitical law. Blood was essential for rites of purification and consecration. But the flow of blood during menstruation or parturition rendered a woman unclean. The circumcision of a male child required 33 days of "blood purification" by the mother.[1] Circumcision, the visible reminder of the covenant of Abraham with Yahweh, was required of newborn Jewish males. It "connote(d) suitability for participation in what God is doing."[2] Hence, free and uncontrolled bleeding of the male child during circumcision, during the ratification of God's covenant, would be noted with awe and concern by those of the Jewish faith. It should not be surprising that the first genetic counseling offered to families with hemophilia is found in the Babylonian Talmud (compilation of Jewish law dated to about the third century AD) and concerns the necessity for circumcision in families with what we would now call hemophilia. It takes no retrodaction of modern genetic concepts into historical texts to demonstrate a sophisticated concept of the inherited basis of bleeding disorders by the 16th century codifier of Jewish law, Rabbi Joseph Karo.[3,4]

This brief historical introduction reminds us that the social costs and burdens of hemophilia render it far more important than it might be judged if only the limited number of affected patients is considered. Certainly, the availability of replacement therapy has had a dramatic effect on the life of the hemophiliac. The median life expectancy for a severe patient was 11 years in the 19th to early 20th centuries. In the seventh decade of this century (and before the advent of AIDS), median life expectancy approached that of normal males, only to decline drastically in the 1980s.[5,6] Technology has removed most or all viral risk factors from currently available coagulation factor preparations but has done nothing to ameliorate the risk of hemophilic patients developing inhibitors (neutralizing antibodies) to F.VIII or F.IX with therapy. Indeed, there are those who feel that the risk of inhibitor formation may be increased. If the social cost of hemophilia is high, the cost of caring for hemophiliacs with inhibitors is higher still. Inhibitor patients can no longer rely on well-characterized factor replacement therapy but must avail themselves of "bypass" therapy, uncertain in efficacy and as yet poorly understood, or must undergo the time-consuming and extremely costly procedure of high-dose tolerance induction.

During November 1983 a symposium organized by Leon Hoyer and Louis Aledort was held on the problem of inhibitors to coagulation factors. In the ten years following that symposium many significant advances have been made in hemophilia research. Pure F.VIII and F.IX preparations produced by monoclonal antibody or recombinant DNA technologies are available to patients. Improved porcine F.VIII preparations, recombinant VIIa, and "activated" prothrombin complex concentrates are available to circumvent preformed inhibitors. High-dose tolerance induction has proven effective in many patients, even those having high-titered inhibitors. But as we stand on the threshold of gene therapy for the hemophilias, we still do not know which patient will develop an inhibitor, how to prevent or ameliorate that occurrence, and, having developed an inhibitor, how best to deal with it.

Considering the advances of the last ten years in the context of future genetic therapies for hemophilia and current unanswered questions, a Second International Symposium on Inhibitors to Coagulation Factors seemed timely and apropos. The Symposium, held in Chapel Hill, North Carolina, November 3-5, 1993, was designed to integrate modern concepts of immunology and hematology toward the design of potential new therapies directed toward the prevention or elimination of antibodies to F.VIII and F.IX. The reader of the following manuscripts, abstracts, and discussion session transcripts can decide how well that goal has been achieved.

I wish to thank my fellow program committee members who generously took time from their pursuits to develop the agenda and plan the meeting: Louis M. Aledort, M.D., Leon W. Hoyer, M.D., Jeanne M. Lusher, M.D., Harold R. Roberts, M.D., and Gilbert C. White II, M.D. My thanks as well to the local committee comprised of Herbert Cooper, M.D., Beth Lubahn, Ph.D., and Gail Macik, M.D. Their input in the early going laid a firm foundation for later planning. It was a pleasure to work with both committees.

The generous financial grants and contributions by federal and state agencies and by industry allowed us to have both a symposium and this volume, a permanent record of the event. This synoptic acknowledgment should encourage the reader to note each individual contributor on the list. A second element, the human one, was also necessary for success. The Office of Continuing Medical Education at the University of North Carolina at Chapel Hill School of Medicine provided constant guidance and logistical support for all phases of the meeting. To name two individuals is to slight many, but the help of William Easterling, Jr., M.D., and Jane Radford is most gratefully acknowledged. Of course, our most profound thanks to Ms. Jaime Welch-Donahue of the Center for Thrombosis and Hemostasis of the UNC School of Medicine. Without her efforts, there would have been no Symposium, without her *persistent* efforts, there would have been no symposium volume.

Howard M. Reisner, Ph.D.
Chapel Hill, North Carolina

References

1. Sperling, S.D. Blood in Freedman DN (ed.): The Anchor Bible Dictionary. New York: Doubleday 1992, vol. 1, p 761-763.
2. Hall, R.G. Circumcision in Freedman DN (ed.):The Anchor Bible Dictionary. New York: Doubleday 1992, vol. 1, p 1025-1031. (The quoted phrase is on 1026 column 2.)
3. Rosner, F. Hemophilia in the Talmud and Rabbinic Writings, Ann Int Med 70:833-837, 1969.
4. Rosendaal, F.R., Smit, C., Briët, E. Hemophilia Treatment in Historical Perspective: A Review of Medical and Social Developments, Ann Hematol 62:5-15, 1991.
5. Larsson, S.A. Life Expectancy of Swedish Haemophiliacs, 1831-1980, Brit J Haematol 59:593-602, 1985.
6. Jones, P.K., Ratnoff, O.D. The Changing Prognosis of Classical Hemophilia (Factor VIII "Deficiency"), Ann Int Med 114:641-648, 1991.

Acknowledgments

The editors wish to thank the following sources for their generous educational grants in support of the symposium:

Alpha Therapeutic Corporation
American Diagnostica Inc.
American Red Cross
Armour Pharmaceutical Company
Baxter Healthcare Corporation, Biotech Group, Hyland Division
Behringwerke AG
Boehringer Mannheim Corporation
Cardiovascular Diagnostics, Inc.
Centers for Disease Control and Prevention
The Coalition for Hemophilia B
COBE BCT, Inc.
Genetics Institute, Inc.
Glaxo Inc. Research Institute
Immuno AG
Immuno-U.S., Inc.
Miles Inc., Pharmaceutical Division, Biological Products
National Heart, Lung, and Blood Institute
North Carolina Biotechnology Center*
Novo Nordisk AS, Biopharmaceuticals Division
Organon Teknika Corporation
Ortho Diagnostic Systems Inc.
Porton Products Limited
Quantum Health Resources

*The proceedings are based upon work supported in whole or in part by the North Carolina Biotechnology Center. Any opinions, findings, conclusions, or recommendations expressed in this publication are those of the authors and do not necessarily reflect the views and policies of the North Carolina Biotechnology Center.

Sponsors

The Second International Symposium on Inhibitors to Coagulation Factors was jointly sponsored by The Center for Thrombosis and Hemostasis and the Office of Continuing Medical Education and Alumni Affairs of the University of North Carolina School of Medicine and the American Red Cross, the Centers for Disease Control and Prevention, the National Heart, Lung, and Blood Institute of the National Institutes of Health, the National Hemophilia Foundation, and The Coalition for Hemophilia B.

Contents

I: Factors VIII, IX and von Willebrand factor, their molecular and antigenic structure: An overview

Ultimately, understanding how inhibitors to coagulation factors develop will require precise knowledge of the structural epitopes in these factors that are recognized by components of the immune system. This is important for understanding both the mechanisms of clotting factor neutralization and those involved in antigen presentation, HLA-restricted antigen processing, and antibody formation. This first section focuses on some of these general structural issues.

In the opening chapter, Lollar discusses the structure and function of factor VIII (F.VII). While the linear sequence of F.VIII has been known since 1984, the functional domains of the molecule remain incompletely defined. Sequences involved in binding phospholipid, von Willebrand factor, and, most recently, factor IXa (J Biol Chem 269, 7150-7155, 1994) have been identified, but the binding site for factor X and the physical relation of these various domains with each other are unknown and are targets of current research efforts.

Next, Antonarakis, Kazazian, Gitschier, and colleagues update current understanding of the molecular defects in patients with hemophilia. Three important features regarding mutations in patients with inhibitors are brought out by these workers. First, of the 31 inhibitor patients with point mutations, the location of the mutation in all 31 is in the carboxyl terminal light chain, suggesting that there may be something unique about this part of the molecule. Second, factors other than, or at least in addition to, the molecular defect play a role in inhibitor formation since among patients with identical defects, some may have an inhibitor and others may not. This observation is consistent with previous findings that family members with the same genetic defect may be discordant for the presence of an inhibitor. Third, the majority of reported inhibitor cases have nonsense mutations or deletions in their F.VIII gene so whatever other factors are involved in inhibitor formation, those patients with severe defects have a greater likelihood of developing inhibitors.

Hoyer presents an elegant analysis of the incidence of inhibitors in hemophilia A, including the relatively novel application of Kaplan-Meier analysis. He compares and contrasts inhibitor prevalence and incidence in studies using new ultrapure forms of F.VIII with older data using intermediate and low purity forms of F.VIII. While it may still be too early for these comparisons, the concepts and methods proposed by Hoyer provide new and potentially important ways of looking at inhibitor occurrence.

One of the most exciting and important areas of inhibitor research is the identification of F.VIII epitopes recognized by inhibitor antibodies. Scandella and Reisner review the current status of work in this area. One important advance is the development by Scandella and coworkers of an assay to identify neutralizing antibodies to F.VIII. Using this assay, she has shown that neutralizing antibodies also appear to be confined to the A2 and/or C2 domains. Attempts to further define inhibitor epitopes using smaller segments of the A2 or

C2 domains have not been successful and may require knowledge of the tertiary structure of F.VIII. Another important observation is the heterogeneity of the antibody response. Reisner shows this at the level of the antibody while Scandella presents evidence from clinical trials with recombinant F.VIII that epitope switching may occur. Reisner also discusses genetic influences on inhibitor formation.

Finally, High and Mannucci and coworkers describe current understanding of inhibitor formation in hemophilia B and von Willebrand's disease, respectively. In both disorders, inhibitor formation is much less prevalent than in hemophilia A, but lessons learned from these inhibitors will be applicable to hemophilia A and should help define a general model for inhibitor formation.

Structure and function of Factor VIII

Pete Lollar

Division of Hematology-Oncology
Department of Medicine
Emory University
Atlanta, Georgia 30322

When the first International Symposium on Inhibitors was held 10 years ago, the existing knowledge of Factor VIII (F.VIII) structure and function was summarized by Fulcher and Zimmerman.[1] F.VIII had been recently purified from bovine, porcine, and human plasma, and investigators interested in the biochemistry of F.VIII were at the threshold of making significant progress. This followed a decade of frustrating work in which advances in protein purification and immunological methods had made the purification of F.VIII a feasible but elusive goal.

From the limited amino acid sequence of purified F.VIII, oligonucleotide probes were made and the human F.VIII cDNA was cloned.[2,3] The deduced amino acid sequence of F.VIII provided the starting point for investigations into F.VIII structure-function correlates. Methods for purifying milligram quantities of plasma-derived and recombinant F.VIII permitted biochemical studies that have clarified some of the mysterious properties of F.VIII that had been noted during early investigations. The expanding knowledge of F.VIII structure and function has provided the background for the investigation of the mechanism of inhibition of F.VIII by alloantibodies and autoantibodies at the molecular level.

Structure of F.VIII

F.VIII is synthesized as a single chain with a polypeptide molecular weight of 265,000. Internal sequence homology defines three types of domains. The term "domain" used in reference to F.VIII structure does not denote an independent folding unit as it commonly does in protein folding studies. The domains are arranged in the sequence A1-A2-B-A3-C1-C2.[4] (See Figure 1.) Due to proteolysis within the B domain and between the A2 and B domains, plasma-derived and recombinant F.VIII are isolated as a heterogeneous population of heterodimers, with either little or no single chain F.VIII present.[5,6] Proteolysis at residue 1648 results in the formation of a subunit called the F.VIII light chain. Addi-

Inhibitors to Coagulation Factors
Edited by Louis M. Aledort *et al.*, Plenum Press, New York, 1995

tionally, proteolysis at undetermined sites within the B domain and between the A2 and B domain at residue 740 results in removal of some or all of the B domain. Since the degree of heavy chain proteolysis varies, heterogeneity in the polypeptide structure of F.VIII is restricted to the heavy chain. F.VIII heterodimers are linked non-covalently in a process that requires divalent metal ions. The protease(s) responsible for

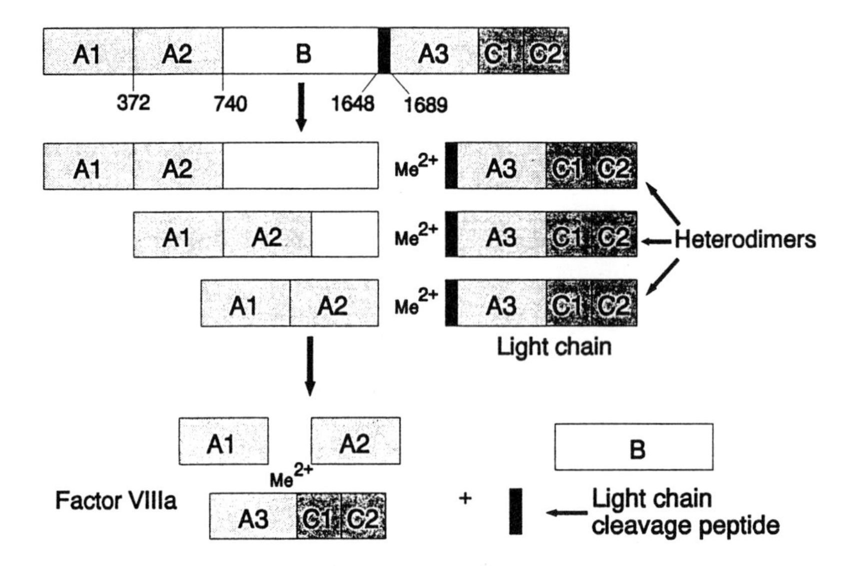

Figure 1. Structure of F.VIII

the cleavage of single chain F.VIII to its heterodimeric forms is/are not known, nor is it known whether F.VIII circulates in single chain form like its homologous procofactor, factor V (F.V),[7] or undergoes proteolysis during the isolation procedure. Since intracellular proteolysis of human recombinant F.VIII to typical heterodimeric forms occurs in heterologous expression systems,[8,9] it is likely that F.VIII circulates predominantly in heterodimeric form.

Originally, domains were designated with respect to the amino acid sequence as A1 (1-329), A2 (380-711), A3 (1649-2032), B (712-1648), C1 (2033-2182), and C2 (2183-2332) based in intron-exon boundaries.[4] Because of the functional importance of the fragments of F.VIII produced by thrombin, domains are more commonly delineated with reference to thrombin cleavage sites at Arg 372, Arg 740, and Arg 1689[10] as A1 (1-372), A2(373-740), B (741-1648), and A3-C1-C2 (1690-2332).[11] The A3-C1-C2 fragment constitutes the thrombin-cleaved light chain, which differs from the light chain in that an acidic, 41-residue fragment corresponding to residues 1649-1689 has been removed. There is an acidic region at the COOH-terminus of the A1 domain that is subject to proteolytic cleavage at Arg 336 by activated protein C and Factor IXa (F.IXa).[10,12-14]

The sequence of mouse F.VIII reveals significantly greater homology to human F.VIII in the A and C domains than in the B domain.[15] A similar observation has been made in comparing human and porcine F.VIII, based on the partial sequence of porcine F.VIII.[2,16] Recombinant, active F.VIII molecules lacking most or all of the B domain have been expressed.[2,11,17] Additionally, a heterodimeric form of F.VIII that appears to lack the entire B domain can be isolated from plasma, and has similar activity to B domain-containing F.VIII.[6,18,19] Scanning transmission electron microscopy[20] and rotary shadowing electron microscopy[21] indicate that the B domain projects as a long thin stalk from the core of the F.VIII molecule. Consistent with the ultrastructural data, removal of the B domain during the activation of F.VIII produces an M_r 160,000 molecule with approximately the same sedimentation coefficient (7 S) as M_r 240,000 B-domain rich F.VIII,[22,23] indicating that the presence of the B domain introduces a large element of asymmetry and an attendant large frictional ratio. The B domain, which contains up to 50% of the mass of F.VIII, has no known function.

Factor V (F.V) is homologous to F.VIII and can be described in terms of a A1-A2-B-A3-C1-C2 domain sequence.[24,25] The A- and C- type domains but not B domains are homologous in factors V and VIII. Of particular interest is the fact that the two acidic regions in F.VIII corresponding to residues 336-372 and 1648-1689 are not found in F.V. The former region is necessary for procoagulant activity of F.VIII and the latter region is necessary for the association of F.VIII with von Willebrand factor (vWf) (see below).

Human F.VIII contains 25 potential N-linked glycosylation sites, 19 of which are in the B domain.[2,4] At least 75% of the sites in the B domain are occupied.[26] The A1 subunit has two potential N-linked glycosylation sites, at least one of which is occupied, and the A2 subunit contains a single unoccupied site.[8] The light chain contains two potential N-linked glycosylation sites, at least one of which is occupied.

There are seven potential tyrosine sulfation sites in F.VIII. Of these, tyrosine sulfate has been identified at Tyr 346, Tyr 1664, and Tyr 1680, and Tyr 718, Tyr 719, and Tyr 723 are probably sulfated.[27] A potential site at Tyr 395 does not appear to be occupied. Inhibition of tyrosine sulfation of recombinant F.VIII expressed in CHO cells by sodium chlorate treatment resulted in the production of a F.VIII molecule with only 16% of the coagulant activity of wild type F.VIII.[27] Thus, tyrosine sulfation appears necessary for the full development of procoagulant activity in an as yet undefined way.

Significant differences between the circular dichroism spectra of human F.VIII and human F.VIIIa are observed below 235 nm that are consistent with less random coil and more beta sheet structure in

F.VIIIa, presumably due to the removal of the thrombin-cleaved B-domain.[28]

Association of F.VIII with von Willebrand factor

F.VIII circulates bound non-covalently to von Willebrand factor (vWf) and this interaction stabilizes F.VIII *in vivo*.[29,30] vWf circulates as a population of multimers composed of identical M_r 270,000 subunits that are covalently linked by disulfide bonds to form a long, linear molecule.[31,32] von Willebrand factor is heterogeneous because intracellular subunit assembly produces a wide range of number of subunits per molecule (from 2 to perhaps greater than 80). (See reference 33 for a review). Ultrastructural studies indicate that the vWf subunit is a long (60 nm) fibrillar structure consisting of an NH_2-terminal G domain and a COOH-terminal R domain.[34,35] The monomers are arranged ...GR-RG-GR-RG.... to form multimers. Electron microscopic and light scattering studies indicate that most multimers are supercoiled in solution in a form perhaps resembling a "ball of yarn."[36,37]

A binding site on the vWf subunit for F.VIII has been localized to the NH_2-terminal 273 residues.[38] More recently, an epitope for a monoclonal antibody to vWf that blocks F.VIII binding was localized to residues 78-96 at the amino terminus of vWf[39] and mutations in the vWf gene corresponding to this region have been identified that produce secondary deficiency of F.VIII. (See reference 40 for a review.)

The heterogeneity of vWf is evident from its broad distribution during centrifugation and gel permeation chromatography. F.VIII associates with all multimeric forms of vWf since F.VIII co-migrates with vWf across this broad distribution.[31,41,42] The fractional saturation of vWf with F.VIII *in vivo* is only 0.01-0.08. However, binding of F.VIII to vWf in plasma indicates that the binding capacity of vWf is considerably higher.[43] Consistent with this, studies using purified F.VIII and vWf indicate that the multimeric vWf can bind one molecule of F.VIII per subunit in solution.[22] However, measurements of F.VIII binding to vWf immobilized on plastic indicate that only 1 in 50 subunits of vWf bind F.VIII.[44] Furthermore, a stoichiometry of 0.1 F.VIII molecule per vWf subunit was inferred from measurements of the inhibition by vWf of F.VIII binding to platelets.[45] Possible reasons for this discrepancy include functional differences between fluid phase and solid phase vWf or methodological differences. For example, the purification of vWf by some methods could lead to a non-native state that either increases or decreases the binding capacity for F.VIII.

von Willebrand factor binds to the light chain of F.VIII.[46-48] A heavy chain determinant on F.VIII for vWf binding has not been definitely identified. The thrombin-cleaved F.VIII light chain does not bind vWf.[47] Antibodies that localize to epitopes between residues 1670-1684 in the acidic NH_2-terminal region of the F.VIII light chain that is cleaved by thrombin have been identified that block the binding of F.VIII to vWf.[46,49,50] Sulfation of F.VIII at Tyr 1680 is necessary for the interaction of F.VIII with vWf.[17] These results indicate that this region of F.VIII contains at least part of the binding site on F.VIII for vWf, or alternatively, stabilizes a conformation of the F.VIII light chain that interacts with vWf. The latter possibility is suggested by the observation that synthetic peptides corresponding to the acidic NH_2-terminal region of F.VIII do not block the binding of F.VIII to vWf.[17,46] Additionally, two inhibitory antibodies have been identified that recognize epitopes in the C2 domain that inhibit binding of F.VIII to vWf.[51] This suggests either that the C2 domain and the acidic NH_2-terminal region are near one another and together make up the binding site, or that the F.VIII binding site on vWf is in the C2 domain and the acidic NH_2-terminal region stables this site allosterically.

The possible interaction of vWf with the C2 domain of F.VIII is important from the standpoint of inhibitor study since one of the two dominant epitopes for human autoantibodies and alloantibodies is in the C2 region and appears to be involved in the recognition site of vWf.[51]

Activation of F.VIII by thrombin and by Factor Xa

F.VIII has no detectable activity prior to proteolytic activation. Thrombin and Factor Xa (F.Xa) are the only two mammalian enzymes that have been shown convincingly to activate F.VIII. The activation of F.VIII is complex since several proteolytic cleavages are involved. Additionally, analysis of F.VIII activation is complicated by the fact that activated F.VIII (F.VIIIa) undergoes non-proteolytic inactivation under certain conditions, including conditions of its plasma concentration ($\approx$ 1 nM) at physiological pH and ionic strength (see below).

The cleavage pattern catalyzed by thrombin and F.Xa differ. Some of the cleavage sites for F.Xa have been identified as Arg 336, Arg 372, Arg 1689, and Arg 1721.[10] Thrombin-activated F.VIII has greater activity than F.Xa-activated F.VIII.[52,53] Thrombin does not inactivate F.VIIIa,[54,55] with the possible exception of very slow, secondary cleavages.[10,23,27] In contrast, F.Xa may inactivate F.VIIIa,[10] although this is controversial.[52]

Exposure of the F.VIII-vWf complex to thrombin results in dissociation of F.VIII from vWf.[56,57] Cleavage of the F.VIII light chain at Arg 1689 appears to be necessary and sufficient for dissociation of F.VIII at physiological ionic strength and pH.[47] Whether cleavage of the F.VIII light chain is necessary to activate the procoagulant function of F.VIII, i.e., its ability to serve as a cofactor for F.IXa during the activation of F.X (see below) has been controversial. A serine protease from *Bothrops jararacussu* venom has been isolated that activates porcine F.VIII to about 50% of the level obtained by thrombin and that cleaves F.VIII at Arg 372 but not at Arg 1689.[58] Additionally, an Arg 1689 Cys mutant form of F.VIII has been identified, F.VIII-East Hartford, that is activated by thrombin in the absence of light chain cleavage.[59] In both of these studies, dissociation of vWf from F.VIII was necessary for the development of procoagulant activity, consistent with the observation that vWf competes for binding of F.VIII to phospholipid membranes (see below). In contrast, another group studying F.VIII derived from another family with an Arg 1689 Cys mutation, were not able to activate the mutant F.VIII with thrombin.[60] Additionally, site-directed mutagenesis of recombinant F.VIII to a produce an Arg 1689 Ile mutation yielded a molecule that could not be activated by thrombin.[61] In this instance, binding of F.VIII to vWf in the culture medium or conformational abnormalities associated with the non-conservative replacement of arginine with isoleucine, may have accounted with the lack of activity.

The rate of activation of F.VIII is not affected when F.VIII is bound to vWf.[62] Thus, dissociation of F.VIII from vWf is not rate-limiting during the activation of the F.VIII-vWf complex by thrombin. Consistent with this, in the presence of vWf, cleavage of the F.VIII light chain by thrombin is accelerated and heavy chain cleavage at Arg 372 becomes the slowest bond cleavage.[62] Heparin is a potent inhibitor of the activation of F.VIII or the F.VIII-vWf complex by thrombin, presumably by competition with F.VIII for binding to an exosite on thrombin.[63] This suggests that the anticoagulant properties of heparin may involve more than antithrombin-dependent reactions.

Thrombin-activated F.VIII is a M_r 160,000 heterotrimer composed of A1, A2, and A3-C1-C2 subunits.[23,64,65] The non-proteolytic decay of F.VIIIa is due to dissociation of the A2 subunit, which is in reversible equilibrium with the A1/A3-C1-C2 dimer.[64] The dissociation constant for the interaction is $\approx$ 0.2 μM at physiological ionic strength at pH 7.4.[66] Thus, the reaction lies far to the right (i.e., toward the dissociated state) at the plasma concentration of F.VIII. The dissociation rate constant is greater for human F.VIIIa than for porcine F.VIIIa, which may explain the superior coagulant activity of porcine F.VIII in standard one-stage coagulation assays.[65,66] This may have implications regarding the use of porcine F.VIII in the treatment of patients with F.VIII inhibitors. Additionally, the inverse correlation between coagulant activity and subunit dissociation rate implies that non-proteolytic loss of F.VIIIa activity by subunit dissociation may participate in the regulation of F.VIIIa activity *in vivo.*

Fully functional porcine F.VIIIa can be isolated that is indefinitely stable at pH 6.0 at 4 °C or room temperature.[23] In contrast, a method for isolating fully functional human F.VIIIa has not been described, although a procedure is available to produce active, heterotrimeric human F.VIIIa that can be stored stably at -80°C.[28] Active human F.VIIIa can be reconstituted from purified preparations of A1/A3-C1-C2 dimer and A2 subunit.[64] A molar excess of A2 subunit is required to achieve optimal activity, suggesting that preparations of A2 subunit are partially denatured. Similarly, pH-induced inactivation of porcine F.VIIIa, which is associated with A2 subunit dissociation, can be only partially reversed by restoring initial conditions.[67] These results indicate that an irreversible folding pathway may be accessible to the A2 subunit.

Function of F.VIIIa

F.VIIIa is a cofactor for F.IXa-catalyzed activation of F.X.[68] The activation of F.X occurs by means of hydrolysis of a single bond at Arg 194-Ile 195. In the absence of F.VIIIa, phospholipid, and Ca^{2+}, the catalytic efficiency of F.IXa toward F.X is very low. Additionally, F.IX is generally poorly reactive toward synthetic substrates and inhibitors, including diisopropylfluorophosphate, *p*-nitrophenyl-*p*'-guanidinobenzoate, N α- benzoyl-L-arginine ethyl ester, and thiobenzyl benzyloxycarbonyl-L-lysinate, which usually are very reactive toward other proteases in this family.

An increase in the catalytic efficiency of F.IXa of several orders of magnitude results from assembly of the intrinsic fXase.[69,70] F.VIIIa apparently alters the active site structure of F.IXa to make it better recognize the rate-limiting transition state for F.X activation, since its dominant kinetic effect is to increase the k_{cat}.[69] The rate-limiting step in this reaction is acyl enzyme formation.[70] A k_{cat} increase due to F.Va also is observed in the homologous prothrombinase complex,[71] and suggests that unlike trypsin and other solution phase serine proteases, the active sites of F.IXa and F.Xa are not fully formed in the absence of cofactors. The 52 residue F.X activation peptide that is removed plays a critical role that is most pronounced in the presence of F.VIIIa.[70]

The formation of the Michaelis complex during assembly of the intrinsic fXase involves six possible binary interactions: F.IXa/F.VIIIa, F.IXa/membrane,F.IXa/F.X, F.VIIIa/membrane, F.VIIIa/F.X, and F.X/membrane. The mechanism by which F.VIII reduces the energy required for conversion of the Michaelis complex to the transition state is unknown. Elucidation of this mechanism will probably require the determination of the three-dimensional structure of F.IXa, perhaps in complex with F.VIIIa or an active fragment of F.VIIIa.

The binding of F.VIIIa to F.IXa has been studied fluorometrically using a derivative of D-phenylalanyl-prolyl-arginyl-F.IXa modified with fluorescein, producing Fl-FPR-FIXa.[72] Titration of Fl-FPR-F.IXa with F.VIIIa at a fixed rate, saturating phospholipid produces a large, saturable increase in fluorescence anisotropy. This signal has been used to determine the apparent dissociation constant for binding (2 nM) and stoichiometry (1 mole F.VIIIa per mole Fl-FPR-F.IXa). Additionally, F.VIII in the presence of phospholipid produces a small, saturable increase in anisotropy in Fl-FPR-F.IXa, followed by a larger increase upon addition of thrombin to activate F.VIII. Thus, somewhat surprisingly, F.VIII binds F.IXa, but proteolytic modification of F.VIII must occur before the complete F.VIIIa-dependent structural change in the active site of F.IXa occurs.

Fluorescence energy transfer experiments indicate that the active site of F.IXa is far from the surface of the phospholipid membrane and that the elongated F.IXa molecule projects approximately perpendicularly from the membrane surface.[73] F.VIIIa does not produce an appreciable change in the distance of the probe from the surface, in contrast to studies with prothrombinase, in which F.Va produces an increase distance of probes in the active site of F.Xa from the membrane surface.[74]

The assembly of intrinsic fXase probably involves the direct binding of F.VIIIa to a membrane surface or component. Synthetic phospholipid membranes containing an acidic phospholipid component provide a surface for efficient assembly of intrinsic fXase. Platelets and monocytes presumably provide the membrane surface for intrinsic fXase assembly *in vivo*. The "receptor" for binding of F.IXa, F.VIIIa, or F.X may consist of phospholipid membrane in which the acidic component is due to phosphatidylserine that has been translocated from the inner membrane leaflet during platelet activation.[75] Whether other components, possibly protein, are involved is one of the great unsolved questions in hemostasis.

It is unlikely that F.VIII *per se* is involved in membrane interactions *in vivo* since binding of F.VIII to vWf inhibits all of its known membrane-dependent interactions, including binding to phospholipid,[76,77] activation by F.Xa,[78] inactivation by activated protein C,[79-81] and binding to platelets.[45] After cleavage of the F.VIII light chain, dissociation from vWf occurs, allowing membrane-dependent interactions.

Both F.VIII[82,83] and F.VIIIa[84] bind with high affinity to negatively charged phospholipids. A specific interaction of F.VIII with phosphatidylserine has been identified, suggesting that the interaction of F.VIII with phospholipid or cellular membranes may not be merely due to a non-specific electrostatic interaction.[85]

A region of F.VIII involved in the binding to phosphatidylserine has been localized to residues 2303-2332 in the C2 domain.[86] Some human inhibitory alloantibodies and autoantibodies bind to the C2 domain[87] and inhibit binding of F.VIII to phosphatidylserine.[88] The epitopes for two inhibitory antibodies that inhibit binding of F.VIII to phosphatidylserine have been localized to residues 2170-2327 and 2248-2312.[51] Interestingly, as discussed above, these antibodies also block binding of F.VIII to vWf, consistent with a common recognition site on F.VIII for vWf and phospholipid.

A direct interaction between F.Va (cofactor) and prothrombin (substrate) components of the prothrombinase complex has been documented.[89,90] Corresponding measurements of binding between F.VIIIa and F.X in the homologous intrinsic fXase complex have not been performed.

Inactivation of F.VIII and F.VIIIa

In addition to the non-proteolytic loss of F.VIIIa activity, both F.VIII and F.VIIIa are proteolytically inactivated by activated protein C and F.IXa. "Inactivation" of F.VIII implies that proteolytically altered F.VIII cannot be activated. Additionally, as mentioned previously, although it activates F.VIII, F.Xa may inactivate F.VIIIa. Other possible inactivating enzymes, such as plasmin, have not been studied using purified proteins. Activated protein C cleaves F.VIII and F.VIIIa at Arg 336 and Arg 562 in the A1 and A2 domains, and possibly at Arg 740 at the A2-B junction.[10,12] Cleavage at Arg 562 correlates most with loss of activity, although cleavage at Arg 336 also probably is involved in the inactivation process, possibly by promoting A2 subunit dissociation.[91] Part of the binding site on F.VIII for activated protein C has been localized near the A3-C1 junction of the light chain.[92]

F.IXa cleaves F.VIII and F.VIIIa at Arg 336 and at Arg 1719.[13,14] Because these cleavages are associated with loss of activity, it has been concluded erroneously that F.IXa does not stabilize F.VIIIa.[14] However, in the presence of phospholipid, addition of F.IXa to purified human or porcine F.VIIIa,[28] or to F.VIII activated *in situ* with thrombin[52,53,55,93] or F.Xa[53] markedly decreases its rate of decay. The stabilization of F.VIIIa by F.IXa is presumably due to slower A2 subunit dissociation while in the intrinsic fXase complex. This is consistent with the observation that F.IXa increases the rate of reconstitution of F.VIIIa activity when purified A2 subunit and A1/A3-C1-C2 dimer are mixed together.[94] Stabilization of F.VIIIa by F.IXa and proteolytic inactivation of F.VIIIa by F.IXa are not mutually exclusive events. Since F.IXa has relatively poor catalytic efficiency toward F.VIIIa, the predominant effect of F.IXa at nanomolar concentrations is stabilization.[28]

Inactivation of F.VIII or F.VIIIa by inhibitory antibodies

Based on the functional properties of F.VIII and F.VIIIa summarized in this chapter, possible mechanisms of inhibition by human autoantibodies or alloantibodies can be specified. Of the known or possible binary interactions of F.VIIIa, binding to F.IXa, F.X, phospholipid, or the putative platelet membrane receptor could produce inhibition. As reviewed elsewhere in the Symposium, it appears very likely that anti-C2 inhibitory antibodies act by blocking membrane binding of F.VIIIa. However, it seems remotely possible that the anticoagulant effect *in vitro* is an artifact of the lipid reagents used, as in the lupus anticoagulant, and instead results from inability of infused F.VIII to bind vWf.

Other possible mechanisms of antibody inhibition include inhibition of proteolytic activation of F.VIII, rate enhancement of A2 subunit dissociation, and interference with formation of the transition state of the F.IXa/F. VIIIa/F.X/membrane complex. This last possibility means that the antibody would not interfere with the formation of the ground state, Michaelis complex, but would somehow interfere with conversion to the transition state. This appears to be how anti-A2 antibodies inhibit F.VIIIa function.[95] Anti-A2 antibodies studied so far are noncompetitive inhibitors of intrinsic fXase. This indicates they either block the binding of F.VIIIa to F.IXa or phospholipid or interfere with the catalytic function of fully assembled intrinsic fXase, but they do not inhibit the binding of the substrate, F.X. A monoclonal anti-A2 antibody, 413, does not inhibit the increase in fluorescence anisotropy that results from the binding of F.VIIIa to Fl-FPR-F.IXa on phospholipid vesicles in the absence of F.X, indicating it does not inhibit assembly of intrinsic fXase. Addition of F.X to Fl-FPR-F.IXa, F.VIIIa, and phospholipid vesicles produces a further increase in fluorescence anisotropy and a decrease in fluorescence intensity. This effect is completely blocked by mAb 413. The effect of mAb 413 on F.X-dependent fluorescence changes indicates that the binding of fX to Fl-FPR-fIXa in the presence of F.VIIIa may induce a partial transition state character to the complex, which is blocked by mAb 413.

Implications for future therapy of patients with F.VIII inhibitors

The mapping of inhibitor epitopes is discussed in detail elsewhere in the Symposium. One of the surprising findings is that, although the large size of the F.VIII molecule represents a vast number of possible epitopes, only two epitopes, one in the A2 domain and the other in the C2 domain, appear to represent the targets for the majority of inhibitory antibodies. Non-inhibitory antibodies have been identified to multiple additional epitopes to hemophilia A plasmas, the clinical significance of which is uncertain.[96] The cross-reactivity of many inhibitory antibodies to porcine F.VIII is sufficiently low that porcine F.VIII concentrate has been used successfully for several years. The subsequent development of anti-porcine antibodies occurs frequently, although the reactive epitopes have not been identified. An attractive possibility for future inhibitor therapy would be to replace the A2 and C2 epitopes in recombinant human F.VIII with homologous porcine sequences, thereby producing hybrid human/porcine F.VIII. Hybrid human/porcine F.VIII might be less immunogenic than porcine F.VIII and the use of a highly purified F.VIII molecule might eliminate untoward side effects associated with infusion of porcine F.VIII concentrate. Additionally, hybrid human/porcine F.VIII molecules containing various degrees of porcine sequence replacement could be useful for epitope mapping studies.

A recombinant hybrid human/porcine F.VIII, designated HP2, has been constructed in which 60% of the amino terminus of the human A2 domain is replaced by the homologous porcine sequence.[97] This region encompasses the A2 inhibitory epitope.[98,99] This construct has been stably expressed in baby hamster kidney cells. The cross-reactivity of HP2 with RC and mAb 413 is less than 1%, indicating that the A2 epitope recognized by these antibodies has been eliminated. Construction of an A2/C2 hybrid human/porcine F.VIII could produce a recombinant F.VIII molecule in which the epitopes recognized by most inhibitory plasmas are essentially eliminated.

References

1. Fulcher, C.A., and T.S. Zimmerman. 1984. Structure and function of factor VIII coagulant protein. In Factor VIII Inhibitors. L.W. Hoyer, editor. Alan R. Liss, New York. 57-72.
2. Toole, J.J., J.L. Knopf, J.M. Wozney, L.A. Sultzman, J.L. Buecker, D.D. Pittman, R.J. Kaufman, E. Brown, C. Shoemaker, E.C. Orr, G.W. Amphlett, W.B. Foster, M.L. Coe, G.J. Knutson, D.N. Fass, and R.M. Hewick. 1984. Molecular cloning of a cDNA encoding human antihaemophilic factor. *Nature* 312:342-347.
3. Gitschier, J., W.I. Wood, T.M. Goralka, K.L. Wion, E.Y. Chen, D.H. Eaton, G.A. Vehar, D.J. Capon, and R.M. Lawn. 1984. Characterization of the human factor VIII gene. *Nature* 312:326-330.
4. Vehar, G.A., B. Keyt, D. Eaton, H. Rodriguez, D.P. O'Brien, F. Rotblat, H. Oppermann, R. Keck, W.I. Wood, R.N. Harkins, E.G.D. Tuddenham, R.M. Lawn, and D.J. Capon. 1984. Structure of human factor VIII. *Nature* 312:337-342.
5. Fass, D.N., G.J. Knutson, and J.A. Katzmann. 1982. Monoclonal antibodies to porcine factor VIII coagulant and their use in the isolation of active coagulant protein. *Blood* 59:594-600.
6. Fay, P.J., M.T. Anderson, S.I. Chavin, and V.J. Marder. 1986. The size of human factor VIII heterodimers and the effects produced by thrombin. *Biochim. Biophys. Acta* 871:268-278.
7. Nesheim, M.E., K.H. Myrmel, L. Hibbard, and K.G. Mann. 1979. Isolation and characterization of single chain bovine factor V. *J. Biol. Chem.* 254:508-517.
8. Kaufman, R.J., L.C. Wasley, and A.J. Dorner. 1988. Synthesis, processing, and secretion of recombinant human factor VIII expressed in mammalian cells. *J. Biol. Chem.* 263:6352-6362.
9. Eaton, D.L., P.E. Hass, L. Riddle, J. Mather, M. Wiebe, T. Gregory, and G.A. Vehar. 1987. Characterization of recombinant human factor VIII. *J. Biol. Chem.* 262:3285-3290.
10. Eaton, D., H. Rodriguez, and G.A. Vehar. 1986. Proteolytic processing of human factor VIII. Correlation of specific cleavages by thrombin, factor Xa, and activated protein C with activation and inactivation of factor VIII coagulant activity. *Biochemistry* 25:505-512.
11. Eaton, D.L., W.I. Wood, D. Eaton, P.E. Hass, P. Hollingshead, K. Wion, J. Mather, R.M. Lawn, G.A. Vehar, and C. Gorman. 1986. Construction and characterization of an active factor VIII variant lacking the central one-third of the molecule. *Biochemistry* 25:8343-8347.
12. Fay, P.J., T.M. Smudzin, and F.J. Walker. 1991. Activated protein C-catalyzed inactivation of human factor VIII and factor VIIIa. Identification of cleavage sites and correlation of proteolysis with cofactor activity. *J. Biol. Chem.* 266:20139-20145.
13. Lamphear, B.J., and P.J. Fay. 1992. Proteolytic interactions of factor IXa with human factor VIII and factor VIIIa. *Blood* 80:3120-3128.
14. O'Brien, D.P., D. Johnson, P. Byfield, and E.G.D. Tuddenham. 1992. Inactivation of factor VIII by factor IXa. *Biochemistry* 31:2805-2812.
15. Elder, B., D. Lakich, and J. Gitschier. 1993. Sequence of the murine factor VIII cDNA. *Genomics* 16:374-379.
16. Toole, J.J., D.D. Pittman, E.C. Orr, P. Murtha, L.C. Wasley, and R.J. Kaufman. 1986. A large region (approximately equal to 95 kDa) of human factor VIII is dispensible for in vitro procoagulant activity. *Proc. Natl. Acad. Sci. USA* 83:5939-5942.
17. Leyte, A., H.B. van Schijndel, C. Neihrs, W.B. Huttner, M.P. Verbeet, K. Mertens, and J.A. van Mourik. 1991. Sulfation of Tyr^{1680} of human blood coagulation factor VIII is essential for the interaction of factor VIII with von Willebrand factor. *J. Biol. Chem.* 266:740-746.

18. Andersson, L.O., N. Forsman, K. Huang, K. Larsen, A. Lundin, B. Pavlu, J. Sandberg, K. Sewerin, and J. Smart. 1986. Isolation and characterization of human factor VIII: molecular forms in commercial factor VIII concentrate, cryoprecipitate, and plasma. *Proc. Natl. Acad. Sci. USA* 83:2979-2983.
19. Lollar, P., C.G. Parker, and R.P. Tracy. 1988. Molecular characterization of commercial porcine factor VIII concentrate. *Blood* 71:137-143.
20. Mosesson, M.W., D.N. Fass, P. Lollar, J.P. DiOrio, C.G. Parker, G.J. Knutson, J.F. Hainfeld, and J.S. Wall. 1990. Structural model of porcine factor VIII and factor VIIIa molecules based on scanning transmission electron microscope (STEM) images and stem mass analysis. *J. Clin. Invest.* 145:1310-1311.
21. Fowler, W.E., P.J. Fay, D.S. Arvan, and V.J. Marder. 1990. Electron microscopy of human factor V and factor VIII: correlation of morphology with domain structure and localization of factor V activation fragments. *Proc. Natl. Acad. Sci. USA* 87:7648-7652.
22. Lollar, P., and C.G. Parker. 1987. Stoichiometry of the porcine factor VIII-von Willebrand factor association. *J. Biol. Chem.* 262:17572-17576.
23. Lollar, P., and C.G. Parker. 1989. Subunit structure of thrombin-activated porcine factor VIII. *Biochemistry* 28:666-674.
24. Jenny, R.J., D.D. Pittman, J.J. Toole, R.W. Kriz, R.A. Aldape, R.M. Hewick, R.J. Kaufman, and K.G. Mann. 1987. Complete cDNA and derived amino acid sequence of human factor V. *Proc. Natl. Acad. Sci. USA* 84:4846-4850.
25. Kane, W.H., and E.W. Davie. 1988. Blood coagulation factors V and VIII: structural and functional similarities and their relationship to hemorrhagic and thrombotic disorders. *Blood* 71:539-555.
26. Dorner, A.J., D.G. Bole, and R.J. Kaufman. 1987. The relationship of N-linked glycosylation and heavy chain-binding protein association with the secretion of glycoproteins. *J. Cell Biol.* 105:2665-2674.
27. Pittman, D.D., J.H. Wang, and R.J. Kaufman. 1992. Identification and functional importance of tyrosine sulfate residues within recombinant human factor VIII. *Biochemistry* 31:3315-3325.
28. Curtis, J.E., S.L. Helgerson, E.T. Parker, and P. Lollar. 1994. Isolation and characterization of thrombin-activated human factor VIII. *J. Biol. Chem.* 269:6246-51.
29. Tuddenham, E.G., R.S. Lane, F. Rotblat, A.J. Johnson, T.J. Snape, S. Middleton, and P.B. Kernoff. 1982. Response to infusions of polyelectrolyte fractionated human factor VIII concentrate in human haemophilia A and von Willebrand's disease. *Br. J. Haematol.* 52:259-267.
30. Brinkhous, K.M., H. Sandberg, J.B. Garris, C. Mattson, M. Palm, T. Griggs, and M.S. Read. 1985. Purified human factor VIII procoagulant protein: comparative hemostatic response after infusion into hemophilic and von Willebrand disease dogs. *Proc. Natl. Acad. Sci. USA* 82:8752-8755.
31. Fass, D.N., G.J. Knutson, and E.J. Bowie. 1978. Porcine Willebrand factor: a population of multimers. *J. Lab. Clin. Med.* 91:307-320.
32. Chopek, M.W., J.P. Girma, K. Fujikawa, E.W. Davie, and K. Titani. 1986. Human von Willebrand factor: a multivalent protein composed of identical subunits. *Biochemistry* 25:3146-3155.
33. Wagner, D.D. 1990. Cell biology of von Willebrand factor. *Ann. Rev. Cell Biol.* 6:217-246.
34. Fretto, L.J., W.E. Fowler, D.R. McCaslin, H.P. Erickson, and P.A. McKee. 1986. Substructure of human von Willebrand factor. Proteolysis by V8 and characterization of two functional domains. *J. Biol. Chem.* 261:15679-15689.
35. Fowler, W.E., L.J. Fretto, K.K. Hamilton, H.P. Erickson, and P.A. McKee. 1985. Substructure of human von Willebrand factor. *J. Clin. Invest.* 76:1491-1500.
36. Ohmori, K., L.J. Fretto, R.L. Harrison, M.E. Switzer, H.P. Erickson, and P.A. McKee. 1982. Electron microscopy of human factor VIII/von Willebrand glycoprotein: effect of reducing reagents on structure and function. *J. Cell Biol.* 95:632-640.
37. Slayter, H., J. Loscalzo, P. Bockenstedt, and R.I. Handin. 1985. Native conformation of human von Willebrand protein. Analysis by electron microscopy and quasi-elastic light scattering. *J. Biol. Chem.* 260:8559-8563.
38. Foster, P.A., C.A. Fulcher, T. Marti, K. Titani, and T.S. Zimmerman. 1987. A major factor VIII binding domain resides within the amino-terminal 272 residues of von Willebrand factor. *J. Biol. Chem.* 262:8443-8446.

39. Bahou, W.F., D. Ginsburg, R. Sikkink, R. Litwiller, and D.N. Fass. 1989. A monoclonal antibody to von Willebrand factor (vWF) inhibits factor VIII binding: Localization of its antigenic determinant to a nonadecapeptide at the aminoterminus of the mature vWF polypeptide. *J. Clin. Invest.* 84:56-61.
40. Ginsburg, D., and E.J.W. Bowie. 1992. Molecular genetics of von Willebrand disease. *Blood* 79:2507-2519.
41. Owen, W.G., and R.H. Wagner. 1972. Antihemophilic factor: separation of an active fragment following dissociation by salts or detergents. *Thromb. Diath. Haemorrh.* 27:502-515.
42. Weiss, H.J., and L.W. Hoyer. 1973. Von Willebrand factor: dissociation from antihemophilic factor activity. *Science* 182:1149-1151.
43. Zucker, M.B., M.E. Soberano, A.J. Johnson, A.J. Fulton, S. Kowalski, and M. Adler. 1983. The in vitro association of antihemophilic factor and von Willebrand factor. *Thromb. Haemostas.* 49:37-41.
44. Leyte, A., M.P. Verbeet, T. Brodniewicz-Proba, J.A. van Mourik, and K. Mertens. 1989. The interaction between human blood-coagulation factor VIII and von Willebrand factor. *Biochem. J.* 257:679-683.
45. Nesheim, M., D.D. Pittman, A.R. Giles, D.N. Fass, J.H. Wang, D. Slonosky, and R.J. Kaufman. 1991. The effect of plasma von Willebrand factor on the binding of human factor VIII to human platelets. *J. Biol. Chem.* 266:17815-17820.
46. Foster, P.A., C.A. Fulcher, R.A. Houghten, and T.S. Zimmerman. 1988. An immunogenic region with residues Val^{1670}-Glu^{1684} of the factor VIII light chain induces antibodies which inhibit binding of factor VIII to von Willebrand factor. *J. Biol. Chem.* 263:5230-5234.
47. Lollar, P., D.C. Hill-Eubanks, and C.G. Parker. 1988. Association of the factor VIII light chain with von Willebrand factor. *J. Biol. Chem.* 263:10451-10455.
48. Hamer, R.J., J.A. Koedam, N.H. Beeser-Visser, and J.J. Sixma. 1987. The effect of thrombin on the complex between factor VIII and von Willebrand factor. *Eur. J. Biochem.* 167:253-259.
49. Leyte, A., K. Mertens, B. Distel, R.F. Evers, M.J.M. De Keyzer-Nellen, M.M.C.L. Groenen-van Dooren, J.D.E. de Bruin, H. Pannekoek, J.A. van Mourik, and M.P. Verbeet. 1989. Inhibition of human coagulation Factor VIII by monoclonal antibodies. Mapping of functional epitopes with the use of recombinant Factor VIII fragments. *Biochem. J.* 263:187-194.
50. Precup, J.W., B.C. Kline, and D.N. Fass. 1991. A monoclonal antibody to factor VIII inhibits von Willebrand factor binding and thrombin cleavage. *Blood* 77:1929-1936.
51. Shima, M., D. Scandella, A. Yoshioka, H. Nakai, I. Tanaka, S. Kamisue, S. Terada, and H. Fukui. 1993. A factor VIII neutralizing monoclonal antibody and a human inhibitor alloantibody recognizing epitopes in the C2 domain inhibit binding to von Willebrand factor and to phosphatidylserine. *Thromb. Haemostas.* 69:240-246.
52. Lollar, P., G.J. Knutson, and D.N. Fass. 1985. Activation of porcine factor VIII:C by thrombin and factor Xa. *Biochemistry* 24:8056-8064.
53. Neuenschwander, P., and J. Jesty. 1992. Thrombin-activated and factor Xa-activated human factor VIII: differences in cofactor activity and decay rate. *Arch. Biochem. Biophys.* 296:426-434.
54. Hultin, M.B., and J. Jesty. 1981. The activation and inactivation of human factor VIII by thrombin: effect of inhibitors of thrombin. *Blood* 57:476-482.
55. Lollar, P., G.J. Knutson, and D.N. Fass. 1984. Stabilization of thrombin-activated porcine factor VIII:C by factor IXa and phospholipid. *Blood* 63:1303-1308.
56. Cooper, H.A., F.F. Reisner, M. Hall, and R.H. Wagner. 1975. Effects of thrombin treatment on preparations of factor VIII and the Ca^{2+}-dissociated small active fragment. *J. Clin. Invest.* 56:751-760.
57. Weiss, H.J., and S. Kochwa. 1970. Molecular forms of antihemophilic globulin in plasma, cryoprecipitate, and after thrombin activation. *Br. J. Haematol.* 18:89-100.
58. Hill-Eubanks, D.C., C.G. Parker, and P. Lollar. 1989. Differential proteolytic activation of factor VIII-von Willebrand factor complex by thrombin. *Proc. Natl. Acad. Sci. USA* 86:6508-6512.
59. Aly, A.M., and L.W. Hoyer. 1992. Factor VIII-East Hartford has procoagulant activity when separated from von Willebrand factor. *J. Clin. Invest.* 89:1382-1387.
60. O'Brien, D.P., J.K. Patinson, and E.G.D. Tuddenham. 1990. Purification and characterization of factor VIII 372-Cys: A hypofunctional cofactor from a patient with moderately severe hemophilia A. *Blood* 75:1664-1672.

61. Pittman, D.D., and R.J. Kaufman. 1988. Proteolytic requirements for thrombin activation of anti-hemophilic factor (factor VIII). *Proc. Natl. Acad. Sci. USA* 85:2429-2433.
62. Hill-Eubanks, D.C., and P. Lollar. 1990. von Willebrand factor is a cofactor for thrombin-catalyzed cleavage of the factor VIII light chain. *J. Biol. Chem.* 265:17854-17858.
63. Barrow, R.T., J.F. Healey, and P. Lollar. 1994. Inhibition by heparin of thrombin-catalyzed activation of the factor VIII-von Willebrand factor complex. *J. Biol. Chem.* 269:593-8.
64. Fay, P.J., P.J. Haidaris, and T.M. Smudzin. 1991. Human factor VIII_a subunit structure: reconstitution of factor VIII_a from the isolated A1/A3-C1-C2 dimer and A2 subunit. *J. Biol. Chem.* 266:8957-8962.
65. Lollar, P., and E.T. Parker. 1991. Structural basis for the decreased procoagulant activity of human factor VIII compared to the porcine homolog. *J. Biol. Chem.* 266:12481-12486.
66. Lollar, P., E.T. Parker, and P.J. Fay. 1992. Coagulant properties of hybrid human/porcine factor VIII molecules. *J. Biol. Chem.* 267:23652-23657.
67. Lollar, P., and C.G. Parker. 1990. pH-dependent denaturation of thrombin-activated porcine factor VIII. *J. Biol. Chem.* 265:1688-1692.
68. Hemker, H.C., and M.J.P. Kahn. 1967. Reaction sequence of blood coagulation. *Nature* 215:1201-1202.
69. van Dieijen, G., G. Tans, J. Rosing, and H.C. Hemker. 1981. The role of phospholipid and factor VIIIa in the activation of bovine factor X. *J. Biol. Chem.* 256:3433-3442.
70. Duffy, E.J., and P. Lollar. 1992. Intrinsic pathway activation of factor X and its activation peptide-deficient derivative, factor $X_{(Des\ 143\text{-}191)}$. *J. Biol. Chem.* 267:7821-7827.
71. Rosing, J., G. Tans, J.W.P. Govers-Riemslag, R.F.A. Zwaal, and H.C. Hemker. 1980. The role of phospholipids and factor Va in the prothrombinase complex. *J. Biol. Chem.* 255:274-283.
72. Duffy, E.J., E.T. Parker, V.P. Mutucumarana, A.E. Johnson, and P. Lollar. 1992. Binding of factor VIIIa and factor VIII to factor IXa on phospholipid vesicles. *J. Biol. Chem.* 267:17006-17011.
73. Mutucumarana, V.P., E.J. Duffy, P. Lollar, and A.E. Johnson. 1992. The active site of factor IXa is located far above the membrane surface and its conformation is altered upon association with factor VIIIa. A fluorescence study. *J. Biol. Chem.* 267:17012-17021.
74. Husten, E.J., C.T. Esmon, and A.E. Johnson. 1987. The active site of blood coagulation factor Xa. Its distance from the phospholipid surface and its conformational sensitivity to components of the prothrombinase complex. *J. Biol. Chem.* 262:12953-12961.
75. Zwaal, R.F.A., P. Comfurius, and L.L.M. van Deenen. 1977. Membrane asymmetry and blood coagulation. *Nature* 268:358-360.
76. Lajmanovich, A., G. Hudry-Clergeon, J.M. Freyssinet, and G. Marguerie. 1981. Human Factor VIII procoagulant activity and phospholipid interactions. *Biochem. Biophys. Acta* 678:132-136.
77. Andersson, L.O., and J.E. Brown. 1981. Interaction of Factor VIII-von Willebrand Factor with phospholipid vesicles. *Biochem. J.* 200:161-167.
78. Koedam, J.A., R.J. Hamer, N.H. Beeser-Visser, B.N. Bouma, and J.J. Sixma. 1990. The effect of von Willebrand factor on activation of factor VIII by factor Xa. *Eur. J. Biochem.* 189:229-234.
79. Koedam, J.A., J.C.M. Meijers, J.J. Sixma, and B.N. Bouma. 1988. Inactivation of human factor-VIII by activated protein-C. Cofactor activity of protein-S and protective effect of von Willebrand factor. *J. Clin. Invest.* 82:1236-1243.
80. Fay, P.J., J.-V. Coumans, and F.J. Walker. 1991. von Willebrand factor mediates protection of factor VIII from activated protein C-catalyzed inactivation. *J. Biol. Chem.* 266:2172-2177.
81. Rick, M.E., N.L. Esmon, and D.M. Krizek. 1990. Factor IXa and von Willebrand factor modify the inactivation of factor VIII by activiated protein C. *J. Lab. Clin. Med.* 115:415-421.
82. Bloom, J.W. 1987. The interaction of rDNA factor VIII, factor VIIIses-797-1562, and factor VIIIses-797-1562-derived peptides with phospholipid. *Thromb. Res.* 48:439-448.

83. Gilbert, G.E., B.C. Furie, and B. Furie. 1990. Binding of human factor VIII to phospholipid vesicles. *J. Biol. Chem.* 265:815-822.
84. Gilbert, G.E., D. Drinkwater, S. Barter, and S.B. Clouse. 1992. Specificity of phosphatidylserine-containing membrane binding sites for factor VIII. *J. Biol. Chem.* 267:15861-15868.
85. Gilbert, G.E., and D. Drinkwater. 1993. Specific membrane binding of factor VIII is mediated by sO-phospho-L-serine, a moiety of phosphatidylserine. *Biochemistry* 32:9577-9585.
86. Foster, P.A., C.A. Fulcher, R.A. Houghten, and T.S. Zimmerman. 1990. Synthetic factor VIII peptides with amino acid sequences contained within the C2 domain of factor VIII inhibit factor VIII binding to phosphatidylserine. *Blood* 75:1999-2004.
87. Scandella, D., S.D. Mahoney, M. Mattingly, D. Roeder, L. Timmons, and C.A. Fulcher. 1988. Epitope mapping of human factor VIII inhibitor antibodies by deletion analysis of factor VIII fragments expressed in Escherichia Coli. *Proc. Natl. Acad. Sci. USA* 85:6152-6156.
88. Arai, M., D. Scandella, and L.W. Hoyer. 1989. Molecular basis of factor-VIII inhibition by human antibodies - antibodies that bind to the factor-VIII light chain prevent the interaction of factor-VIII with phospholipid. *J. Clin. Invest.* 83:1978-1984.
89. Guinto, E.R., and C.T. Esmon. 1984. Loss of prothrombin and of factor Xa-factor Va interactions upon inactivation of factor Va by activated protein C. *J. Biol. Chem.* 259:13986-13992.
90. Luckow, E.A., D.A. Lyons, T.M. Ridgeway, C.T. Esmon, and T.M. Laue. 1989. Interaction of clotting factor V heavy chain with prothrombin and prethrombin 1 and role of activated protein C in regulating this interaction: analysis by analytical ultracentrifugation. *Biochemistry* 28:2348-2353.
91. Fay, P.J., P.J. Haidaris, and C.F. Huggins. 1993. Role of the COOH-terminal acidic region of A1 subunit in A2 subunit retention in human factor VIIIa. *J. Biol. Chem.* 268:17861-17866.
92. Walker, F.J., D. Scandella, and P.J. Fay. 1990. Identification of the binding site for activated protein C on the light chain of factors V and VIII. *J. Biol. Chem.* 265:1484-1489.
93. Lollar, P., and D.N. Fass. 1984. Inhibition of activated porcine factor IX by dansyl-glutamyl-glycyl-arginyl-chloromethylketone. *Arch. Biochem. Biophys.* 233:438-446.
94. Lamphear, B.J., and P.J. Fay. 1992. Factor IXa enhances reconstitution of factor VIIIa from isolated A2 subunit and A1/A3-C1-C2 dimer. *J. Biol. Chem.* 267:3725-3730.
95. Lollar, P., E.T. Parker, J.E. Curtis, S.L. Helgerson, L.W. Hoyer, M.E. Scott, and D. Scandella. 1994. Inhibition of human factor VIIIa by anti-A2 subunit antibodies. *J. Clin. Invest.* submitted:
96. Gilles, J.G.G., J. Arnout, J. Vermylen, and J.-M.R. Saint-Remy. 1993. Anti-factor VIII antibodies of hemophiliac patients are frequently directed towards nonfunctional determinants and do not exhibit isotypic restriction. *Blood* 82:2452-2461.
97. Lubin, I.M., J.F. Healey, P. Lollar, D. Scandella, and M.S. Runge. 1993. Expression of a recombinant hybrid human/porcine factor VIII molecule with elimination of reactivity toward an inhibitory anti-human A2 domain antibody. *Blood* 82a:in press.
98. Ware, J., M.J. MacDonald, M. Lo, S. de Graaf, and C.A. Fulcher. 1992. Epitope mapping of human factor VIII inhibitor antibodies by site-directed mutagenesis of a factor VIII polypeptide. *Blood Coagul. Fibrinolysis.* 3:703-716.
99. Scandella, D., M. Mattingly, and R. Prescott. 1993. A recombinant factor VIII A2 domain polypeptide quantitatively neutralizes human inhibitor antibodies which bind to A2. *Blood* 82:1767-1775.

Molecular etiology of factor VIII deficiency in hemophilia A

Stylianos E. Antonarakis,[1,2] Haig H. Kazazian,[2] Jane Gitschier,[3] Pierre Hutter,[1] Philippe de Moerloose,[4] Michael A. Morris[1]

[1]Division of Medical Genetics, University of Geneva Medical School and Cantonal Hospital, Geneva, Switzerland; [2]Center for Medical Genetics, The Johns Hopkins University School of Medicine, Baltimore, Maryland; [3]Howard Hughes Medical Institute and University of California, San Francisco, California; [4]Hemostasis Unit, Cantonal Hospital of Geneva, Switzerland

The gene for factor VIII (F.VIII) is located on the human X chromosome; and, consequently, hemophilia A is a classic example of X-linked recessive inheritance. It occurs almost exclusively in males who have only one X chromosome; females with one abnormal F.VIII gene are asymptomatic carriers because the other X chromosome contains a normal gene. The frequency of the disorder is 1-2 in 10,000 male births in all ethnic groups. The severity and frequency of bleeding in the patients correlates with the F.VIII activity in plasma (Rizza et al., 1983; Hoyer, 1987).

About 50% of patients have severe hemophilia A with F.VIII activity less than 1% of normal; they have frequent spontaneous bleeding into joints, muscles, and internal organs. Moderately severe hemophilia A occurs in about 10% of patients; the F.VIII activity is 2-5% of normal, and there is bleeding after minor trauma. Mild hemophilia which occurs in 30-40% of patients, is associated with F.VIII activity of 5-30% and there is bleeding only after significant trauma or surgery. Of particular interest for the understanding of the function of F.VIII is a category of patients who have considerable amount of F.VIII protein in their plasma (at least 30% of normal), but the protein is non-functional, i.e., the F.VIII activity is much less than the F.VIII plasma level. Approximately 5% of patients belong to this CRM (cross-reacting material) positive category (Lazarchick & Hoyer, 1978). CRM-reduced is another category in which the F.VIII antigen and activity are reduced to approximately the same level.

Before the introduction of modern treatment, severe hemophilia A

Inhibitors to Coagulation Factors
Edited by Louis M. Aledort *et al.*, Plenum Press, New York, 1995

was a genetically lethal disease, and Haldane in 1935 predicted that about one-third of cases would be the result of novel mutations. This prediction was proven correct, since many patients have been identified who carry a de novo mutation which is not present in the X chromosomes of their mothers.

F.VIII gene and deduced protein structure

The gene for F.VIII was cloned in 1984 (Gitschier et al., 1984; Wood et al., 1984; Toole et al., 1984). Factor VIII genomic DNA is 186 kb long (approximately 0.1 % of the DNA of the X chromosome) and contains 26 exons. The nucleotide sequences of the exons, intron-exon boundaries, and 5′ and 3′ untranslated regions have been determined (Gitschier et al., 1984; Wood et al., 1984; Toole et al., 1984). The exon length varies from 69 to 262 nucleotides except for exon 14 which is 3106 nucleotides long and exon 26 which has 1958 nucleotides . There are 6 large introns of more than 14 kb, such as IVS22 which is 32 kb long. (See Figure 1.)

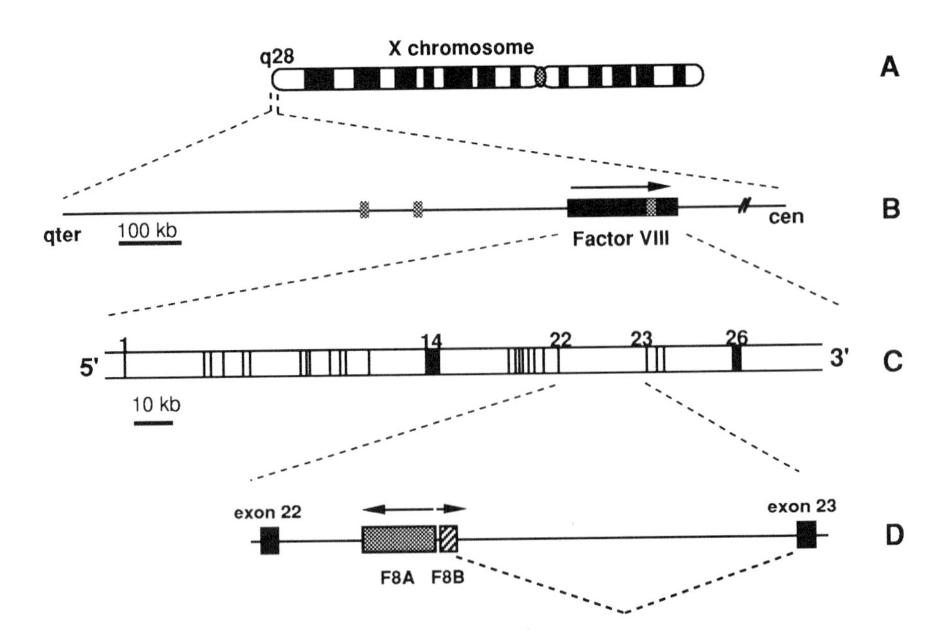

Figure 1. Schematic representation of the chromosomal localization and structure of the F.VIII gene. The gene is located about 1000 kb from the Xqter (panel B). It is 186 kb long and contains 26 exons (panel C). The large IVS22 contains two nested genes, the intronless F8A and F8B which utilizes the exon 23 of F.VIII gene as its second exon (panel D). There are two further copies of sequences homologous to F8A on Xq28 as shown by the gray boxes of panel B.

The F.VIII gene maps on the most distal band (Xq28) of long arm of the X chromosome. (See Figure 1.) Pulsed field gel electrophoresis and physical mapping of Xq28 using Yeast Artificial Chromosomes suggested that the F.VIII gene mapped distal to G6PD and the distance from F.VIII gene to the Xq telomere is approximately 1 Mb. (Poustka et al., 1991; Freije and Schlessinger, 1992). The order of these loci and the direction of transcription is Xcen-G6PD-3'F8-5'F8-Xqter (Freije und Schlessinger, 1992; Migeon et al., 1993).

The normal F.VIII mRNA is approximately 9 kb, of which the coding sequence is 7053 nucleotides. There is a CpG island within IVS22 which is associated with 2 additional transcripts. One transcript of 1.8 kb, termed F8A, is produced abundantly in a wide variety of cells. It is oriented opposite to that of F.VIII, contains no introns (Levinson et al., 1990; see Figure 1) and is conserved in the mouse (Levinson et al., 1992a). The second transcript of 2.5 kb, termed F8B, is transcribed in the same direction as F.VIII and after a short private exon, it utilizes exons 23 to 26 of the F.VIII gene (Levinson et al., 1992b). The two transcripts F8A and F8B originate within 122 nucleotides of each other. The sequences of F8A and several kilobases of surrounding DNA are also present in two other areas of the X chromosome, approximately 400 kb telomeric to the F.VIII gene (Levinson et al., 1990; Freije und Schlessinger, 1992). The function of the F8A and F8B transcripts and their potential protein products are unknown.

Factor VIII is expressed in the liver, spleen, lymph nodes, and a variety of human tissues but not in bone marrow, peripheral blood lymphocytes, or endothelial cells (Wion et al., 1985). In the liver, which is the major tissue of expression, the hepatocyte is the predominant cell that synthesizes F.VIII RNA (Zelechowska et al., 1985; Wion et al., 1985).

The gene for F.VIII encodes a precursor protein of 2351 amino acids. The first 19 comprise the leader peptide (Vehar et al., 1984). The mature protein contains 2332 amino acids and the calculated molecular weight is 264763 Da. There are several domains of F.VIII with internal homology. The first three are the A1 (residues 1-329), A2 (380-711), and A3 (1649-2019) domains and have amino acid sequence homology of approximately 30%. The A domains are homologous to domains of ferroxidase, ceruloplasmin, and factor V (F.V) (Vehar et al., 1984, Koschinsky et al. 1986, Kane et al., 1986). The A2 and A3 domains are separated by a region of 983 amino acids, the B domain, which contains 19 of the 25 potential N-glycosylation sites. The B domain has no known function, is encoded by nearly all of the large exon 14 and is not conserved between F.VIII and F.V (Kane et al. 1987). At the carboxyl-terminus of the mature protein, there are two homologous domains: C1 and C2. These domains are also homologous to the A, C, and D chains of discoidin I, the C domains of F.V and mouse milk fat globular membrane protein (Vehar et al. 1984, Stubbs et al. 1990). Domains C1 and C2 extend from amino acid residues 2020 to 2172 and 2173 to 2332, respectively. The arrangement of the six domains described above from the amino to the carboxyl-terminus is A1-A2-B-A3-C1-C2 (see Figure 3). In addition, there are two acidic regions rich in aspartic and glutamic acid residues. One is located between the A1 and A2 domains (residues 331 to 372) and the second between residues 1648 to 1689. Factor VIII contains 23 cysteine residues of which 14 are conserved between F.V and F.VIII, suggesting that these proteins might be folded in a similar manner. The murine F.VIII cDNA has been recently cloned and sequenced (Elder et al., 1993). The mouse cDNA encodes a protein of 2319 amino acids with overall 74% identity to the human sequence. The amino acid identity in the A and C domains is 84 to 93%, whereas the B domain and the two acidic regions are more divergent, since the identity in these regions is 42-70%. All thrombin cleavage sites and sulfated tyrosine residues are conserved. Five of 6 potential N-glycosylation sites in the non-B domain parts of F.VIII are also conserved. In addition, the B domain of the murine deduced protein sequence contains 19 potential N-glycosylation sites; the human sequence contains the same number of potential N-glycosylation sites but in different positions. This suggests that the glycosylation of the B domain is important for the biosynthesis of F.VIII.

F.VIII gene defects

Since the cloning of the F.VIII gene, the DNA of more than 1000 hemophilia A patients has been examined for molecular defects. Initially, Southern blotting, cloning, and sequencing were the methods used. The introduction of the polymerase chain reaction (PCR) amplification from genomic DNA or from RNA (RT-PCR) has revolutionized the mutation detection protocols. Several screening methods for recognition of mutations have been employed, including denaturing gradient gel electrophoresis, single-stranded conformational analysis, RNA cleavage analysis, and subsequent direct sequencing of PCR products. A database of mutations in the F.VIII gene has been published (Tuddenham et al., 1991) and has been recently updated (November 1993) by S.E.A. This hemophilia A mutation database is termed HAMD-Nov93 throughout this paper. Table 1 depicts the different kinds of mutations found in patients with severe or mild to moderate hemophilia A from the studies of Higuchi et al., 1991a, 1991b; Naylor et al., 1993, as updated after the identification of the common inversion by Lakich et al., 1993. These studies have been selected because all the mutations have been identified in a given sample.

Gross DNA rearrangements

Large deletions. In about 5% of the patients with hemophilia A there are large (more than 50 nucleotides) deletions in the F.VIII gene

(Antonarakis & Kazazian, 1988). The mutation database of Tuddenham et al., 1991, contains 59 different deletions. Southern blot analysis indicated that there were no two unrelated patients with the same breakpoints suggesting that F.VIII gene does not contain sequences that are prone to become deletion breakpoints. Deletions almost always produce severe hemophilia A; however, a deletion of exon 22 was associated with moderate disease probably because of in-frame splicing of exons 21 and 23 and production of a protein lacking 52 amino acids (Youssoufian et al., 1987). Few deletion breakpoints have been characterized at the nucleotide sequence level. The majority of the breakpoints examined do not occur in repetitive elements such as Alu sequences. There is usually 2-4 nucleotide homology at the junction point and the deletion mechanism is probably via non-homologous recombination (Woods-Samuels et al., 1991).

Table 1. Hemophilia A: Categories of mutations in patients with severe or mild to moderate disease (data from two selected studies by Higuchi et al. 1991a,1991b, and Naylor et al. 1993)

	Severe	Mild/Moderate
Point mutations	24	48
Missense	8	46
Frameshift	9	1
Nonsense	7	
Splicing mutations		1
Gross Rearrangements	26	
Partial inversion	22	
Large deletion	4	
Unknown	2	5
Total	52	53

The common partial inversion

The efforts to characterize all the mutations in the F.VIII gene in a defined sample of hemophilia A patients revealed a surprising and unexpected finding. After scanning all the exons of F.VIII gene using denaturing gradient gel electrophoresis, Higuchi et al. found the causative mutations in about 90% of patients with mild to moderate hemophilia A (Higuchi et al., 1991a); however, when severely affected patients were similarly extensively studied, the causative mutation was found in only about 50% (Higuchi et al., 1991b). The cause of the remaining 50%

of severe hemophilia A remained elusive. Subsequently, Naylor et al., using RT-PCR of illegitimate transcription of the F.VIII gene, found that in about 40% of severely affected patients no RT-PCR amplification was possible between exons 22 and 23 of the gene (Naylor et al., 1992). Lakich et al. recently found that these patients have a partial inversion of the F.VIII gene (the gene up to and including exon 22 is inverted), due to homologous recombination between the region that includes the F8A gene in intron 22 and one of the two other homologous regions located more than 400 kb 5′ (telomeric) to the F.VIII gene (Lakich et al., 1993). (See Figure 2.) The discovery of this mutation and the subsequent development of a simple diagnostic test using Southern blot is of considerable clinical significance since this mutation mechanism may account for about 25% of all patients with hemophilia A and certainly for more than 40% of those with severe disease.

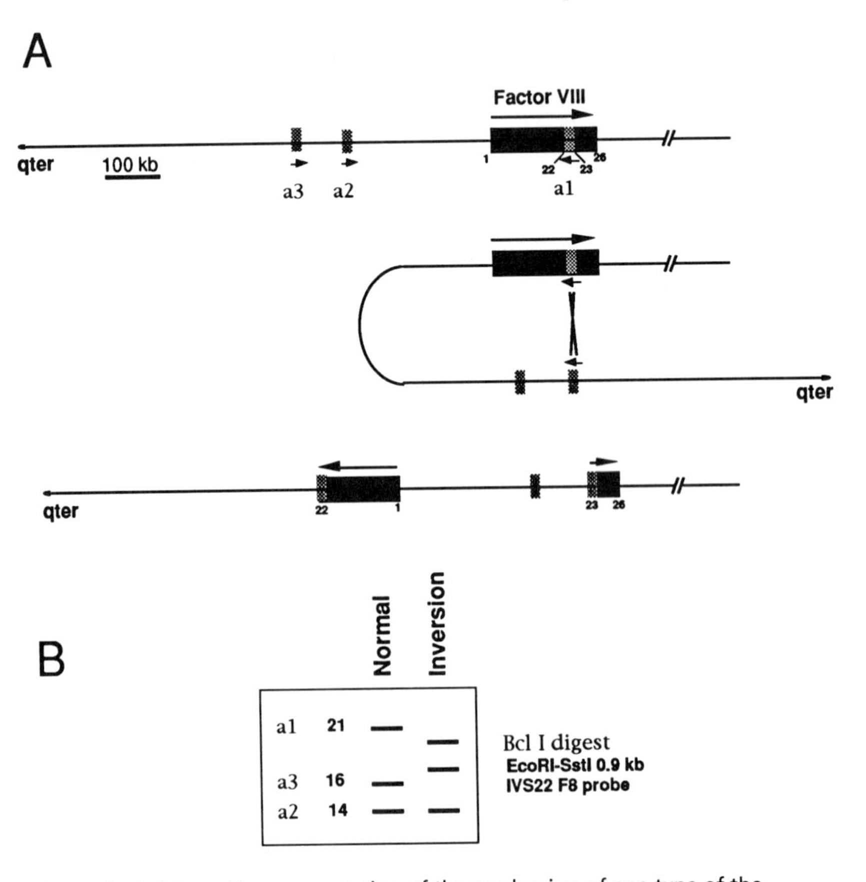

Figure 2. A. Schematic representation of the mechanism of one type of the common inversion of the F.VIII gene (as proposed by Lakich et al., 1993) due to intrachromosomal crossing-over between the homologous sequences A1 and A3. The partial inversion includes exons 1 to 22 of the gene. B. Diagnostic Southern blot analysis for the recognition of this type of inversion of the F.VIII gene.

Insertion of retrotransposons

De novo insertion of LINE retrotransposons in the human genome was first reported in the F.VIII gene in two cases of severe de novo hemophilia A (Kazazian et al., 1988). In one case a 3.8 kb portion of a LINE element was inserted in exon 14. The inserted DNA had a poly(A) tail and produced a target site duplication of an adenine rich sequence. In the second case a 2.1 kb portion of a LINE element was inserted in a different site of exon 14 with all the characteristics of retrotransposition. LINE elements compose about 5% of the human genome and there are approximately 10^5 copies (Fanning & Singer, 1987). The full length of the element is 6.1 kb and most of the copies in the human genome are partial and defective. The consensus sequence of LINE element contains two open reading frames, the second of which predicts a polypeptide with homology to viral reverse transcriptase. About 3000 LINE copies are full-length but only a few, perhaps those with open reading frames, can produce a new insertion through an RNA intermediate. They are probably transcribed into DNA and then reinserted as double stranded DNA into a new genomic site (Kazazian & Scott, 1993). The full-length "active" LINE element responsible for the insertion in the first hemophilia A patient was cloned and characterized. It maps on chromosome 22 and encodes a peptide that has reverse transcriptase activity (Dombroski et al., 1991; Mathias et al., 1991). A third LINE element insertion in intron 10 has also been observed, but since it did not co-segregate with hemophilia A, it represents a recently established private polymorphism (Woods-Samuels et al., 1989). There are about a dozen additional reports for LINE and Alu insertions in other genes (Kazazian & Scott, 1993).

Duplications

There are two such lesions described in the F.VIII gene. In one there was a duplication of 23 kb of IVS22 inserted between exons 23 and 25 (Gitschier, 1993). This rearrangement found in two female siblings was apparently unstable and led to deletion of exons 23-25 in the male offspring of one of the females. In the second case, there was an in-frame duplication of exon 13 in a patient with mild hemophilia A (Murru et al., 1990).

Point mutations and small (<50 bp) deletion/insertions

Nucleotide substitutions and insertion/deletions of up to 50 nucleotides are discussed under point mutations. Table 2 contains the different kinds of point mutations included in the HAMD-Nov93.

Small deletion/insertions

Small deletions or insertions in the coding region of F.VIII gene that result in frameshifts and cause severe hemophilia A have been reported.

The HAMD-Nov93 contains 22 small deletions (of 1, 2, 4, 11, and 23 nucleotides) among 259 independent mutations recorded (8.5%). There are 7 small insertions (1 and 10 nucleotides) (2.7% of the total point mutations). About half of the small deletions (11 of 22) or insertions (4 of 7) were found in exon 14.

Nonsense mutations

There are 63 independent nonsense mutations in 21 different codons that are included in the point mutation database comprising 25% of the total number of point mutations. This percentage is perhaps biased since many investigators have used restriction digestion analysis with Taq I to detect CG to TG mutations and in particular CGA to TGA (Arg to Stop) substitutions (Youssoufian et al., 1986). In two samples of 52 severe hemophilia A patients of table 1 in which all point mutations have been characterized, the number of nonsense codons were 7, i.e., 13.5% (7 of 52) of the total severe mutations or 29% (7 of 24) of the point mutations.

Table 2. Hemophilia A; point mutations in the Factor VIII gene (data from the HAMD-Nov93)

Patients with known point mutations	259
Number of different point mutations	149
Nucleotide substitutions	120
Missense	93
Nonsense	21
Small insertions	7
Small deletions	22
CRM positive	13
Splicing mutations	7
Different CG to TG mutations	33
Patients with CG to TG mutations	119
Missense polymorphism	3
Silent substitutions in exons	5

CpG dinucleotide hypermutability

The study of point mutations in F.VIII gene uncovered two general lessons concerning human mutations. The first was the discovery of the mutation hotspot at CpG dinucleotides in which there is a common substitution CG to TG if the mutation occurs in the sense strand, or CG to CA if the mutation occurs in the antisense strand (Youssoufian et al., 1986). The mutations probably occur because, in mammalian DNA, the majority of CpG dinucleotides are methylated (at the 5 carbon of cytosine) by methyltransferase; the subsequent spontaneous deamination of the 5-methylcytosine produces a TpG dinucleotide. The mutation usually occurs in tissues in which the gene of interest is not expressed (Cooper & Youssoufian, 1988). There are 119 independent mutations that conform to the CG to TG rule in the point mutation database (46%). This high proportion of CpG mutations in the database is probably due to the deliberate screening by restriction analysis. There are 24 sites in which recurrent mutation at CpG dinucleotides have occurred. An unbiased estimate of the frequency of CG to TG mutation may be obtained from studies in which all point mutations have been characterized in a given sample of patients (Higuchi 1991 a, b; Diamond et al.., 1992; Naylor et al., 1993). In these studies a total of 84 point mutations have been characterized and 32 were of the CG to TG rule (38%). It has been estimated that in the F.VIII gene CG to TG or CA mutations are 10-20 times more frequent than mutations of CG to any other dinucleotide (Youssoufian et al., 1988). The mutation hotspot has subsequently been observed in a wide variety of other human genes related to disease phenotypes.

Exon skipping due to nonsense mutations

An important observation concerning the pathophysiology of nonsense mutations has recently been made in the F.VIII gene and independently in the fibrillin and OAT genes using RT-PCR for mutation detection (Naylor et al., 1993; Dietz et al., 1993). In some cases a nonsense codon mutation can lead to abnormal RNA processing where the exon containing the mutation is skipped. In one case of Glu1987Stop mutation in exon 19 all detectable RNA lacked the sequences of this exon. In the second case of Arg2116Stop mutation of exon 22 there was about 50% of RNA without the sequences of exon 22 while the remaining 50% of RNA was of normal size. It is of interest that the junctions of exons 18 to 20 and 21 to 23 do not result in translational frameshift. The mechanism, significance, and frequency of exon skipping due to nonsense mutations are presently unknown.

Missense mutations

The study of missense mutations, i.e., nucleotide substitutions that result in amino acid substitutions, is important for the understanding of the function of the protein and the pathophysiology of the disease. A total of 96 mutations leading to amino acid substitutions (including three rare normal variants) have been described and presented in Figure 3. These mutations are spread throughout the different domains of the gene except for exon 14 (that encodes the B domain) which is devoid of amino acid substitutions that cause hemophilia A. The mode of action of the majority of these missense mutations in reducing F.VIII activity in plasma is unknown.

Mutations at thrombin cleavage sites

There are mutations in patients with CRM positive hemophilia A that affect the thrombin cleavage needed for activation of the molecule. Mutations Arg372His and Arg372Cys have been shown to abolish the normal cleavage by thrombin in the heavy chain (Gitschier et al., 1988; Arai et al., 1989; O'Brien et al., 1990). It is not clear if the Ser373Leu mutation has an effect in thrombin cleavage. Mutations Arg1689Cys and Arg1689His abolish thrombin cleavage at the light chain (O'Brien et al., 1989; Arai et al., 1990).

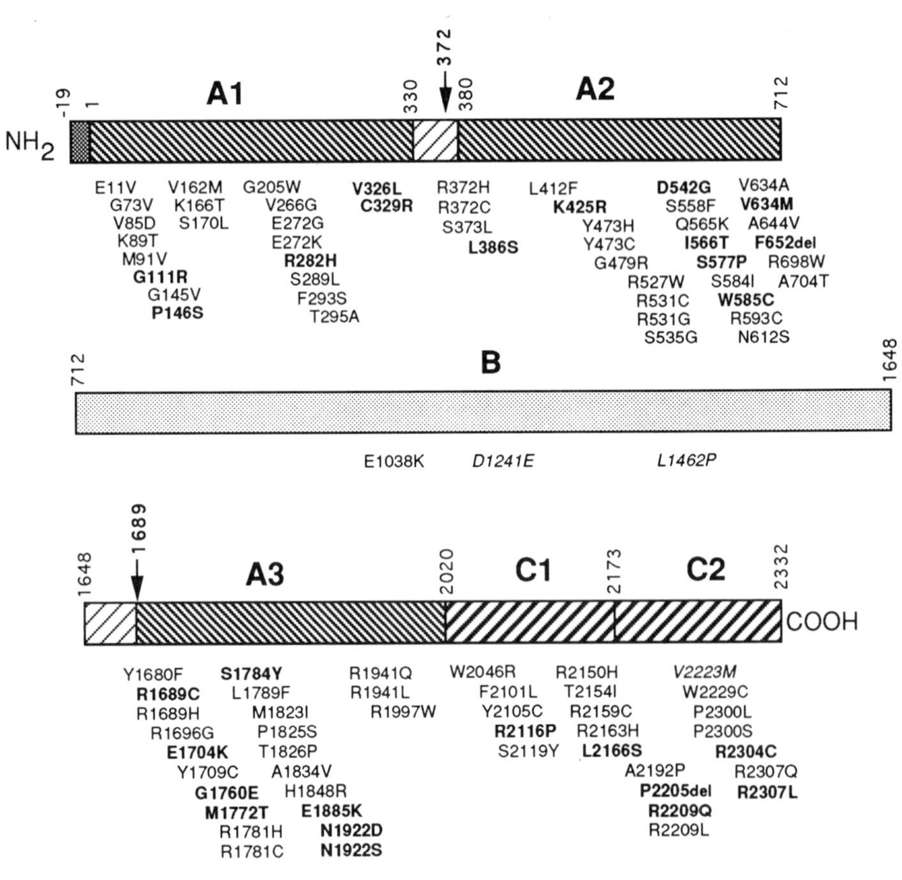

Figure 3. Missense mutations and single amino acid deletions in the F.VIII protein. The structural domains of the protein are shown and the amino acid substitutions are depicted using the one letter code. For example E11V indicates a Glu to Val substitution at residue 11. Mutations in boldface letters were found in patients of the CRM positive category. Mutations in *italics* are rare normal variants.

Mutations at N-glycosylation sites

Two other CRM positive mutations produce severe hemophilia A by creating new N-glycosylation sites in the protein (Aly et al., 1992). The first, Ile566Thr, creates a new site at Asn564 in the A2 domain of the heavy chain. The second new site is at Asn1770 of the A3 domain of the light chain (the mutation is Met1772Thr). When the plasma of either patient is deglycosylated, F.VIII activity is restored to a significant degree. The significance of the Ser577Pro mutation that in theory eliminates a potential N-glycosylation site at Asn575 is unknown.

Mutations at the von Willebrand factor binding site

There are two sulfated tyrosine residues Tyr1664 and Tyr1680 in the region of F.VIII between amino acids Lys1673 and Glu1684 in which a vWf binding site has been localized. A mutation Tyr1680Phe has been observed in patients with moderate, CRM reduced hemophilia A (Higuchi et al., 1990). Site-directed mutagenesis of Tyr1680Phe results in a molecule that has lost high-affinity binding to vWF presumably because the phenylalanine residue cannot be sulfated (Leyte et al., 1991).

Study of mutations in CRM-positive and CRM-reduced patients

The elucidation of mutations in these patients is important in understanding the significance of specific amino acid residues. A small number of such mutations have been described (see Figure 3 and McGinniss et al., 1993). Since about 40% of CRM positive mutations occur in the A2 domain, which consists of 228 amino acids or about 10% of the coding region of F.VIII, this region must be important in procoagulant activity. The majority of mutations, however, are CRM negative and probably affect the folding and stability of the protein. Since these mutations result in the absence of secreted F.VIII and the *in vitro* functional studies depend on the analysis of the protein produced in eukaryotic cells after transfection with F.VIII cDNA, the mechanisms of action of these mutants will be difficult to elucidate.

Other missense mutations of interest

In the next few years an effort may be made to elucidate the function of missense mutations of interest. Amino acid substitutions that lead to severe phenotype with F.VIII activity <1% are of importance; to date, for the majority of these mutations the level of F.VIII antigen is not known and therefore they cannot be categorized in the CRM positive group. Cases with two different mutations in the same amino acid are also of interest. (See Figure 3.) These are at Glu272, Tyr473, Arg531, Val634, Arg1781, Asn1922, Arg1941, Arg2209, Pro2300, Arg2307. Mutations in the last 30 amino acids of F.VIII (C2 domain) may cause reduced phospholipid binding. Candidates are Arg2304Leu, Arg2307Leu, Arg2307Gln. The domains of F.VIII for binding to factors IX, X, and others have not yet been clearly defined. As our understanding of the binding domains is improved, missense mutations in the respective domains will become candidates for specific aberrant and testable function.

Splicing errors

A small number of potential splicing errors have been identified (HAMD-Nov93). It seems that in spite of the presence of 50 splice junctions in the F.VIII gene, splicing mutants do not account for a sizeable fraction of hemophilia patients. There are two mutations in the invariant AG of the acceptor splice site in introns 5 and 6 associated as expected with severe hemophilia A. There are four mutations in the donor splice site consensus and two in cryptic splice sites. No formal proof that the mutations cause abnormal splicing has been obtained.

Promoter mutations

No examples of mutations in the 5' untranslated region of F.VIII gene have been reported to date. If such mutations do occur, they are probably infrequent since no substitutions have been found in 530 nucleotides of the 5' flanking region of F.VIII in aproximately 230 patients with hemophilia A analyzed (Higuchi et al., 1990; Gitschier et al., unpublished). Notably, however, the cis-regulatory elements for F.VIII gene expression are either unknown or poorly understood; therefore, intelligent search for promoter mutations has not yet been performed.

Mutations in patients with Factor VIII inhibitors

Approximately 10-20% of patients with severe hemophilia A develop antibodies (known as inhibitors) to F.VIII after treatment with exogenous F.VIII (Sultan, 1992). The etiology of development of inhibitors is unknown. Epitope mapping has shown specificities against the heavy or light chain or both in different patients (Fulcher et al., 1987; Scandella et al., 1989). The analysis of many F.VIII mutations and their association with inhibitor development may uncover some rules, if such exist, concerning the contribution of the nature of the mutations to inhibitor formation. Table 3 shows the mutations associated with inhibitors from the HAMD-Nov93. This table is incomplete, since the status on inhibitors is not available for all mutations. In addition, the inhibitor titer and the duration of the presence of the antibodies is not reported. Finally, the frequency of inhibitors in patients with the partial inversion has not yet been extensively studied; in the Naylor et al. 1993 study, 2 of 10 such patients had inhibitors.

The majority of the reported inhibitor cases have nonsense mutations or deletions in their F.VIII gene (35 of 39; 90%). There are, however, four missense mutations (see Table 3) associated with low levels of inhibitors. Perhaps, these mutations create local structural modifications which lead to the creation of new immunogenic epitopes. Among the nonsense mutations, Arg1941Stop is associated with inhibitors in 5/7 cases; Arg2147Stop in 3/5 and Arg2209Stop in 3/7, Lys1827Stop in 2/2 cases. Other nonsense mutations, however, (for example, 6 cases with Arg336Stop) are never associated with inhibitors. In these cases, exon skipping may provide a protein lacking 20-50 amino acids. Gross deletions of F.VIII result in 4.5-fold increased incidence of inhibitors compared to patients without detectable deletions (Millar et al., 1990). However, no clear picture has emerged as to the correlation between the size or the breakpoints of the deletions and the development of inhibitors. The information on the frequency of inhibitors in patients with the partial inversion of F.VIII gene is limited (Naylor et al., 1993).

Table 3. Hemophilia A; molecular defects in patients with inhibitors (from HAMD-Nov93)

Patients reported in the point mutation database			21 of 259
Nonsense			17
	Arg1696Stop	1 of 1	
	Gln1796Stop	1 of 1	
	Lys1827Stop	2 of 2	
	Gln1874Stop	1 of 1	
	Arg1941Stop	5 of 7	
	Arg1966Stop	1 of 3	
	Arg2147Stop	3 of 5	
	Arg2209Stop	3 of 7	
Missense			4
	Asn1922Asp	1 of 1	
	Tyr2105Cys	1 of 1	
	Arg2209Glu	1 of 10	
	Trp2229Cys	1 of 1	
Patients reported in the deletion database			18 of 49
Patients with Factor VIII inversion			2 of 10

DNA polymorphisms in the F.VIII gene

A number of DNA polymorphisms have been reported in the F.VIII gene and can be used for the indirect detection of defective genes. Carrier detection and prenatal diagnosis can be provided in informative pedigrees. The detailed description of polymorphisms is beyond the scope of this short chapter. For more information, see Peake et al., 1993, Antonarakis and Kazazian, 1992.

Acknowledgments

The Hemophilia A project of S.E.A. and H.H.K. is supported by grant HL38165 and that of J.G. by grant HL42968 from the NIH. The authors thank many physicians and patients for their collaboration during the different studies.

References

Aly AM, Higuchi M, Kasper CK, Kazazian HH Jr, Antonarakis SE, Hoyer LW. Hemophilia A due to mutations that create new N-glycosylation sites. Proc Natl Acad Sci USA 89: 4933-4937, 1992.

Antonarakis SE, Kazazian HH Jr. The molecular basis of Hemophilia A in man. Trends Genet 4: 233-237, 1988.

Antonarakis SE, Kazazian HH Jr. Hemophilias. in Brock DJH, Rodeck CH, Ferguson-Smith MA eds. "Prenatal Diagnosis and Screening" Churchill Livingstone, Edinburgh, pp 477-481, 1992.

Arai M, Higuchi M, Antonarakis SE, Kazazian HH, Phillips JA, Janco RL, Hoyer LW. Characterization of a thrombin cleavage site mutation (arg 1689 to cys) in the Factor VIII gene of two unrelated patients with cross-reacting material positive Hemophilia A. Blood 75: 384-389, 1990.

Arai M, Inaba H, Higuchi M, Antonarakis SE, Kazazian HH, Fujimaki M, Hoyer LW. Detection of a mutation altering a thrombin cleavage site (arginine-372-histidine). Proc Natl Acad Sci USA 86: 4277-4281, 1989.

Barker D, Schafer M, White R. Restriction sites containing CpG show a higher frequency of polymorphism in human DNA. Cell 36: 343-348, 1984.

Cooper DN, Youssoufian H. The CpG dinucleotide and human genetic disease. Hum Genet 74: 151-155, 1988.

Diamond C, Kogan S, Levinson B, Gitschier J. Amino acid substitutions in conserved domains of factor VIII and related proteins: study of patients with mild to moderately severe Hemophilia A. Hum Mut 1: 248-257, 1992.

Dietz HC, Valle D, Francomano CA, Kendzior RJ Jr, Pyeritz RE, Cutting GR. The skipping of constitutive exons in vivo induced by nonsense mutations. Science 259: 680-682, 1993.

Dombroski BA, Mathias SL, Nanthakumar E, Scott AF, Kazazian HH Jr. Isolation of a human transposable element. Science 254: 1808-1910, 1991.

Elder B, Lakich D, Gitschier J. Sequence of the murine Factor VIII cDNA. Genomics 16: 374-379 ,1993.

Fanning TG, Singer MF. LINE-1: a mammalian transposable element. Biochem Biophys Acta 910: 203-212, 1987.

Freije D, Schlessinger D. A 1.6 nb contig of yeast artificial chromosomes around the human factor VIII gene reveals three regions homologous to probes for the DXS115 locus and two for the DXYS64 locus. Am J Hum Genet 51: 66-80, 1992.

Fulcher CA, de Graff S, Mahoney S, Zimmerman TS. FVIII inhibitor IgG subclass and FVIII polypeptide specificity determined by immunoblotting. Blood 69: 1475-1480, 1987.

Gitschier J, Wood WI, Goralka TM, Wion KL, Chen EY, Eaton DE, Vehar GA, Capon DJ, Lawn RM. Characterization of the human factor VIII gene. Nature 312: 326-330, 1984.

Gitschier J. Maternal duplication associated with gene deletion in sporadic Hemophilia. Am J Hum Genet 43: 274-279, 1988.

Gitschier J, Drayna D, Tuddenham EGD, White RL, Lawn RM. Genetic mapping and diagnosis of haemophilia A achieved through a BclI polymorphism in the factor VIII gene. Nature (London) 314: 738-740, 1988.

Gitschier J, Kogan S, Levinson B, Tuddenham EGD. Mutations of factor VIII cleavage sites in Hemophilia A. Blood 72: 1022-1028, 1988.

Haldane JBS. The rate of spontaneous mutation of a human gene. J Genet 31: 317-326, 1935.

Higuchi M, Antonarakis SE, Kasch L, Oldenburg J, Economou-Petersen E, Olek K, Inaba H, Kazazian HH. Towards a complete characterization of mild to moderate Hemophilia A: detection of the molecular defect in 25 of 29 patients by denaturing gradient gel electrophoresis. Proc Natl Acad Sci USA 88: 8307-8311, 1991a.

Higuchi M, Kazazian HH, Kasch L, Warren TC, McGinniss MJ, Phillips JA, Kasper C, Janco R, Antonarakis SE. Molecular characterization of severe Hemophilia A suggests that about half the mutations are not within the coding region and splice junctions of the factor VIII gene. Proc Natl Acad Sci USA 88: 7405-7409, 1991b.

Higuchi M, Wong C, Kochhan L, Olek K, Aronis S, Kasper CK, Kazazian HH, Antonarakis SE. Characterization of mutations in the factor VIII gene by direct sequencing of amplified genomic DNA. Genomics 6: 65-71, 1990.

Hoyer LW. Molecular pathology and immunology of factor VIII. Hum Pathol 18: 153-161, 1987.

Kane WH, Davie EW. Cloning of a cDNA coding for human factor V, a blood coagulation factor homologous to factor VIII and ceruloplasmin. Proc Natl Aca Sci USA 83 : 800-6804 , 1986.

Kane WH, Ichinose A, Hagen FS, Davie EW. Cloning of cDNAs coding for the heavy chain region and connecting region of human factor V , a blood coagulation factor with four types of repeats. Biochemistry 26: 6508-6514 , 1987.

Kazazian HH Jr, Scott AF. "Copy and paste" transposable elements in the human genome. J Clin Invest 91: 1859-1860, 1993.

Kazazian HH Jr, Wong C, Youssoufian HG, Scott AF, Phillips D, Antonarakis SE. A novel mechanism of mutation in man: Hemophilia A due to de novo insertion of L1 sequences. Nature (London) 332: 164-166, 1988.

Koschinsky ML, Funk WD, van Ooar BA, MacGillivray RTA. Complete cDNA sequence of human preceruloplasmin. Proc Natl Acad Sci USA 83: 5086-5090, 1986.

Lakich D, Kazazian HH, Antonarakis SE, Gitschier J. Inversions disrupting the factor VIII gene as a common cause of severe Hemophilia A. Nature Genet 5: 236-241, 1993.

Lalloz MR, McVey JH, Pattinson JK, Tuddenham EGD. Haemophilia A diagnosis by analysis of a hypervariable dinucleotide repeat within the factor VIII gene. Lancet 338: 207-211, 1991.

Lazarchick J, Hoyer LW. Immunoradiometric measurement of the factor VIII procoagulant antigen. J Clin Invest 62: 1048-1052, 1976.

Levinson B, Kenwrick S, Lakich D, Hammonds G, Gitschier J. A transcribed gene in an intron of the human factor VIII gene. Genomics 7: 1-11, 1990.

Levinson B, Bermingham JR, Metzenberg A, Kenwrick S, Chapman V, Gitschier J. Sequence of the human factor VIII-associated gene is conserved in mouse. Genomics 13 : 862-865 , 1992a.

Levinson B, Kenwrick S, Gamel P, Fisher K, Gitschier J. Evidence for a third transcript from the human factor VIII gene. Genomics 14: 585-589, 1992b.

Leyte A, van Schijndel HB, Niehrs C, Huttner WB, Verbeet MP et al. Sulfation of Tyr1[680] of human blood coagulation factor VIII is essential for the interaction of factor VIII with von Willebrand factor. J Biol Chem 266: 17572-17576, 1991.

Mathias SL, Scott AF, Kazazian HH Jr, Boeke JD, Gabriel A. Reverse transcriptase encoded by a human transposable element. Science 254: 1808-1810, 1991.

McGinniss MJ, Kazazian HH Jr, Hoyer LW, Bi L, Inaba H, Antonarakis SE. Spectrum of mutations in CRM-positive and CRM-reduced Hemophilia A. Genomics 15: 392-398, 1993.

Migeon BR, McGinnis MJ, Antonarakis SE, Axelman J, Stasiowski B, Youssoufian H, Keams WG, Chung A, Pearson PL, Kazazian HH Jr, Muneer RS. Severe Hemophilia A in a female by cryptic translocation: order and orientation of factor VIII within Xq28. Genomics 16: 20-25, 1993.

Murru S, Casula L, Pecorara M, Mori P, Cao A, Pirastu M. Illegitimate recombination produced a duplication within the factor VIII gene in a patient with mild Hemophilia A. Genomics 7: 115-118, 1990.

Naylor JA, Green PM, Rizza CR, Gianelli F. Factor VIII gene explains all cases of haemophilia A. Lancet 340: 1066-1067, 1992.

Naylor JA, Green PM, Rizza CR, Gianelli F. Analysis of factor VIII mRNA reveals defects in every one of 28 haemophilia A patients. Hum Mol Genet 2: 11-17, 1993.

O'Brien D, Pattinson JK, Tuddenham EGD. Purification and characterization of factor VIII 3372-cys: a hypofunctional cofactor from a patient with moderately severe Hemophilia A. Blood 75: 1664-1672, 1990.

O'Brien DP, Tuddenham EGD. Purification and characterization of factor VIII 1689Cys: A non-functional cofactor occurring in a patient with severe haemophilia A. Blood 73: 2117-2122, 1989.

Peake IR, Lillicrap DP, Boulyjenkov V, Briët E, Chan V, Ginter EK, Kraus EM, Ljung R, Mannucci PM, Nicolaides K, Tuddenham EGD. Report of a joint WHO/WFH meeting on the control of haemophilia: carrier detection and prenatal diagnosis. Blood Coagulation and Fibrinolysis 4: 313-344, 1993.

Poustka A, Detrich A, Langenstein G, Toniolo D, Warren ST, Lehrach H. Physical map of Xq27-Xqter: localizing the region of the fragile X mutation. Proc Natl Acad Sci USA 88: 8302-8306, 1991.

Rizza CR, Spooner RJD. Treatment of haemophilia and related disorders in Britain and Northern Ireland during 1976-80: Report on behalf of the Directors of Haemophilia Centres in the United Kingdom. Brit Med J 286: 929-933, 1983.

Scandella D, Mattingly M, de Graaf S, Fulcher CA. Localization of epitopes for human factor VIII inhibitor antibodies by immunoblotting and antibody neutralization. Blood 74: 1618-1626, 1990.

Stubbs JD, Lekutis C, Singer KL, Bui A, Yuzuki D, Srinivasan U, Parry G. cDNA cloning of a mouse mammary epithelial cell surface protein reveals the existence of epidermal growth factor-like sequences. Proc Natl Acad Sci USA 87: 8417-8421, 1990.

Sultan Y. Prevalence of inhibitors in a population of 3435 Hemophilia patients in France. Thromb Haemostas 67: 600-602, 1992.

Toole JJ, Knopf JL, Wozney JM, Sultzman LA, Buecker JL, Pittman DD, Kaufman RJ, Brown E, Shoemaker C, Orr EC, Amphlett GW, Foster WB, Coe ML, Knutson GJ, Fass DN, Hewick RM. Molecular cloning of a cDNA encoding human antihaemophilic factor. Nature 312: 342-347, 1984.

Tuddenham EGD, Cooper DN, Gitschier J, Higuchi M, Hoyer L, Yoshioka A, Peaker I, Schwaab R, Olek K, Kazazian H, Lavergne J-M, Giannelli F, Antonarakis S. Haemophilia A: database of nucleotide substitutions, deletions, insertions and rearrangements of the factor VIII gene. Nucl Acid Res 19: 4821-4833, 1991.

Vehar GA, Keyt B, Eaton D, Rodriguez H, O'Brien DP, Rotblat F, Oppermann H, Keck R, Wood WI, Harkins RN, Tuddenham EGD, Lawn RM, Capon DJ. Structure of human factor VIII. Nature 312: 337-342, 1984.

Wion KL, Kelly D, Summerfield JA, Tuddenham EGD, Lawn RM. Distribution of factor VIII mRNA and antigen in human liver and other tissues. Nature 317: 726-729, 1985.

Wood WI, Capon DJ, Simonsen CC, Eaton DL, Gitschier J, Keyt B, Seeburg PH, Smith DL, Hollingshead P, Wion KL, Delwart E, Tuddenham EGD, Vehar GA, Lawn RM. Expression of active human factor VIII from recombinant DNA clones. Nature 312: 330-337, 1984.

Woods-Samuels P, Kazazian HH Jr, Antonarakis SE. Nonhomologous recombination in the human genome: Deletions in the human factor VIII gene. Genomics 10: 94-101, 1991.

Woods-Samuels P, Wong C, Mathias SL, Scott AF, Kazazian HH Jr, Antonarakis SE. Characterization of a non-deleterious L1 insertion in an intron of the human factor VIII gene and evidence for open reading frames in functional L1 elements. Genomics 4: 290-296, 1989.

Youssoufian H, Antonarakis SE, Aronis S, Triftis G, Phillips DG, Kazazian HH Jr. Characterization of five partial deletions of the factor VIII gene. Proc Natl Acad Sci USA 84: 3772-3776, 1987.

Youssoufian H, Antonarakis SE, Bell W, Griffin AM, Kazazian HH Jr. Nonsense and missense mutation in Hemophilia A: estimate of the relative mutation rate at CpG dinucleotides. Am J Hum Genet 42: 718-725, 1988.

Youssoufian H, Kazazian HH Jr, Phillips DG, Aronis S, Tsiftis G, Brown VA, Antonarakis SE. Recurrent mutations in haemophilia A give evidence for CpG mutations hotspots. Nature (London) 324: 380-382, 1986.

Zelechowska MG, van Mourik JA, Brodniewicz-Proba T. Ultrastructural localization of factor VIII procoagulant antigen in human liver hepatocytes. Nature 317 : 729-731 , 1985.

The incidence of factor VIII inhibitors in patients with severe hemophilia A

Leon W. Hoyer

Holland Laboratory
American Red Cross
Rockville, Maryland 20855

Inhibitor development is a serious complication of hemophilia A treatment. Although studies of inhibitor prevalence had been published earlier,[1-3] the possibility that there is increased risk of inhibitor formation with different Factor VIII (F.VIII)-containing products was first raised when heat-treated concentrates were introduced. However, this concern was quickly dropped when it was apparent that heat treatment was necessary to inactivate HIV. As a result, no systematic study during the 1980s directly compared heat treated and non-heat treated F.VIII concentrates.

When more highly purified concentrates were developed, usually using an immunoaffinity step, the question was raised again. Identification of an inhibitor in a very young patient,[4] as well as in an adult patient previously treated for many years with other products,[5] suggested that the new process might cause an increase in the incidence of inhibitors. Unfortunately, no large prospective studies have specifically addressed this question for the concentrates that contain highly purified plasma F.VIII. While several studies have summarized patient outcomes after treatment with these products, they have emphasized virus safety issues rather than inhibitor development.[6,7] In this regard it is important to note the recent reports from Belgium and the Netherlands that document an unexpected increase in inhibitor development that appeared to be related to a change in the F.VIII manufacturing process.[8,9] Thus, some F.VIII preparations are more immunogenic than others, and a concern about increased inhibitor development is entirely appropriate.

A possible relationship between the type of therapeutic product and inhibitor incidence has also been examined during the evaluation of F.VIII prepared by recombinant DNA technology. It is recognized that recombinant products might have small differences from plasma-derived F.VIII, e.g., in post-translational modifications. However, initial studies demonstrated no unusual inhibitor frequency in patients who had previously received plasma-derived F.VIII for many years.[10-12] In addition,

Inhibitors to Coagulation Factors
Edited by Louis M. Aledort *et al.*, Plenum Press, New York, 1995

large prospective studies have been undertaken to determine the incidence of F.VIII inhibitors in patients who had not been previously exposed to F.VIII. While some physicians have considered the inhibitor incidence data in preliminary reports to be higher than expected, this contention has been very controversial. This chapter will address this concern by reviewing the published data describing the incidence and prevalence of inhibitors to F.VIII in patients with severe hemophilia A.

Variables that affect the development of F.VIII inhibitors

If different publications are to be compared, it is important at the outset to identify those variables that affect inhibitor development. (See Table 1.)

Table 1. Variables that affect F.VIII inhibitor incidence

I. The Patient

- Molecular defect
- Immunologic response characteristics
- Other immunologic challenges

II. The Treatment

- Type of F.VIII product
- Number of exposures
- Pattern of exposures
- Cumulative exposure
- Effect of exposure to several different products

III. The Evaluation

- Frequency
- Sensitivity

It is essential that the patient populations be equivalent in susceptibility to inhibitor development if different reports are to be compared. Specifically, it is clear that patients with severe hemophilia A (undetectable F.VIII, reported in most series as <1% or <2% of normal) are at a much higher risk than patients with moderate or mild hemophilia A.[3] In addition, a registry of molecular defects suggests that patients with gene deletions or nonsense mutations are at higher risk than individuals with missense mutations.[13] The importance of disease severity and of the specific molecular defect is undoubtedly at least part of the reason why there is a familial tendency to inhibitor formation.[14] However, although related patients obviously have the same molecular defect, they do not consistently develop or fail to develop an inhibitor. While HLA characterizations have not been helpful to date,[15-17] it is still likely that specific immunologic response capabilities modulate inhibitor development. In addition, the immunologic response to F.VIII may be affected by concurrent exposures to other immunologic challenges that might deflect or enhance F.VIII antibody formation. For example, patients first exposed to F.VIII at the time of an infection might respond somewhat differently than patients who are first transfused under different circumstances. In fact, there is a report of two pairs of identical twins with severe hemophilia who differ in their inhibitor status.[18]

The patient treatment schedule is also important, and there are a number of variables in addition to the type of F.VIII concentrate. The number of exposures and their pattern may be as important as the cumulative amount of F.VIII infused. Although there are no definitive studies, at least one report suggests that patients exposed to a single product may be less likely to develop an inhibitor than patients treated with several different concentrates.[19]

The patient evaluation schedule will also influence the number of inhibitors that are recognized. Until recently, inhibitor testing was done only for patients who appeared to have become resistant to treatment. This clinical evaluation will only detect inhibitors that are of a sufficiently high titer to affect F.VIII replacement therapy. In contrast, prospective studies with regular screening assays have identified a higher incidence of inhibitors. Patients with low titer inhibitors will certainly be detected earlier when assays are done every three to six months–or after a defined number of F.VIII treatments. In fact, transient inhibitors may only be identified if routine screening tests are carried out. For example, a large prospective study detected a rather high frequency of transient antibodies, 12 of 31, and 7 of these patients had only a single positive sample.[3]

Definitions

Before evaluating the published studies of F.VIII inhibitor development, it is also important to point out that even though "incidence" and "prevalence" have often been used interchangeably, they need to be clearly distinguished. The prevalence of inhibitors is that percentage of hemophilic patients who have an inhibitor at any specific point in time. As Rasi and Ikkala pointed out, this measure is very much affected by blood product use and inhibitor patient mortality.[20] Because these factors are rarely described in detail for retrospective studies, it is very difficult to compare prevalence estimates. For this reason, inhibitor incidence has been emphasized in most recent reports. This is the occurrence of an inhibitor in a defined population during a specific period of time. While incidence has been calculated in some reports as the number of new inhibitors per patient year of observation, this measure is unsatisfactory. It is a useful measure only when the likelihood of inhibitor development is independent of the duration of exposure. However, hemophilia A patients with a long history of F.VIII exposure have a very low likelihood of subsequent inhibitor formation. Thus, this statistic is only useful for patient populations that have comparable F.VIII exposure at the outset of the observation period–and the same follow-up period. The duration of the study is important since inhibitors usually develop after a relatively limited number of F.VIII infusions. Thus, the same population followed twice as long will appear to have a lower incidence of inhibitors per patient year.

A better approach is the description of the cumulative risk of inhibitor development, and the method of Kaplan and Meier is suitable for nonparametric estimations from incomplete observations.[21] In these analyses, data are censored for patients who are followed for periods of time that are shorter than those of the remaining study population.[22]

Cumulative risk analysis of F.VIII inhibitor incidence

Although few published studies have included a cumulative risk analysis of inhibitor formation, this evaluation is quite easy if the needed data are available for all of the patients in the study. To establish the most important, measurable risk factor for inhibitor formation, Lee and colleagues evaluated the effect of different variables using stepwise logistic regression analysis. They determined that the risk factors that best correlated with inhibitor development were the number of infusions and the number of F.VIII exposure days.[22] Unfortunately, most published studies have not reported these data, and information about these measures for both noninhibitor and inhibitor patients was available for only two groups of patients treated with plasma-derived F.VIII.[2,23] More studies provide data for inhibitor development as a function of patient age, but it must be emphasized that this is a less satisfactory parameter since the intensity of F.VIII treatment for comparable bleeding episodes will affect the exposure history of patients at any given age.

Table 2 summarizes data for six groups of patients with severe (or, in one series, severe and moderate[24]) hemophilia A treated primarily with cryoprecipitate and/or intermediate purity F.VIII concentrates. Some of the patients received high purity F.VIII concentrates as well. When data for these groups are compared by cumulative risk analysis, there is great heterogeneity of inhibitor development and cumulative incidence (see Figure 1).

Table 2. Inhibitor incidence as a function of patient age

Study	Number of patients	Number of inhibitors: Total	High responders	Transient responders	Median age: At inhibitor development	Of noninhibitor patients*	Frequency of inhibitor evaluation
Patients treated with cryoprecipitate, intermediate purity, or high purity F.VIII concentrates							
Strauss[2]	78	17	13	N.S.**	8	13	Clinical
Rasi et al.[25]	60	9	9	0	7.9	16.8	Clinical
Ehrenforth et al.[24]	46***	15	12	0	2	(~7)	Every 20 exposures
de Biasi[26]	48	11	9	0	3.3	12.3	Clinical
Peerlinck[19]	48	3	1	2	2	12	Yearly
Addiego et al.[23]	89	25	21	2	1.75	10	Yearly
Patients treated with recombinant F.VIII							
Bray et al.[27]	72	17	8	5	1.3	2.4	Every 3 months
Lusher et al.[28]	59	14	6	4	1.4	2.3	Every 3 months

* at the time the study was published

** not stated

*** This study includes patients with both severe and moderate hemophilia A (F.VIII <5% normal)and estimates of median age were made from the published figure. For the other series, raw data were available for both the inhibitor and noninhibitor groups.

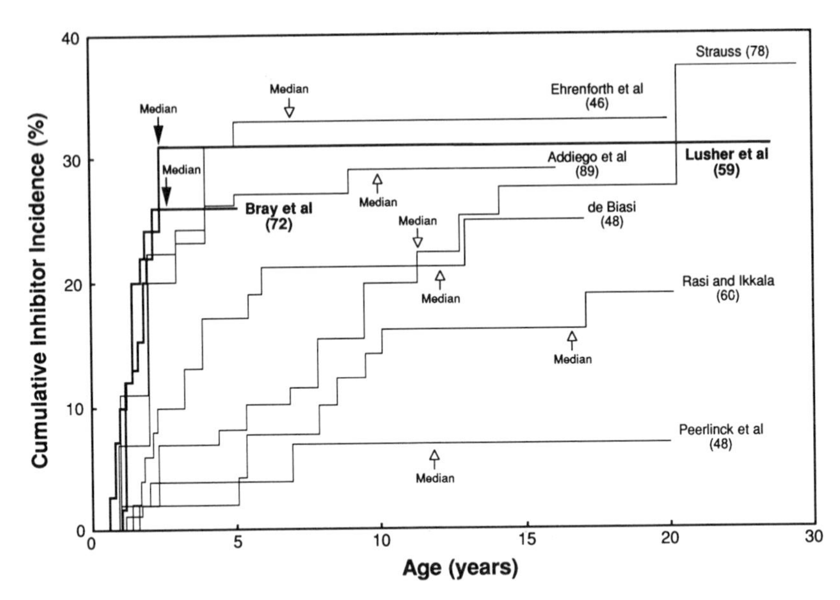

Figure 1. Cumulative inhibitor incidence as a function of patient age data for the 8 studies listed in Table 2. The recombinant F.VIII studies are plotted in heavier lines and have filled arrows at the median age for patients who had not developed an inhibitor at the time the study was analyzed. The lighter lines and open arrows are for data from studies of plasma-derived F.VIII. The number of patients in each study is indicated in parenthesis.

Inhibitor development appears to be less frequent in some series in which the patients received lyophilized cryoprecipitate,[19,25] but not in one large series in which plasma and cryoprecipitate were the only available products.[2] Data for patients treated with recombinant F.VIII have a cumulative incidence pattern like that of two of the series in which plasma-derived F.VIII was used.[23,24] It must be noted, however, that monitoring for inhibitor development was more frequent for these patients.

Data have been obtained from the investigators as well as from the publications for three groups that collected data for the F.VIII exposure days of all study patients. They have been compared to the observations of an earlier report that included this information (see Table 3). While F.VIII inhibitors were recognized after fewer exposures in the recombinant trials (see Figure 2), the cumulative risk of inhibitor development appears to be comparable. The shorter observation period for the recombinant trials is striking, however, and any interpretations must be tentative until the median treatment with recombinant F.VIII is at least 30 exposure days. Again, the different intensities of inhibitor monitoring must be kept in mind when comparing data for different studies.

Table 3. Inhibitor incidence as a function of exposure days in patients with severe hemophilia A

Study	Number of patients	Total number of inhibitors	Median exposure days (Range) At inhibitor development	Median exposure days (Range) Of noninhibitor patients*	Frequency of inhibitor evaluation
Addiego et al.[23]	89	25	10 (2-30)	35 (8 - >100)	Yearly
Strauss[2]	48**	16	41 (21-260)	57 (5- >250)	Clinical
Bray et al.[27]	72	17	9 (3-45)	10 (1-129)	Every 3 months
Lusher et al.[28]	59	14	9 (3-18)	16 (1-196)	Every 3 months

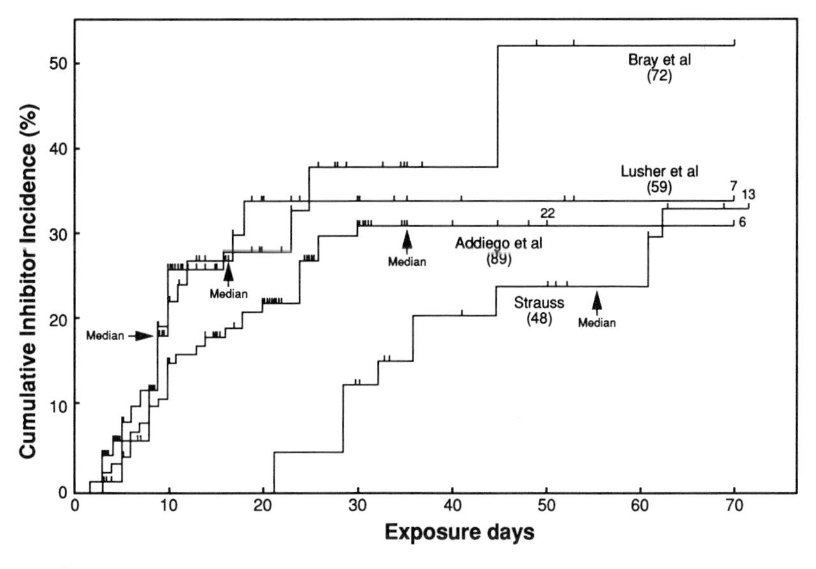

Figure 2. Cumulative inhibitor incidence as a function of F.VIII exposure days. The arrows identify the median exposure days of patients who had not developed an inhibitor at the time the study was analyzed. Tick marks represent individual patients free of an inhibitor and the numbers at the far right of lines indicate the number of noninhibitor patients with at least that many F.VIII exposure days.

The effect of a single patient on a Kaplan-Meier plot can be very great when only a few individuals are at risk in the late phase of the study. The rise in inhibitor incidence after 45 exposure days for the Bray et al.[27] study patients in Figure 2 identifies inhibitor development in one of the four patients followed that long. Because the duration of exposure is still so limited for patients in these recombinant trials, it is not yet clear whether or not there will be significant differences between these groups with longer follow-up.

Inhibitor persistence

Although the cumulative inhibitor incidence figures provide better comparisons than do simple calculations of the percentage of patients who have developed an inhibitor, they tend to overstate the problem. Kaplan-Meier plots do not reflect the heterogeneous nature of the F.VIII inhibitors, and transient or low level inhibitors are not differentiated from high titer, high responder inhibitors that seriously affect patient management. For this reason, additional information is needed if there is to be a meaningful comparison of how inhibitor incidence affects different patient populations. One aspect of this analysis is inhibitor persistence. As Tables 2 and 3 indicate, there have been different proportions of high responder patients and low responder patients in the

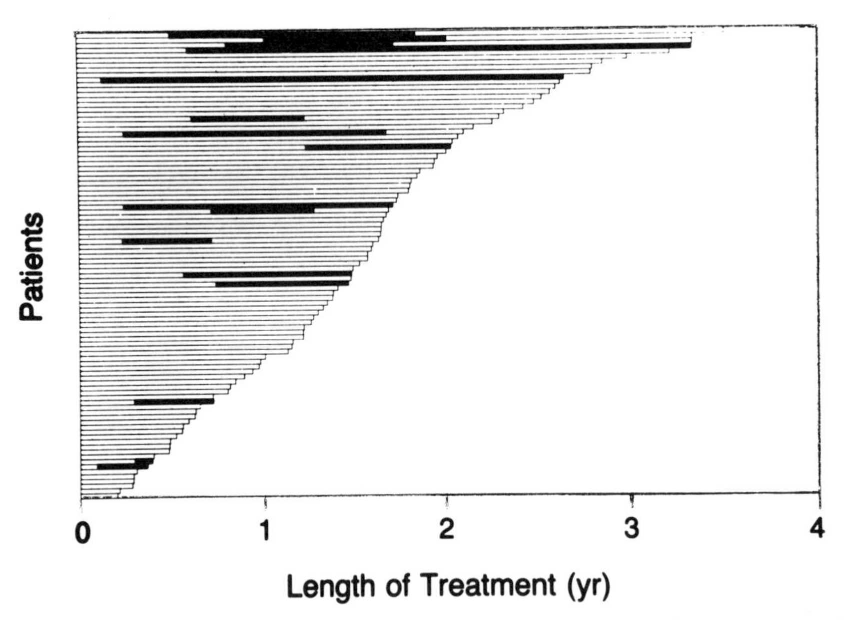

Figure 3. Courses of treatment for 95 patients with hemophilia A who were treated with recombinant F.VIII. Each horizontal bar represents a single patient. Time 0 denotes the beginning of F.VIII treatment. The solid and crosshatched bars denote the presence of an inhibitor in 16 patients. The inhibitor disappeared in 4 of 15 patients who continued to receive episodic treatment and in 3 others following induction of immune tolerance. One patient (crosshatched bar) withdrew from the study shortly after the inhibitor was detected. From Lusher et al.[28] with permission.

different series. Moreover, the "most recent inhibitor titer" for many of these patients has been less than 5 Bethesda units–and is often zero. While this fall may be due to avoidance of F.VIII treatment, or the converse, vigorous F.VIII treatment as a part of an immune tolerance protocol, the fall in titer simply reflects a transient antibody for many patients. As Figure 3 illustrates, the impact of inhibitor development may be quite different from that suggested by cumulative risk analysis.

Although data like that of Figure 3 are not available for other studies, there is some information about the clinical course for many of the inhibitor patients (see Table 4). While the cumulative risk varies from 7 to 37%, the percentage of patients whose most recent inhibitor titer was above 5 Bethesda units ranged from 0 to 9%.

Table 4. Persistence of F.VIII inhibitors

	# Patients	# Inhibitors	Cumulative risk* (%)	Most recent titer at the time the study was reported: 0	< 5 BU	> 5 BU	Prevalence of > 5 BU (%)
Peerlinck et al.[19]	48	3	7	3	3	0	0
de Biasi et al.[26]	48	11	25	0	8	3	6
Ehrenforth et al.[24]	46*	15	33	11	11	4	9
Addiego et al.[23]	89	25	32	6	12	6	7
Bray et al.[27]	72	17	37	5	11	6	8
Lusher et al.[28]	77	16	35	6	19	4	5

* From Figures 1 (at 15 years)[19,24,26] and 2 (at 40 exposure days)[23,27,28]

Interpretation

The observations summarized here suggest that the ways in which we have been describing inhibitor risk are inadequate, especially if exposure to F.VIII generates an immune response to F.VIII in most patients with severe hemophilia A. The recent recombinant studies suggest that frequent patient monitoring identifies transient inhibitors that are followed by "natural" tolerance to F.VIII. Many low responder patients also appear to develop tolerance after continued episodic F.VIII treatment. If these speculations are correct, the detection of a transient or low responder inhibitor during routine surveillance has a very different meaning from the identification of an inhibitor that rises in an amnestic response after each F.VIII exposure. Thus, the critical issue for comparing F.VIII products is not simply cumulative incidence, whether this be a calculation of percentage of treated patients that develop an inhibitor or a Kaplan-Meier plot of cumulative incidence, but the prevalence of inhibitors that *complicate therapy* after a comparable number of F.VIII infusions. Useful comparison of studies will require data that permit this kind of analysis. While these graphs would be quite different from those in figures 1 and 2, the prevalence of inhibitors after comparable exposure is the most clinically relevant measure of patient risk. Prospective trials are needed to obtain this information.

Summary

1. Many factors must be considered when retrospective studies are compared, for the intensity of F.VIII treatment and the frequency of inhibitor evaluation have a marked effect on inhibitor incidence.

2. The incidence of F.VIII inhibitors in patients treated with cryoprecipitate and/or intermediate purity concentrates varies greatly in different studies, with cumulative risks of 7-32% after 10-15 years of follow-up.

3. The rate of inhibitor development appears to be higher for patients treated with recombinant F.VIII, but the cumulative incidence is not greater. This may change, however, when there has been an equivalent follow-up period for the patients receiving recombinant F.VIII.

4. Cumulative risk figures may inappropriately overstate the magnitude of the problem since some inhibitors disappear after short periods of time.

5. Only prospective, long-term studies will provide more satisfactory incidence and prevalence data.

Acknowledgments

Drs. Joseph Addiego, Ruffaello de Biasi, Gordon Bray, Deborah Hurst, Kathelijne Peerlinck, and Vesa Rasi and Ms. Suzanne Courter kindly made primary patient data available so that the Kaplan-Meier plots could be prepared. I also want to recognize the expert assistance of Debbie Wilder in the preparation of the manuscript and of Kris Thompson in the preparation of the figures.

References

1. C.K. Kasper. Incidence and course of inhibitors among patients with classic hemophilia, *Thromb.et Diath.Haem.* 30:263 (1973).
2. H.S. Strauss. Acquired circulating anticoagulants in hemophilia A, *N.Engl.J.Med.* 281:866 (1969).
3. C.W. McMillan, S.S. Shapiro, D. Whitehurst, et al. The natural history of factor VIII:C inhibitors in patients with hemophilia A: A national cooperative study. II. Observations on the initial development of factor VIII:C inhibitors, *Blood* 71:344 (1988).
4. B.A. Bell, E.M. Kurczynski and G. Bergman. Inhibitors to monoclonal antibody purified factor VIII, *Lancet* 336:638 (1990).
5. C.M. Kessler and K. Sachse. Factor VIII:C inhibitor associated with monoclonal-antibody purified FVIII concentrate, *Lancet* 335:1403 (1990).

6. J.E. Addiego, Jr., E.D. Gomperts, S.-L. Liu, et al. Treatment of hemophilia A with a highly purified factor VIII concentrate prepared by anti-FVIIIc immunoaffinity chromatography, *Thromb.Haem.* 67:19 (1992).
7. J.M. Lusher, P.M. Salzman and Monoclate Study Group. Viral safety and inhibitor development associated with factor VIIIC ultra-purified from plasma in hemophiliacs previously unexposed to Factor VIIIC concentrates, *Semin.Hematol.* 27:1 (1990).
8. K. Peerlinck, J. Arnout, J.G. Gilles, J.M. Saint-Remy and J. Vermylen. A higher than expected incidence of factor VIII inhibitors in multitransfused haemophilia A patients treated with an intermediate purity pasteurized factor VIII concentrate, *Thromb.Haem.* 69:115 (1993).
9. F.R. Rosendaal, H.K. Nieuwenhuis, H.M. van den Berg, et al. A sudden increase in factor VIII inhibitor development in multitransfused hemophilia A patients in the Netherlands, *Blood* 8:2180 (1993).
10. R.S. Schwartz, C.F. Abildgaard, L.M. Aledort, et al. Human recombinant DNA-derived antihemophilic factor (Factor VIII) in the treatment of hemophilia A, *N.Engl.J.Med.* 323:1800 (1990).
11. G.C. White, C.W. McMillan, H.S. Kingdon and C.B. Shoemaker. Use of recombinant antihemophilic factor in the treatment of two patients with classic hemophilia, *N.Engl.J.Med.* 320:166 (1989).
12. G.L. Bray. Current status of clinical studies of recombinant factor VIII (Recombinate®) in patients with hemophilia A, *Trans.Med.Rev.* VI:252 (1992).
13. E.G.D. Tuddenham, D.N. Cooper, J. Gitschier, et al. Haemophilia A: database of nucleotide substitutions, deletions, insertions and rearrangements of the factor VIII gene, *Nucl.Acids Res.* 19:4821 (1991).
14. R. Pozzatti, J. Vogel and G. Jay. The human lymphotropic T virus type I taxgene can cooperate with the RAS oncogene to induce neoplastic transformation of cells, *Mol.Cell.Biol.* 10:413 (1990).
15. L.E. Lippert, L.M. Fisher and L.B. Schook. Relationship of major histocompatibility complex class II genes to inhibitor antibody formation in hemophilia A, *Thromb.Haem.* 64:564 (1990).
16. A.M. Aly, L.M. Aledort, T.D. Lee and L.W. Hoyer. Histocompatibility antigen patterns in hemophilic patients with factor VIII antibodies, *Br.J.Haematol.* 76:238 (1990).
17. D. Frommel, J.P. Allain, E. Saint-Paul, et al. HLA antigens and factor VIII antibody in classic hemophilia. European study group of factor VIII antibody, *Thromb.Haem.* 46:687 (1981).
18. European Study Group of Factor VIII Antibody. Development of factor VIII antibody in haemophilic monozygotic twins, *Scand.J.Haemat.* 23:64 (1979).
19. K. Peerlinck, F.R. Rosendaal and J. Vermylen. Incidence of inhibitor development in a group of young hemophilia A patients treated exclusively with lyophilized cryoprecipitate, *Blood* 81:3332 (1993).
20. R. Schwaab, M. Ludwig, J. Oldenburg, et al. Identical point mutations in the factor VIII gene that have different clinical manifestations of hemophilia A, *Am.J.Hum.Genet.* 47:743 (1990).
21. E.L. Kaplan and P. Meier. Nonparametric estimation from incomplete observations, *J.Am.Statist.Assn.* 53:457 (1958).
22. M.L. Lee, S. Liu-Maruya, S. Courter and E. Gomperts. Determining inhibitor-development risk in previously untreated hemophilia A patients treated (PUPs) with recombinant factor VIII (recombinate™), *Blood* 80:491a (1992).
23. J. Addiego, C. Kasper, C. Abildgaard, et al. Frequency of inhibitor development in haemophiliacs treated with low-purity factor VIII, *Lancet* 342:462 (1993).
24. S. Ehrenforth, W. Kreuz, I. Scharrer, et al. Incidence of development of factor VIII and factor IX inhibitors in haemophiliacs, *Lancet* 339:594 (1992).
25. V. Rasi and E. Ikkala. Haemophiliacs with factor VIII inhibitors in Finland:prevalence, incidence and outcome, *Br.J.Haematol.* 76:369 (1990).
26. R. De Biasi, A. Rocino, M.L. Papa and E. Salerno. Incidence of inhibitor development in hemophilia A patients, *Thromb.Haem.* 69:1103 (1993).
27. G.L. Bray, E.D. Gomperts, S. Courter, et al. A multi-center study of recombinant factor VIII (Recombinate): Safety, efficacy and inhibitor risk in previously untreated patients with hemophilia A, *Blood* (1993).
28. J.M. Lusher, S. Arkin, C.F. Abildgaard, R.S. Schwartz and Kogenate. Previously Untreated Patient Study Group. Recombinant factor VIII for the treatment of previously untreated patients with hemophilia A, *N.Engl.J.Med.* 328:453 (1993).

Epitope specificity and functional characterization of factor VIII inhibitors

Dorothea Scandella,[1] Craig Kessler,[2] Pamela Esmon,[3] Deborah Hurst,[3] Suzanne Courter,[4] Edward Gomperts,[4] Matthew Felch,[1] Richard Prescott,[1] and the Recombinate and Kogenate Study groups

[1]*Holland Laboratory, American Red Cross, Rockville, Maryland 20855*
[2]*George Washington University School of Medicine, Washington, DC 20037*
[3]*Miles Inc., Berkeley, California 94701*
[4]*Baxter/Hyland, Glendale, California 91203*

Alloantibodies that inactivate factor VIII (F.VIII) develop in up to 24% of hemophilia A patients given therapeutic infusions of F.VIII. The probability of inhibitor formation is greatest in individuals with severe hemophilia (<2% F.VIII antigen), but they may also occur in moderate or mild hemophiliacs.[1] Autoantibodies with similar properties appear rarely in individuals with normal F.VIII levels. In both cases bleeding episodes are difficult to control.

Inhibitor epitopes were initially localized to the A2 domain or to the light chain of thrombin cleaved F.VIII by immunoblotting analysis.[2] Sixty-eight of 76 inhibitor plasmas tested in this assay contained IgG antibodies which bound to F.VIII. Of these antibodies, 46% bound only to the light chain, 26% bound only to the A2 domain and 32% bound to both. Rare plasmas in which antibodies bound to A1 and A2 occurred in 3% of the patients.[3] More detailed epitope localization studies utilizing immunoblotting of F.VIII deletion polypeptides constructed from the F.VIII cDNA and expressed in E. coli revealed that the major inhibitor epitopes were contained within amino acid regions 373-546 of the A2 domain[4,5] and within the C2 domain. Rare plasmas containing antibodies that bound to the light chain within the A3 or C1 domains or to the heavy chain acidic region between A1 and A2 were also detected.[4] No significant differences in the epitope specificity of allo- or autoantibody inhibitors were seen in these assays.

A plasmid encoding the cDNA of the A2 domain was used to express A2 as a soluble, secreted protein in insect cells. Antibody binding to radiolabeled A2 was analyzed by immunoprecipitation assays. The

Inhibitors to Coagulation Factors
Edited by Louis M. Aledort *et al.*, Plenum Press, New York, 1995

results revealed that a number of plasmas contained antibodies that bind to the soluble form of A2 but not to A2 in immunoblotting assays. Most plasmas (>50%) contained both anti-A2 and anti-C2 antibodies by immunoprecipitation assay.[6]

In this study we compare the epitope specificity of 28 allo- and autoantibody inhibitors by immunoblotting and immunoprecipitation assays. These assays measure antibody binding to F.VIII fragments but not their effect on F.VIII function. In order to determine the relative contribution of anti-A2, anti-C2 or other antibodies to the plasma inhibitor activity, we also studied the ability of the A2 domain and in some cases the C2 domain or the light chain to neutralize the inhibitor activity in each plasma.

Materials and methods

Inhibitors

The inhibitor plasmas used in this study are referred to by the same initials used in previous studies.[4,6] Purified total inhibitor IgG was prepared by caprylic acid precipitation[7] or by affinity chromatography on protein G Sepharose.[8] The inhibitor titer was determined in the Bethesda assay.[9]

Immunoblotting assay

Details have been published in previous studies.[4,10] Briefly, deletions were constructed from the F.VIII cDNA, and the deletion polypeptides were expressed in E. coli.[10] A more extensive series of C2 domain deletions was constructed as described[11] and also expressed in E. coli. Bacterial extracts containing these polypeptides were separated by SDS-polyacrylamide gel electrophoresis and electroblotted to nitrocellulose paper. Incubation with inhibitor plasma at 10-20 Bethesda units/ml was followed by monoclonal anti-human IgG4 antibody, biotinylated goat anti-mouse IgG, and streptavidin-alkaline phosphatase. The bound phosphatase was measured colorimetrically. When biotinylated goat anti-mouse antibody was used as the secondary antibody, similar results were obtained.

Expression of F.VIII fragments in eukaryotic cells and in E. coli

The cDNA encoding the A2 domain or the light chain were altered by site-specific mutagenesis to contain the signal peptide for β-interferon,[6] and the C2 cDNA was similarly altered to contain the signal peptide for tissue plasminogen activator. The A2 domain was expressed from a baculovirus vector in insect cells. Forty-eight hours after viral infection, the cells producing A2 were pulse labeled for 4 hours with ^{35}S-methionine. The C2 domain and the light chain DNA were transfected into COS-1 cells by the DEAE-dextran method and pulse labeled after 48 hours. The extracellular medium was used for all immunoprecipitation experiments. When unlabeled A2 was used for inhibitor neutralizations, it was first precipitated with 45% ammonium sulfate and dialyzed.[6]

Plasmid pGEX-2T DNA[12] (Pharmacia LKB Biotechnology) was digested with restriction endonucleases BamHI and XmaI. A DNA fragment encoding F.VIII amino acid residues 2173-2332, the C2 domain, and bounded by the same restriction sites was generated by the polymerase chain reaction. After digestion with BamHI and XmaI, the fragment was cloned into the corresponding sites of pGEX-2T so that it was fused in frame 3' to the DNA encoding the glutathione S-transferase. Fusion protein expression was induced during log phase growth, and it was affinity purified from the bacterial extract after detergent solubilization by passage over a reduced glutathione column.[12]

Immunoprecipitation assay

These assays employed soluble, secreted F.VIII A2 and C2 domain or light chain polypeptides biosynthetically labeled with ^{35}S-methionine.[8] Inhibitor plasmas or IgG at 50-100 Bethesda units/ml or 100 ml plasma for titers <20 BU/ml were mixed with 100-200 ml ^{35}S-labeled F.VIII domain 18 hours and the immune complexes were isolated on 50ml protein G Sepharose. The beads were washed extensively, and the labeled polypeptides were eluted and analyzed by SDS-PAGE and autoradiography. Quantitative immunoprecipitation was performed with highly purified ^{125}I-light chain. Varying concentrations of inhibitor plasma or IgG were incubated with the same concentration of ^{125}I-light chain. The immune complexes were precipitated after incubation with protein G Sepharose, washed extensively, and the remaining radiolabel was determined in a gamma counter.

Inhibitor neutralization assay

The same concentration of inhibitor plasma was mixed with an equal volume of varying concentrations of partially purified A2 domain, purified light chain, or purified glutathione S-transferase-C2 fusion protein (GST-C2). The amount of inhibitor was adjusted to correspond to that which inhibited approximately 90% of plasma F.VIII. After 2 hours incubation the remaining inhibitor titer was measured in the Bethesda assay. The percentage neutralization was calculated relative to an inhibitor control without an F.VIII polypeptide. The F.VIII light chain was obtained from highly purified recombinant F.VIII, which was dissociated and separated from the heavy chain on S-Sepharose (Pharmacia LKB Biotechnology) according to conditions used previously.[13,14] The final purity was >90%. A similar purity was achieved for GST-C2 by affinity chromatography on glutathione Sepharose.

Competitive enzyme-linked immunosorbent assays

The details for these assays have been previously described.[6] Briefly, partially purified A2 domain was immobilized on Dynatech removawell microtiter plates with MAb 8 (epitope 606-740 of A2). Equal concentrations of affinity purified inhibitor ^{125}I-CC Fab′ were incubated with increasing concentrations of unlabeled Fab′ from other inhibitors. After 4 hours incubation at 37°C, the plates were extensively washed and the wells were counted in an LKB gamma counter. The percentage inhibition of ^{125}I-CC binding was calculated relative to a control with no competitor. Competition for antibody binding to the C2 domain was tested by immobilizing recombinant F.VIII on microtiter plates with MAb 413, epitope 373-605 of A2. Unlabeled inhibitor IgGs competed for F.VIII binding of biotinylated MAb ESH8 (epitope 2248-2285 in C2). Bound biotin-ESH8 was detected with streptavidin-alkaline phosphatase and a colorimetric phosphatase assay.

Results

Epitope localization by immunoblotting assays

A group of 28 inhibitor plasmas were tested for epitope specificity by immunoblotting assay. Thirteen of these were from hemophilia A patients and the remaining 15 from spontaneous inhibitor patients. The results for 21 of the inhibitors has been previously described,[4] and the remainder were analyzed in this study by similar methods. The epitope distribution for these inhibitors is shown in Table 1.

Table 1. Epitope specificity of inhibitors by immunoblotting assay

Patients	Number plasmas with antibody binding to F.VIII domains					
	A2	A2 + A1-A2	C2	A2 + C2	A2 + C2 + A3 or C1	Total
Hemophilia A	2	1	6	3	1	13
Spontaneous	6	0	8	1	0	15

Patients in each group are as follows: hemophilia A, A2 only (CHA, RM), A2 + A1-A2 (WD), C2 only (MP, GK1824, L, RI, MU, HG); A2 + C2 (CC, JR, GK1831), A2 + C2 + A3 or C1 (WG); spontaneous, A2 only (RC, JM, SC, FM, EM, NS), C2 only (D, SLC, DP, F, WT, MR, WC, 1801), A2 + C2 (CHI). A1-A2 is the region of amino acid residues 336-372 between the 2 domains.

The epitope for all inhibitors that bound to A2 was localized to amino acid residues 373-740 by deletion mapping.[6] Ware et al.[15,16] used additional F.VIII deletions to localize the epitope for one inhibitor to residues 345-536. Since this epitope had been previously shown to have an amino terminal limit at residue 373, the epitope would appear to be 373-

536. Since the C2 domain epitope had not been further defined, a series of amino- and carboxy-terminal deletions of C2 were constructed and used for immunoblotting as described[11] with 9 inhibitor plasmas and the IgG from F.VIII inhibitory monoclonal antibody (MAb) ESH8. The epitopes varied in length among the inhibitors as shown in Table 2. The C2 domain consists of F.VIII residues 2173-2332.

Table 2. Epitopes within the C2 domain

Antibody source	Amino acid residues comprising epitope
MP	2170-2332
MR, GK1824	2170-2312
RI	2203-2312
SLC, CC, F, DP	2218-2312
MU, WG	2248-2312
MAb ESH8	2248-2285

The epitope regions defined within the A2 and C2 domains are large, suggesting that these epitopes are discontinuous. Based on high resolution structural information available for some antigen-antibody complexes (see review[17]), most amino acid residues within the F.VIII epitope regions would be expected to function in providing the correct three-dimensional structure for antibody recognition. Only a few residues that are adjacent in the native C2 structure would bind directly to the anti-C2 antibody. In contrast, inhibitor WD has been shown by synthetic peptide inhibition studies to bind to residues 354-62.[18] It is not known if there are additional residues which contribute to maximal WD binding to this region. The epitope(s) within the light chain domains A3 or C1 has not been further localized.

Antibody competition to determine epitope overlap within the A2 and C2 domains

The epitope localization experiments indicated that the major epitopes within the A2 and C2 domains are shared by a number of different inhibitor antibodies. To test this more directly, the following inhibitor competition assays were done. The soluble A2 domain was captured onto microtiter plates with MAb 8 which recognizes an epitope within residues 606-740 of A2 outside the inhibitor epitope. Constant concentrations of affinity purified ^{125}I-labeled inhibitor CC Fab′ and varying concentrations of unlabeled competitor Fab′ were then added. Following incubation and washing, the ^{125}I-CC remaining in each well was determined in a gamma counter. The percentage remaining label was calculated from a control of ^{125}I-CC with no competing inhibitor. The results for several inhibitors are shown in Figure 1.

The IgG and Fab' of each human inhibitor shown (CC, RC, RI, GK1831) and MAb 413 competed with ^{125}I-CC for A2 binding. IgG of the other inhibitors tested (EM, WC, JM, NS, and FM) also competed for ^{125}I-CC binding (not shown). This assay was specific since unlabeled CC was a competitor of CC binding but MP, which does not bind to A2[6], had no effect. When MAb 413 was similarly labeled and used for competition assays, its binding to A2 was prevented by inhibitors FM, CC, and RC (not shown).

Possible epitope overlap within the C2 domain was tested by competition of biotinylated MAb ESH8 by unlabeled inhibitor IgG for F.VIII binding by ELISA. According to the immunoblotting results these epitopes overlapped within residues 2248-2285 (see Table 2). As shown in Figure 2, the six unlabeled inhibitors tested (MU, CC, F, MR, RI, MP) competed for biotin-ESH8 binding as did unlabeled ESH8 (not shown), whereas inhibitor RC, which does not bind to C2, did not. The effect of Fab' fragments was not tested.

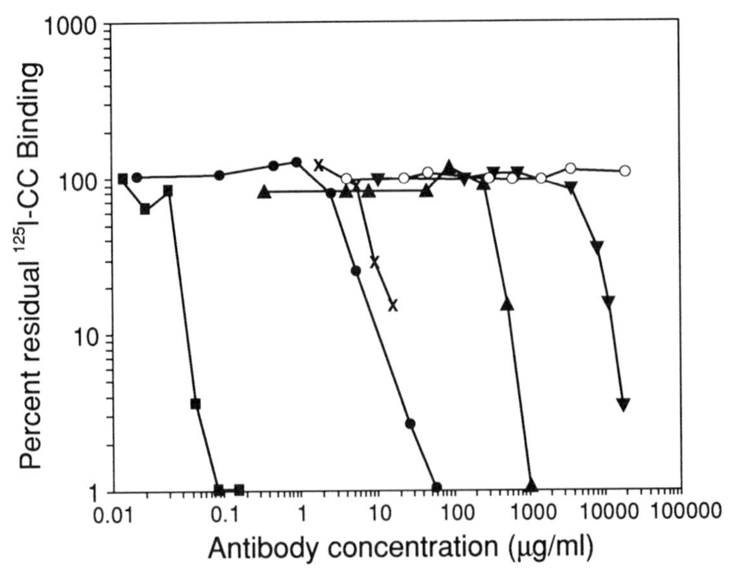

Figure 1. Competition of inhibitor Fab' with ^{125}I-inhibitor CC for binding to the A2 domain. Fab' fragments were used for all inhibitors except MAb 413 for which IgG was used. The figure was modified from Scandella et al.[6] Symbols represent the following inhibitors left to right: MAb 413 (-■-), CC (-•-), RC (-x-), GK1831 (-▲-), RI (-▼-), and MP (-0-).

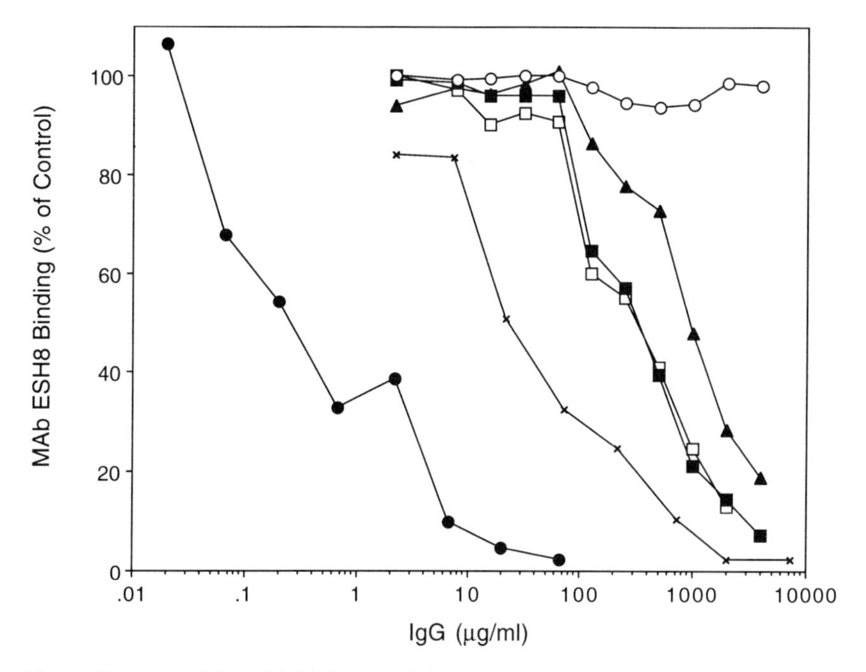

Figure 2. Competition of inhibitors with biotinylated MAb ESH8 for binding to recombinant F.VIII. The data were calculated as for Figure 1. IgG from each inhibitor plasma was used. The symbols represent inhibitors MU (-•-), F (-x-), MC (-□-), RI (-■-), MP (-▲-), and RC (-0-).

For both the A2 and C2 domains, the inhibitor competition data suggest that all of the inhibitors we have tested recognize overlapping or very closely spaced epitopes.

Epitope localization by immunoprecipitation assays

The cDNA encoding the F.VIII A2 or C2 domains or the light chain was modified to contain a heterologous signal peptide so that these polypeptides might be secreted from the host cell. A2 was expressed from a baculovirus vector in Sf9 insect cells, and C2 and the light chain were expressed in COS-1 cells following transfection. Each of the F.VIII polypeptides was secreted.[6,8]

After intrinsic labeling of these F.VIII domains with ^{35}S-methionine, the growth medium was used for further immunoprecipitation analysis. An example of the binding of 100 μl of various hemophilic inhibitor plasmas with the A2 domain is shown in Figure 3, lanes 1-7. A negative control of a normal plasma did not bind to A2 (lane 8). The positive control was MAb 413 IgG (lane 9) which has an epitope within amino acid residues 373-606 of A2.

The epitope binding characteristics were determined by immunoprecipitation assay for the same 28 plasmas studied by immunoblotting. Although the binding of different plasmas to A2 varied considerably (as shown by the different band intensities, Figure 3), and the same results were obtained for C2 binding (not shown), the following summary of the epitope specificity determined by this method lists results only as positive or negative. (See Table 3.)

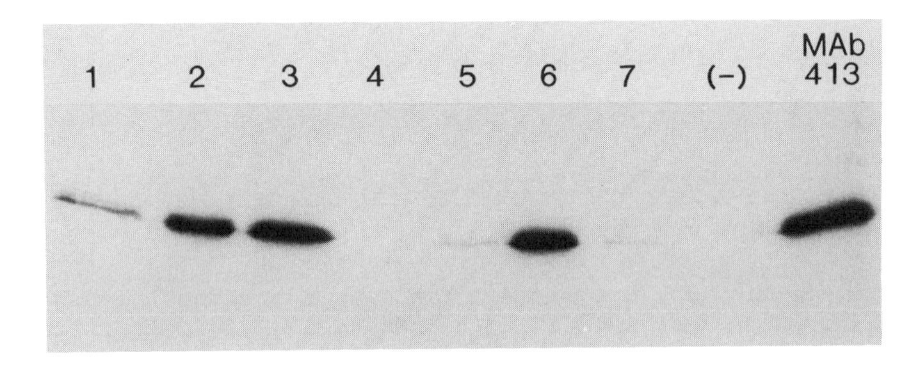

Figure 3. Immunoprecipitation assay of inhibitor binding to the F.VIII A2 domain. Patient plasmas are shown in lanes 1-7. Lane 8 contains a normal plasma, and lane 9 is the positive control, MAb 413 anti-A2 monoclonal antibody.

Table 3. Inhibitor epitope specificity determined by immunoprecipitation

Inhibitors	Number Binding to F.VIII Domains			
	A2	C2	A2 + C2	Total
Hemophilic	1	2	10	13
Spontaneous	2	4	9	15

Inhibitors binding to each region: hemophilic, A2 (RM), C2 (MP, L), A2 + C2 (CHA, WD, CC, WG, JR, GK1831, GK1824, RI, MU, HG); spontaneous, A2 (RC, FM), C2 (D, SLC, MR, MSI1801), A2 + C2 (JM, SC, EM, NS, CHI, DP, F, WT, WC).

Compared to the epitope specificity determined by immunoblotting in which 18% of the 28 inhibitor patients tested had antibodies that bound to A2 and C2, the immunoprecipitation results show that 68% had both types of antibodies. Both anti-A2 and anti-C2 antibodies were detected with higher frequency by the immunoprecipitation assay. Such antibodies may be detected by immunoprecipitation because it is a more sensitive assay or because some antibodies cannot bind to the denatured, immobilized F.VIII fragments used for immunoblotting. The latter was shown for inhibitor RI[6]. As seen for immunoblotting, the epitope specificity of hemophilic and spontaneous inhibitors with regard to the A2 and C2 domains was similar in the immunoprecipitation assay.

Neutralization of inhibitors by the A2 or C2 domains or the light chain

The most clinically relevant antibodies are likely to be those that bind to F.VIII in a way that disrupts its function. The possible effect of antibodies that bind to sites not involved in F.VIII function is not presently known. In order to distinguish functional inhibitors from other antibodies, we have developed inhibitor neutralization assays. A constant concentration of inhibitor was incubated with increasing concentrations of soluble F.VIII domain. After 2 hours incubation at 37°C, the remaining inhibitor titer was measured in the Bethesda assay. The percentage neutralization was calculated from a control with no F.VIII polypeptide. Figures 4 and 5 show results for neutralization of three inhibitors by the A2 domain and the light chain, and the results for the remainder of the 20 inhibitors tested are summarized in tables 4, 5, 6, and 7. In the tables, the inhibitors are divided into groups based on their reactivity with F.VIII by immunoblotting.

Table 4. Neutralization of inhibitors that bind only to the A2 domain by immunoblotting

Name	Type	Binding to F.VIII Domain by Immunoprecipitation			% Neutralization by		
		A2	C2	LC	A2	LC	HC
CHA	HA	+	-	-	>90	<10	
RM	HA	+	-	-	>90	<10	
RC	HA	+	-	-	>90	<10	
JM	SP	+	+(w)	+	>90	<10	
SC	SP	+	+(w)	+	42	46	
FM	SP	+	-	+	70	<10	70
EM	SP	+	+	+	>90	<10	
NS	SP	+	+(w)	+	>90	<10	

The symbol +(w) means barely detectable but distinguishable from background control of single donor plasma. HA, hemophilia A; SP, spontaneous inhibitors. Tests not done are represented by a blank space. This table was modified from Scandella et al.[8]

Inhibitors RC and NS that bound to the A2 domain by immunoblotting and immunoprecipitation assays were >90% neutralized by A2 but not by the light chain (<10%, Figure 4A, Table 4). However, inhibitor MP which did not bind to A2 in these assays was neutralized by the light chain (>90%, Figure 4B) but not by A2 (<10%, Figure

4A). In contrast, inhibitor SC was partially neutralized by A2 and by the light chain (Figure 5A) and completely neutralized when both of them were added simultaneously. This inhibitor bound only weakly to C2 by immunoprecipitation, suggesting that part of its inhibitor binds to an epitope within A3 or C1 of the light chain. Inhibitor FM was partially neutralized by A2 (see Figure 5B) and by the heavy chain (see Table 4) but not by the light chain. The reasons for the lack of complete neutralization of FM by any fragment or combination of fragments is not known.

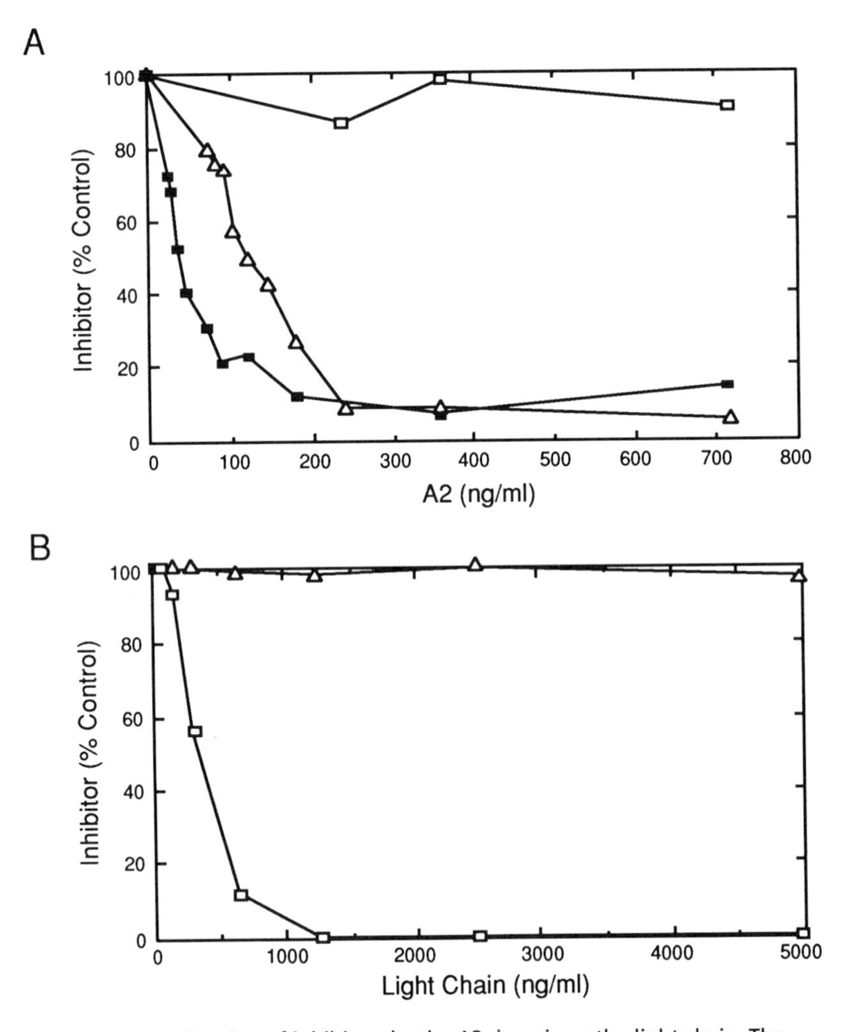

Figure 4. Neutralization of inhibitors by the A2 domain or the light chain. The residual inhibitor titer of samples that were incubated with increasing concentrations of A2 or light chain is expressed as a percentage of the inhibitor titer in the absence of A2. (A) Assay of inhibitors MP (-□-), RC (-Δ-), and NS (-■-) with the A2 domain. (B) Assay of inhibitors MP (-□-) and RC (-Δ-) with the light chain. This figure was reproduced from Scandella et al.[8] by permission of W.B. Saunders.

Among the plasmas from spontaneous inhibitors with antibodies that bound to A2 as well as to C2 by immunoprecipitation, several (JM, SC, FM, EM, NS) showed very weak binding to C2 (similar to lane 7, Figure 3) relative to a MAb control but binding to the light chain equal to that of the control. In order to determine the relative concentration of anti-light antibody within the total IgG, the binding of these antibodies to purified ^{125}I-labeled light chain was measured in a quantitative immunoprecipitation assay, and the results are shown in Figure 6.

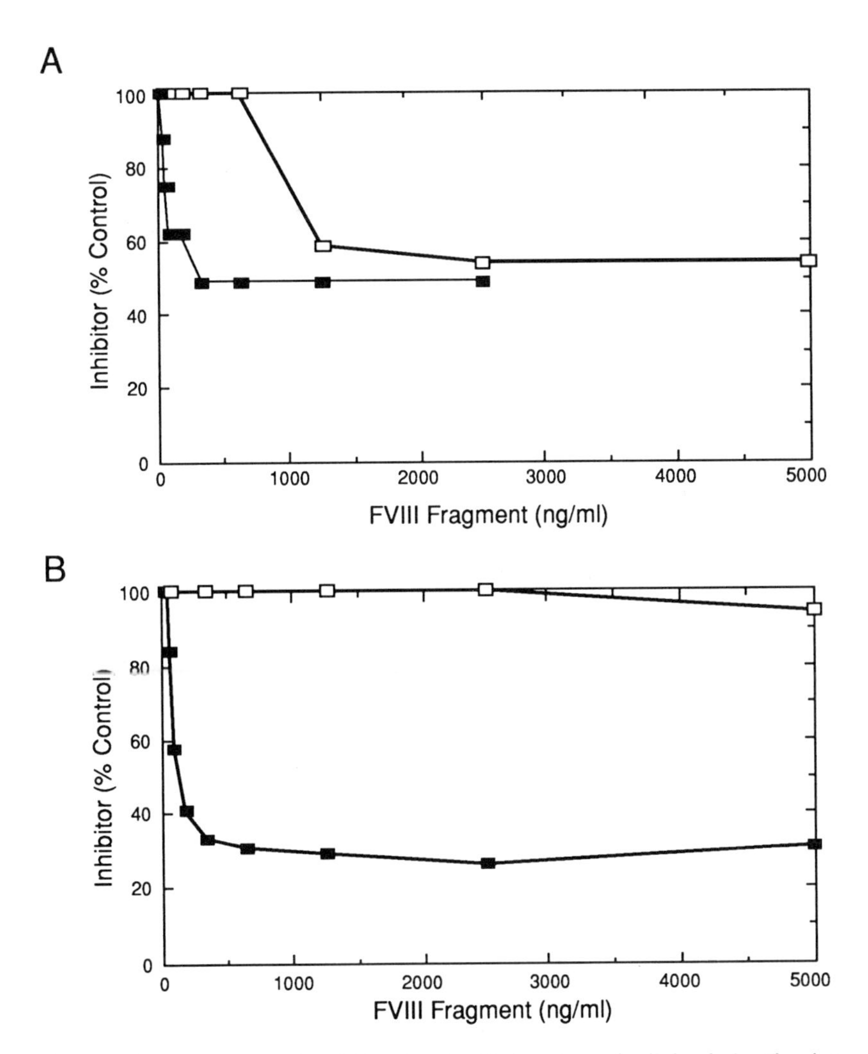

Figure 5. Neutralization of inhibitors by the A2 domain or the light chain. The data were calculated as described in Figure 4. Assay of inhibitors SC (A) or FM (B) with the A2 domain (-■-) or the light chain (-□-). This figure was reproduced from Scandella et al.[8] by permission of W.B. Saunders.

There was up to a 4 log difference in binding among the inhibitors tested although the curves showed similar slopes, suggesting that the antibodies bound with similar affinity. The differences in light chain binding within the total inhibitor IgG would thus appear to reflect differences in anti-light chain antibody concentration. Two inhibitors with anti-C2 antibodies but very low or no anti-A2 antibodies, MU and MP, showed 50% maximal binding, respectively, at 0.08 μg/ml and 3 μg/ml total IgG. Inhibitors FM, SC, and NS had corresponding values of 0.5, 9, and 10 mg/ml although all of them bound minimally to the C2 domain by immunoprecipitation. These results indicated that inhibitors with anti-A2 antibodies may also contain significant concentrations of anti-light chain antibodies, which are not primarily anti-C2. In the case of inhibitor SC, we have shown that these anti-light chain antibodies have F.VIII inhibitory activity, and they thus appear to define a new inhibitor epitope. SC is a spontaneous inhibitor, but we have also detected such anti-light chain antibodies in a hemophilic inhibitor (shown below). Others such as FM may have anti-light chain antibodies which are not inhibitory since there was no neutralization of the plasma by the light chain.

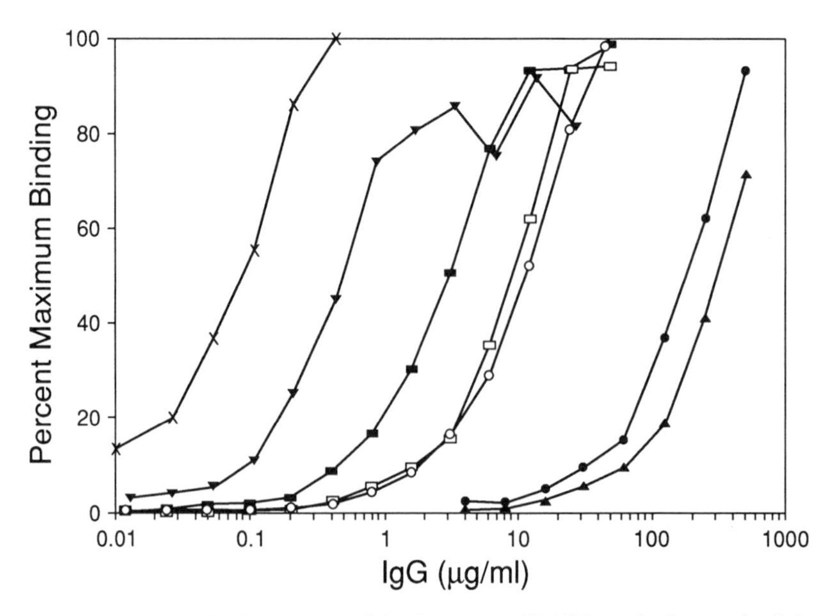

Figure 6. Quantitative immunoprecipitation assay of inhibitor binding to the light chain. The figure was modified from Scandella et al.[8]. The amount of bound ^{125}I-light chain is expressed as a percentage of that for a control containing excess anti-light chain monoclonal antibody ESH 8 (10 μg/μl). Symbols left to right: MU (-x-), FM (-▼-), MP (-■-), SC (-□-), NS (-0-), EM (-•-), and JM (-▲-).

Tables 5-7 summarize neutralization data for the remaining inhibitors tested. The concentration of soluble C2 in the COS cell extract, 30 ng/ml, is not high enough for neutralization assays; therefore, these inhibitors have been neutralized only with A2.

Table 5. Characteristics of an inhibitor that binds to A2 and the region between A1 and A2 by immunoblotting

Name	Type	Binding to F.VIII domain by immunoprecipitation A2	C2	% Neutralization by A2	Peptide 341-63
WD	HA	+	+	40	60

The GST-C2 fusion protein gave variable neutralization results for MU, MP, and MR although these inhibitors were completely neutralized by the light chain (not shown). The 80% neutralization of MR by GST-C2 shows directly that some anti-C2 antibodies are inhibitors. The differences in neutralization of MR, MP, and MU, which all bound to soluble C2 by immunoprecipitation assay, could be due to the presence of as yet unidentified inhibitors in addition to the anti-C2 antibodies. Alternatively, these antibodies may each bind to different residues within the epitope region, not all of which are similarly displayed in the structure of the fusion protein. GST-C2 was thus not used further.

Table 6. Characteristics of inhibitors which bind only to C2 by immunoblotting

Name	Type	Binding to F.VIII domain by immunoprecipitation		% Neutralization by	
		A2	C2	A2	GST-C2
MP	HA	-	+	<10	<10
RI	HA	+	+	<10	
MU	HA	+(w)	+	<10	50
F	SP	+(w)	+	<10	
MR	SP	-	+	<10	80
WC	SP	+	+	<10	

Table 7. Characteristics of inhibitors that bind to A2 and to C2 by immunoblotting

Name	Type	Binding to F.VIII domain by immunoprecipitation		% Neutralization by
		A2	C2	A2
WG	HA	+	+	17
CC	HA	+	+	55
JR	HA	+	+	<10
GK1831	HA	+	+	<10
CHI	SP	+	+	<10

Among the plasmas tested for inhibitor neutralization only 2 (WD and SC) of 10, which had sufficient data to give definitive results, contained 2 different inhibitors which happened to be present at roughly equal concentrations. The one predominant inhibitor in the others was either an anti-A2 (CHA, RM, RC, JM, FM, EM, NS) or an anti-C2 antibody (MR).

Epitope switching and transient inhibitors

Fulcher et al.[19] demonstrated that during long-term F.VIII replacement therapy, the epitope specificity of about 50% of inhibitors (6 of 14), as measured by immunoblotting, can change in response to, or independent of, recent F.VIII infusion. Using immunoprecipitation assays with the A2 and C2 domains and the light chain, we have studied serial plasma samples from two infants, RU2214 and RU1911, treated only with the recombinant F.VIII product Recombinate. The autoradiograms were scanned with a laser densitometer so that a semi-quantitative evaluation could be made of inhibitor binding relative to controls of a saturating concentration of a MAb.

The 3-16-92 sample of RU2214 was the first in which a measurable inhibitor of 4 BU/ml was detected. There was no antibody binding to either the A2 or C2 domains by immunoprecipitation, but there was binding to the light chain. This light chain epitope specificity resembled that of inhibitor SC (see Table 4), and it demonstrated that both hemophilic and spontaneous inhibitors may have light chain inhibitor epitopes that are not within the C2 domain. It is not yet known if these light chain epitopes in the two patients are identical. The presence of the "new" light chain epitope has yet to be directly confirmed. The F.VIII A3 and C1 domains are poorly expressed in heterologous mammalian cells, thus they are not available for immunoprecipitation assays. Six weeks later RU2214 antibodies which bind to A2 and C2 were evident, and by September 1992, the inhibitor titer had dropped to zero.

Plasmas from patient RU7611 treated only with Recombinate were tested in a similar manner. Inhibitors first appeared in September 1990, and the four samples assayed by immunoprecipitation were from February 1991 to January 1992. During this time period the concentration of anti-A2 antibodies rose slowly from 46-94% of the control in parallel with the inhibitor titer (5-33 BU/ml), whereas the anti-C2 antibodies were high in the first three time points (132-165% control), and they dropped to very low levels (7%) by the fourth sample. In April 1991, the anti-A2 and anti-C2 antibody levels were 32% and 177%, respectively, by immunoprecipitation assay, but the inhibitor activity was not neutralized by A2 (10%), presumably because it could be accounted for by the anti-C2 antibodies. In January 1992, the anti-A2 and anti-C2 antibody levels had changed to 95% and 7%, respectively, and the anti-A2 antibodies made up 36% of the total inhibitor activity. The finding that the anti-A2 antibody was not responsible for 100% of the later inhibitor titer could mean that anti-C2 antibodies are more inhibitory to F.VIII activity on a molar basis or that the antibody quantitation by immunoprecipitation is not precise enough.

Several inhibitor patients among a group of hemophiliacs treated with either Recombinate (RU) or Kogenate (K) recombinant F.VIII preparations lost their inhibitor titers spontaneously or after undergoing treatment for immune tolerance induction. Samples from these patients were tested for epitope specificity to determine if the loss of inhibitor titer corresponded to the loss of anti-F.VIII antibodies, as expected. The plasmas from patients who lost their inhibitor spontaneously contained very low levels (RU2112) or no (K5067061 and K5183084) anti-F.VIII antibodies. However, the patients treated in an immune tolerance induction protocol (K561070 and K5271110) both had significant levels of anti-F.VIII antibodies. The F.VIII patient K5271110 showed normal F.VIII recovery up to 4 hours but the 24 hour sample had 50% of the expected F.VIII level. F.VIII recovery for K561070 was normal.

Discussion

We have used soluble, radiolabeled F.VIII domains in immunoprecipitation assays to demonstrate that 75% of 28 patients with inhibitors have multiple anti-F.VIII antibodies. This assay is more sensitive and detects a larger spectrum of antibodies than immunoblotting assays. The predominant epitopes are located within the A2 and C2 domains, and rare epitopes are located in the region between the A1 and A2 domains and at unspecificied locations within the A3 or C1 domains.

Although these assays yielded only semi-quantitative data, it was clear that the levels of anti-A2 or anti-C2 antibodies appeared to vary widely within an inhibitor plasma or among different plasmas. A quantitative immunoprecipitation assay with ^{125}I-labeled highly purified light chain, however, allows the determination of the concentration of total IgG from an inhibitor plasma that is required to precipitate 50% of a standard amount of the ^{125}I-light chain. A 10,000 fold range of anti-light chain antibody concentrations were detected among different inhibitors in this assay. The use of other purified F.VIII domains or fragments in a similar assay should make it possible to determine the concentrations of the full range of anti-F.VIII antibodies that are present in inhibitor plasmas.

Antibody binding assays such as immunoblotting or immunoprecipitation are valuable in determining the antibody spectrum within a plasma. However, identification of the clinical relevance of specific antibodies requires an assay that detects those antibodies with F.VIII inhibitor activity. To this end, we have used the same soluble F.VIII fragments described above to neutralize inhibitors that bind to a particular region of F.VIII. For those plasmas that have been assayed with several fragments, 8 of 10 were fully neutralized by only one of them, and 2 were neutralized only with a combination of 2 fragments. This result was surprising because most of these plasmas contain more than one type of anti-F.VIII antibody.

The epitope specificity of patients treated over long periods of time with F.VIII may change unpredictably, as demonstrated by Fulcher et al.[19] We have shown in this study that this is also true for hemophilic infants who had been treated less than six months with F.VIII. In one patient, RU2214, the initial inhibitor containing plasma showed binding to the light chain but not to the C2 domain, suggesting the presence of a new rare epitope. In another sample six months later, the plasma contained antibodies which bound to the A2 and C2 domains. A second patient, RU7611, initially had low levels of anti-A2 antibody and high levels of anti-C2 antibody, but 11 months later the proportions of these antibodies were reversed. When the two samples were tested for neutralization by the A2 domain, the former was not neutralized (<10%), and the latter was 36% neutralized. This demonstrates that epitope specificity is changing not only in the binding assays but also at the functional level. These results point out that the epitope specificity for an inhibitor patient must be related to a particular plasma sample and it is not a permanent classification. Individuals may switch from anti-A2 to anti-C2 inhibitor specificity or vice versa or to another of the rare specificities (between A1 and A2 or within A3 or C1).

Three patients who lost their inhibitor titer spontaneously no longer had antibodies which bound to F.VIII domains by immunoprecipitation. However, two patients who were treated in an immune tolerance induction protocol to eradicate their inhibitor still had anti-C2 antibodies and one of them had anti-A2 antibodies as well. Previously, Nilsson et al.[20] described the appearance of new non-inhibitory antibodies after tolerance induction. The significance of such antibodies is presently not understood. In addition, the mechanisms of spontaneous or induced loss of inhibitor antibodies remain to be determined.

References

1. S. Ehrenforth, W. Kreuz, I. Scharrer, R. Linde, M. Funk, T. Güngör, B. Krackhardt, and B. Kornhuber, Incidence of development of factor VIII and factor IX inhibitors in haemophiliacs, Lancet 339:594-598 (1992).
2. C.A. Fulcher, S. de Graaf Mahoney, J.R. Roberts, C.K. Kasper, and T.S. Zimmerman, Localization of human factor fVIII inhibitor epitopes to two polypeptide fragments, Proc. Natl. Acad. Sci. USA 82:7728-7732 (1985).
3. C.A. Fulcher, S. de Graaf Mahoney, and T.S. Zimmerman, FVIII inhibitor IgG subclass and FVIII polypeptide specificity determined by immunoblotting, Blood 69:1475-1480 (1987).
4. D. Scandella, M. Mattingly, S. de Graaf, and C.A. Fulcher, Localization of epitopes for human factor VIII inhibitor antibodies by immunoblotting and antibody neutralization, Blood 74:1618-1626 (1989).
5. J. Ware, M. Lo, S. de Graaf, and C. Fulcher, Epitope mapping of human factor VIII inhibitor antibodies by site-directed mutagenesis of a factor VIII polypeptide, Blood Coag. Fibrinol. 3:703-716 (1992).
6. D. Scandella, L. Timmons, M. Mattingly, N. Trabold, and L.W. Hoyer, A soluble recombinant factor VIII fragment containing the A2 domain binds to some human anti-factor VIII antibodies that are not detected by immunoblotting, Thromb. Haemostas. 67:665-671 (1992).
7. M. Steinbuch and R. Audran, The isolation of IgG from mammalian sera with the aid of caprylic acid, Arch. Biochem. Biophys. 134:279-284 (1969).
8. D. Scandella, M. Mattingly, and R. Prescott, A recombinant factor VIII A2 domain polypeptide quantitatively neutralizes human inhibitor antibodies which bind to A2, Blood 82:1767-1775 (1993).
9. C.K. Kasper, L.M. Aledort, R.B. Counts, J.R. Edson, J. Fratantoni, D. Green, J.W. Hampton, M.W. Hilgartner, J. Lazerson, P.H. Levine, C.W. McMillan, J.G. Pool, S.S. Shapiro, N.R. Shulman, and J. van Eys, A more uniform measurement of factor VIII inhibitors, Thrombos. Diathes. haemorrh. 34:869-872 (1975).
10. D. Scandella, S. deGraaf Mahoney, M. Mattingly, D. Roeder, L. Timmons, and C.A. Fulcher, Epitope mapping of human factor VIII inhibitor antibodies by deletion analysis of factor VIII fragments expressed in Escherichia coli, Proc. Natl. Acad. Sci. USA 85:6152-6156 (1988).
11. M. Shima, D. Scandella, A. Yoshioka, H. Nakai, I. Tanaka, S. Kamisue, S. Terada, and H. Fukui, A factor VIII neutralizing monoclonal antibody and a human inhibitor alloantibody recognizing epitopes in the C2 domain inhibit factor VIII binding to von Willebrand factor and to phosphatidylserine, Thromb. Haemostas. 69:240-246 (1993.)
12. D.B. Smith and K.S. Johnson, Single-step purification of polypeptides expressed in Esherichia coli as fusions with glutathione S-transferase, Gene 67:31-40 (1988).
13. P. Lollar, D.C. Hill-Eubanks, and C.G. Parker, Association of the factor VIII light chain with von Willebrand factor, J. Biol. Chem. 263:10451-10455 (1988).
14. P.J. Fay and T.M. Smudzin, Intersubunit fluorescence energy transfer in human factor VIII, J. Biol. Chem. 264:14005-14010 (1989).
15. J. Gitschier, W.I. Wood, M.A. Shuman, and R.M. Lawn, Identification of a missense mutation in the factor VIII gene of a mild hemophiliac, Science 232:1415-1416 (1986).
16. M. Arai, M. Higuchi, S.E. Antonarakis, H.H.Jr. Kazazian, J.A.III Phillips, R.L. Janco, and L.W. Hoyer, Characterization of a thrombin cleavage site mutation (Arg 1689 to Cys) in the factor VIII gene of two unrelated patients with cross-reacting material-positive hemophilia A, Blood 75:384-389 (1990).
17. E.A. Padlan and E.A. Kabat, Modeling of antibody combining sites, Methods Enzymol. 203:3-21 (1991).
18. P.A. Foster, C.A. Fulcher, R.A. Houghten, S. de Graaf Mahoney, and T.S. Zimmerman, A murine monoclonal anti-factor VIII inhibitory antibody and two human factor VIII inhibitors bind to different areas within a twenty amino acid segment of the acidic region of factor VIII heavy chain, Blood Coag. Fibrinol. 1:9-15 (1990).
19. C.A. Fulcher, K. Lechner, and S. de G.Mahoney, Immunoblot analysis shows changes in factor VIII inhibitor chain specificity in factor VIII inhibitor patients over time, Blood 72:1348-1356 (1988).
20. I.M. Nilsson, E. Berntorp, O. Zettervall, and B. Dahlback, Noncoagulation inhibitory factor VIII antibodies after induction of tolerance to factor VIII in hemophilia A patients, Blood 75:378-383 (1990).

Immunogenetics of the human immune response to factor VIII

Howard M. Reisner,[1] Adella Clark,[1] and Linda Levin [1,2]

[1]Department of Pathology, School of Medicine
[2]Department of Endodontics, School of Dentistry
University of North Carolina at Chapel Hill
Chapel Hill, North Carolina 27599-7525

The anti-Factor VIII (F.VIII) antibodies produced by hemophilic patients have the potential to be directed against many different epitopes on the F.VIII molecule. Although first recognized as inhibitors of coagulation factor activity, these antibodies may or may not interfere with F.VIII bioactivity.[1,2] Indeed, as has been suggested by Nilsson and co-workers, antibodies without inhibitory reactivity may play a critical role in the generation and/or maintenance of tolerance in hemophiliacs.[3] At the same time, many individuals with severe hemophilia A may never develop inhibiting antibodies (inhibitors), even after hundreds of days of replacement therapy. What factors are important in determining which individuals with hemophilia A will have an immune response to F.VIII and become refractory to continued replacement therapy? To date, there are no clear answers. It is clear that inherited factors both at the F.VIII locus and at other loci, possibly associated with the immune response, play a role in inhibitor production. This brief review will try to summarize some of the more recent information on how 1) the structure of the F.VIII protein, 2) the nature of F.VIII mutation that results in hemophilia A, 3) additional genetic factors (at the HLA complex locus) may affect the probability of an individual with severe hemophilia (F.VIII:C >1%) developing antibodies to F.VIII, and 4) how the clonality of the immune response to F.VIII varies with time.

Certain localized factors in a protein's tertiary structure can affect the probability that the region will be antigenic in the humoral immune response, i.e., be recognized as a B-cell epitope. Hence, variation in the sequence and/or conformation of the F.VIII might be expected to change the epitopic structure of the molecule. The determination of antigenicity based on primary sequence data, which is both controversial and complex, has been reviewed in some detail during this meeting. (See "B-

Inhibitors to Coagulation Factors
Edited by Louis M. Aledort *et al.*, Plenum Press, New York, 1995

Cell Epitopes: Fact and Fiction," D.C. Benjamin). Primary sequences that are likely to be found on the surface of proteins, which also contain certain other presumptive structural features, are often described in the literature as potential antigenic areas. However, this contention is based only on a priori assumptions or statistical analysis of primary sequence data.[4,5] Computer programs that generate such "antigenicity maps" are widely available and frequently used (see Figure 1[6]), but great skepticism has been voiced about the value of such predictive techniques.

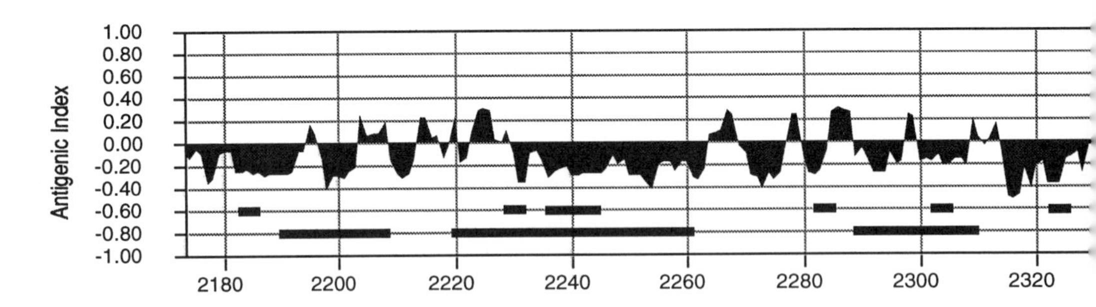

Figure 1. F.VIII C2 Domain: T & B Cell Epitopes. B cell epitopes determined as in (6). T cell epitopes determined as by (7) in 1; as by (8) in 2.

Certain areas of the F.VIII molecule may indeed be "hot spots" containing commonly recognized sets of epitopes such as regions in the A2 and C2 domains discussed in detail in this volume ("Epitope Specificity and Functional Characterization of Factor VIII Inhibitors," D. Scandella and see also references 9-12). The role of structural features intrinsic to the F.VIII protein in generating these "hot spots" of response is still unclear. One epitope on F.VIII detected by inhibitors (albeit rarely) is associated with residues 338-362 of the acidic 1 inter-domain.[13] This region is highly hydrophilic with a high antigenic index (see Figure 2) and hence, would have been predicted to be antigenic.

However, the previously discussed potential B-cell epitopes tell only part of the story. The human immune response to F.VIII has many of the hallmarks of a T-cell dependent response. IgM antibodies are rarely if ever observed, the IgG4 isotype is common in the inhibitor response (suggesting that class switching in response to chronic challenge has occurred), and strong anamnesis in response to repeated challenge is often seen in patients.[14] Because helper T-cell responses are likely to be important in the response to F.VIII, recognition of epitopes presented in concert with MHC class II antigens by TCRs must also be considered. Structural correlates of T-cell epitopes have also been clarified and related to primary amino acid sequence either by a priori statistical methods[7] or by algorithms that predict amphipathic helices.[8] Figures 1 and 2 also indicate areas of F.VIII sequence that are reasonable candidates as T-cell epitopes. The presence of a likely T-cell epitope adjacent to the B-cell epitope in the acidic 1 inter-domain may be significant in the immune response to this region (as will be discussed later).

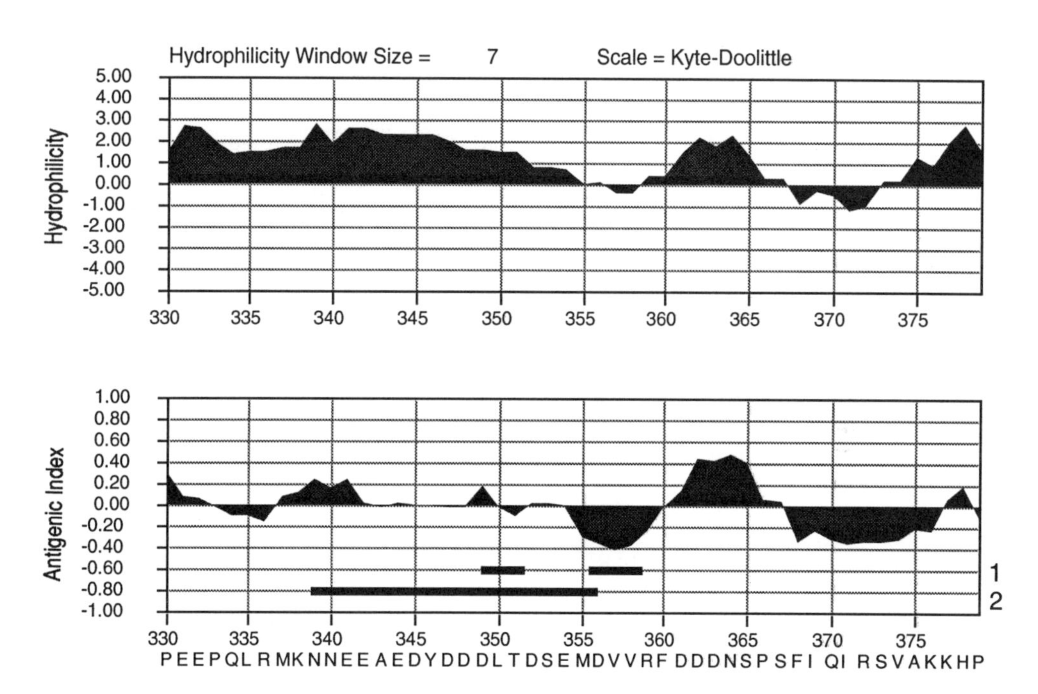

Figure 2. F.VIII Acidic 1 Inter-domain: T & B Cell Epitopes. B cell epitopes determined as in (6). T cell epitopes determined as by (7) in 1; as by (8) in 2.

Excluding possible unidentified sequence polymorphisms, the F.VIII molecule is essentially the same in all individuals. Even if sequence polymorphisms do occur, hemophilic patients have been commonly treated with F.VIII derived from pools of thousands of donors—prior to the use of recombinant protein (rF.VIII). Hence, differences in the structure of the infused F.VIII molecules are unlikely to account for differences in the inhibitor response from patient to patient. The case may be different in previously untreated patients (PUPS) given rF.VIII. An apparent racial difference in the incidence of inhibitors in white and black PUPs could be explained by inherited sequence variation in the F.VIII molecule (reference 15 and unpublished observations) in different racial groups.

Changes in the F.VIII molecule during viral inactivation procedures are capable of rendering therapeutic concentrates uniquely antigenic but this is likely the result of the generation of conformational neoantigens by inadvertent protein denaturation.[16] There is concern that recombinant rF.VIII therapy may result in an increased frequency of inhibitors (reviewed in "The Incidence of Factor VIII Inhibitors," L.

Hoyer, this volume). Should such be the case, and it is far from certain, subtle differences in the structure of rF.VIII, related to glycosylation or other factors,[17] and the potential for sequence polymorphism mentioned above might alter the nature of the immune response to the protein.

Since differences in the structure of infused F.VIII are unlikely to be the major factor in determining patient immune response, a genetic factor or factors intrinsic to the patient, most obviously the nature of the F.VIII mutation resulting in hemophilia, is likely to be responsible. Therefore, an obvious approach is to examine the concordance of related individuals in inhibitor status. A significant tendency toward concordance has been affirmed in brother pair studies, most notably those of Allain and Frommel in France.[18-20] Table 1 presents locally generated data.

Table 1. Distribution of Inhibitors in Hemophilic Brother Pairs

Inhibitor Combination	Observed Occurrence	Expected Occurrence
Inhibitor/inhibitor	6	2.3
Inhibitor/noninhibitor	5	12.3
Noninhibitor/noninhibitor	20	16.3

This population has been treated at the UNC Comprehensive Hemophilia Diagnostic and Treatment Center. Data is from (21).

Clearly the frequency of concordant pairs is unexpectedly high. Why should this be? Hemophilic brother pairs are expected to have the same F.VIII mutation. If certain mutations are associated with inhibitor production, an excess of concordance will be observed in brother pairs. Deletions at the F.IX locus are associated with an increased incidence of inhibitor production to F.IX (see Table 2).

Table 2. Gene Deletions and Inhibitors: Hemophilia A and B

- Hemophilia B
  - Inhibitors Present in F. IX Deletions:
    - 10/15 (67%) Total
    - 5/12 (42%) Partial
    - 56% of Detectable Deletions
    - About 14% of Mutations Are Deletions
- Hemophilia A
  - Inhibitors Present in F. VIII Deletions:
    - 13/33 Detectable Deletions
    - About 1% of Mutations Are Deletions

Data from (22,23)

Point mutations, small additions and deletions that lead to early chain termination, are also associated with inhibitor production (at least at the F.IX locus). About half of F.IX inhibitors not associated with a major deletion occur in individuals with mutations that should lead to truncation of F.IX synthesis within about the first 50 amino acid residues (see Table 3). The question is less clear for F.VIII mutations. Some

Table 3. Point mutations associated with inhibitors in hemophilias A and B

Name	IX:C (%)	IX:Ag (%)	Residue	Mutation
UK 12	<1	?	6	frameshift
Bonn 2	<1	<1	9	frameshift
Madrid 3	<1	<1	9	frameshift
Malmo 4	<1	<1	29	stop
Seattle F	<1	<1	29	stop
HB 7	<1	<1	39	frameshift
HB 5	<1	<1	191	stop (&-739)
Malmo 5	<1	<1	194	stop
HB 6	<1	<1	exon h	acceptor splice
Malmo 3	<1	<1	248	stop
Malmo 1	<1	<1	277	frameshift
Varel	<1	89	365	Ser to Gly (active site)

Name (Number)[1]	VII:C (%)	VIII:Ag (%)	Residue	Mutation
OX 23	<1	?	1696	stop
TWN 93	<1	?	1796	stop
TWN 13	<1	?	1827	stop
JH 51 (2)	<1	?	1922	Asn-Asp
JH 2 (5)	<1	?	1941	stop
Ox 10	<1	?	1966	stop
Ox 14 (3)	14	?	2105	Tyr-Cys
JH 14	<1	<1	2147	stop
H 2 (3)	<1	?	2209	stop
HP 16	7	"CRM+"	2209	Arg-Gln
Anon.	3	?	2229	Trp-Cys

[1] Where more than one inhibitor with an identical mutation is known the number is given.
Data was kindly provided by Drs. High and Atonarakis.

compilations of data confirm an increased incidence of inhibitors in individuals with major gene deletions of F.VIII (although the point remains quite controversial) (see Table 2). Mutations leading to **early** chain termination in F.VIII synthesis do not appear to be associated with excess inhibitor production in hemophilia A (unlike the case with hemophilia B) (see Table 3).

Thirteen of twenty mutations associated with inhibitor production in hemophilia A lead to stops having no obvious association with any region of the molecule. Although the relation of F.VIII mutation to inhibitor production is still in doubt, it is clear that severe hemophilia A is associated with a higher risk of inhibitor formation than mild or moderate forms of the disease.[1,24,25] Even low amounts of circulating F.VIII appear to allow for self-recognition of the protein. Likewise, most inhibitors occur in patients who have no detectable circulating F.VIII antigen (see Table 4). The protection afforded by circulating F.VIII is not absolute. Inhibitors have been seen in individuals with moderate to mild hemophilia A (see Table 3 and 4 for examples). Some of these inhibitors are likely to be related to autoantibodies seen in acquired hemophilia A and are transient (see Table 4). Although the etiology of autoimmune hemophilia A is unrelated to that seen in the genetic form of the disease, the specificity of anti-F.VIII antibodies is similar in both diseases (D. Scandella, this volume).

Table 4. Inhibitors and F.VIII:Ag

	Inhibitor[1]		Inhibitor[2]	
VIII:Ag	Present (%)	Absent	Present (%)	Absent
Present[3]	2 (1.3)	44	3 (2.7)	21
Absent	30 (20.1)	73	19 (17.1)	68

[1] Relationship between presence of inhibitor and level of F.VIII:Ag. F.VIII:Ag was measured using an alloantibody **in the presence of inhibitor**.

[2] F.VIII:Ag status was **inferred in inhibitor patients** as follows: (1) from samples of patient plasma available before inhibitor developed or (2) from non-inhibitor hemophilic members of patient's kindred.

[3] F.VIII:Ag scored as present if equal or greater than 1%.

Brother pairs tend to be concordant in inhibitor response because they are concordant in the disease-causing mutation, but clearly not all brother pairs are concordant. Can loci other than those causing the disease influence the inhibitor response? Considering what we know about the human immune response and the many loci that interact in antigen recognition and response at both the T and B cell level, the answer is undoubtedly yes. Nevertheless, it has proven difficult to demonstrate. Over the last several years, Adella Clark, Emily Reisner, and I have looked for factors at the HLA class I and II loci that might be associated with inhibitor responsiveness. This work is still in progress and is moving into DNA-based analysis of variation. Preliminary findings in which serological typing is exclusively used have proven interesting. There is a highly significant association of DR1 and DR5 with our hemophilic population taken as a whole. Three previous studies have noted either a significant or suggestive increase in the DR5 antigen in hemophilia A populations.[26-28] Our local population of hemophiliacs shows a significant **decrease** in DR5 and a significant **increase** in DR1. Although this finding has little relation to inhibitor development it is quite interesting and difficult to explain. (See Table 5.) These associations, noted independently by several groups, are worthy of some consideration. The association of hemophiliacs as a group with DR5 and DR1 does not appear to involve HIV serological status (unpublished observation).

When inhibitor positive and inhibitor negative patient populations were examined for serological HLA class I and II markers and for immunoglobulin Gm allotypes no significant associations were found. Initial analysis of an incomplete data set did indicate a possible association of inhibitor status and the G1m[1] allotype which was confirmed by blind sample retyping.[29] Addition of all locally available inhibitor patients to the study removed the significant association. Although the two sets of inhibitor patients studied are believed to be homogeneous it is possible that subtle differences do exist–particularly in regard to patient age. This is a point worthy of further study with a larger data set.

Table 5. Selected HLA Associations

HLA Antigen	Association with	Probability
DR1	Hemophilia A (increase)	0.00001[1]
Dr5	Hemophilia A (decrease)	0.00055[2]
Cw5	Antibody Activity[2]	0.028[3]

[1] Significant after correction for multiple comparisons.
[2] Anti-F.VIII heavy and light chain reactivity detected by Western immunoblots.
[3] Suggestive (but not significant after correction for multiple comparisons).
Data is taken from (26).

Reasoning that the multiple epitopes recognized by hemophiliacs could obscure significant associations, we attempted to classify inhibitor patients by the relatively crude measure of which thrombin cleaved F.VIII derived chains reacted with patient antibodies.[26] Hoyer and coworkers had previously found an association of the lack of Cw5 with inhibitor development. To our surprise, the only potentially significant association we could discern was with the presence of either heavy or light chain immunoreactivity with the presence of Cw5 (see Table 5). Although inhibitor status itself was not significantly associated with Cw5, the subgroup of patients with predominantly high inhibitor titer that did react in immunoblotting showed a positive association with this antigen. The probability of two groups finding these associations independently–although seemingly in the opposite direction would appear to be very low indeed.

Because no strong associations were found between inhibitor status or F.VIII chain specificity (as defined by immunoblotting) and HLA or Gm allotype in our population of inhibitor patients, we hypothesized that brother pair studies might be more sensitive in detecting additional genetic factors important in the inhibitor response. In this study, ten brother pairs and one triplet were analyzed for concordance or discordance in the thrombin-derived F.VIII fragments detected. Eight pairs were concordant, two were discordant, and the triplet had a concordant pair and discordant singleton (see Table 6).

Table 6. Pattern of Immunoreactivity to F.VIII Peptides in Sibships

Fragment Specificity (Thrombin Cleaved F.VIII)	Number in Sibship with Given Specificity										
	P1	P2	P3	P4	P5	P6	P7	P8	P9	P10	T11
73,50,42								2	1		
73,43		2	2	2	2	2			1	1	1
73	2						2			1	2

The statistical analysis of this data is quite complex and depends on assumptions as to the true population frequency of patient reactivity with the thrombin-produced fragments. Population estimates done locally or by Fulcher et al. do significantly differ (see Table 6). [21,30] Depending on the exact set of presumptions accepted about the data, the level of concordance in fragment specificity in sibships is significant (P=0.01-0.05)–albeit not strongly so. Considering the many confounding factors in this type of a study and the crude nature of the immunoblotting assay, we feel these results are more than merely suggestive.

If sibships show increased concordance in the area of F.VIII toward which inhibitors are directed, more exact mapping of epitopes combined with family studies are likely to yield stronger associations. For example, only three of 44 inhibitors tested reacted with the 50kd heavy chain derived peptide of F.VIII. Two of these samples represented an uncle-nephew pair. The two samples had identical immunoreactivity and both recognized the 50 and 43kD heavy chain-derived thrombin peptides. Only the uncle-nephew pair (and not the third unrelated patient)

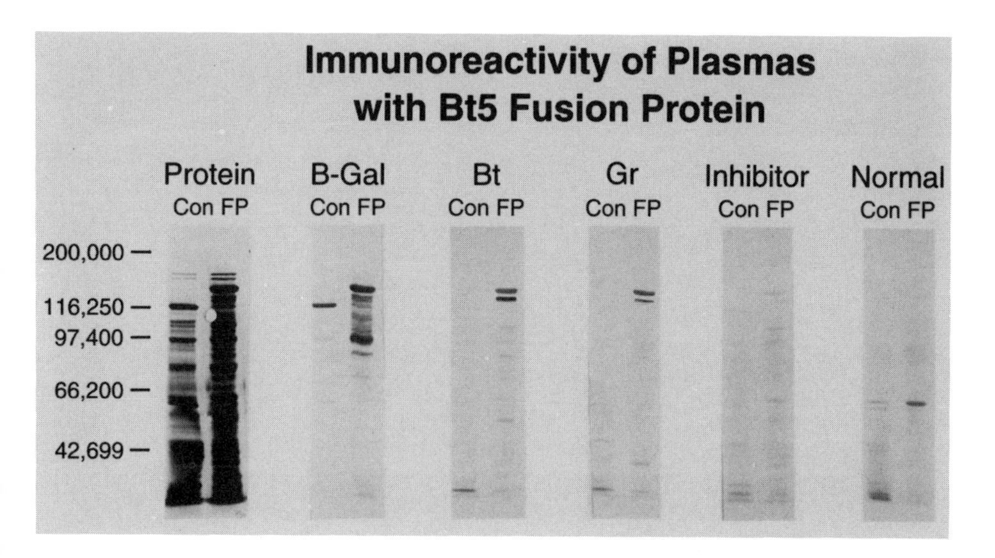

Figure 3. Con = Control *E.coli* lysate, FP = lysate containing Bt5 fusion protein Protein lanes stained for total protein, B-Gal lanes stained with enzyme conjugated antibody to beta galactosidase, other lanes immunostained as indicated. Bt & Gr are sera from uncle, nephew pair. Inhibitor and normal sera show no reaction with fusion protein.

reacted with a fusion protein (ßt5) which defines an epitope subtending 338-362 in the acidic 1 inter-domain (see Figure 3). In Figure 3, only Bt and Gr react with the characteristic doublet pattern indicative of binding to the fusion protein. The additional inhibitor patient and normal sample react with other E. coli derived proteins as these samples have not been pre-adsorbed with bacterial products. Immunostaining for beta-gal in control lysates spiked with beta gal or in lysates of the transformed line demonstrates the higher molecular weight beta gal ßt5 fusion protein as well as what may be proteolytic breakdown products of the fusion protein.

We are intrigued that the ßt5 reactive uncle-nephew pair share the DR6, DRw52, DQw1 haplotype. Inhibitor reactivity with epitopes in the 50kD peptide are very uncommon,[21,30] hence the finding of two related individuals both reacting with the same limited region of this peptide is unlikely to be a chance occurrence. It is recognized that these studies come no where near the exactitude with which particular amino acid variants within class II antigen binding sites have been related to specific auto immune responses, nevertheless we believe they demonstrate the strong potential of looking at the immune response to F.VIII on an epitope by epitope basis. It may well be that the response to the ßt-5 epitope is restricted by a particular set of class II antigens; hence, the inhibitor response to this epitope may depend on HLA class II haplotype.

One criticism that may be voiced of the above studies is that they treat the immune response of hemophiliacs as time invariant. How much variation with time occurs in the clonal nature of the immune response in inhibitor patients? Certainly, chain specificity of inhibitors can vary with time (D. Scandella, this volume and see also reference 31). A second question to be addressed is the complexity of the immune response to F.VIII in terms of the number of B-cell clones that play a role. We have looked at these questions by examining the isoelectrophoretic spectrum of antibodies to F.VIII and comparing these results with thrombin-fragment specificity over time and course of therapy (see Figure 4). Initial studies looked at two patients over about a two year period. Patient 1 exhibited an extraordinary anamnestic response to F.VIII replacement therapy which declined with time. Patient 2 was undergoing successful high dose tolerance induction prior to elective surgery (see Figure 5). In this study, radiolabelled recombinant F.VIII was used to probe blots of isoelectrofocused patient plasma to determine spectrotype pattern. Isoelectric focusing reveals an extremely

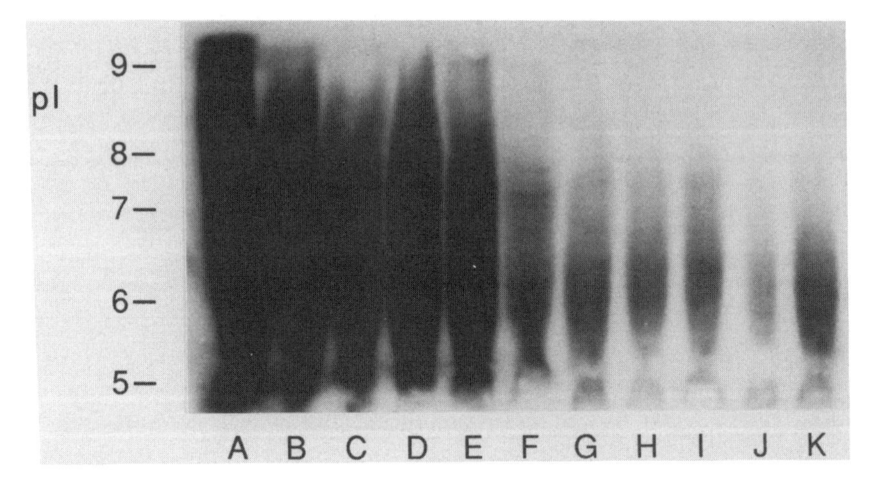

Figure 4. Patient 1: Isoelectric Focusing of Antibodies. Samples A-K span a 120-week time period during which the inhibitor titer in patient 1 fell from 30,000 to 500 Bethesda units.

polyclonal response to F.VIII. Individual clonal components are discernible through the latter part of the response and remain quite constant over about a period of a year (see Figure 4). The patient reacted with the 73kD and 43kD light and heavy chain derived peptides during the

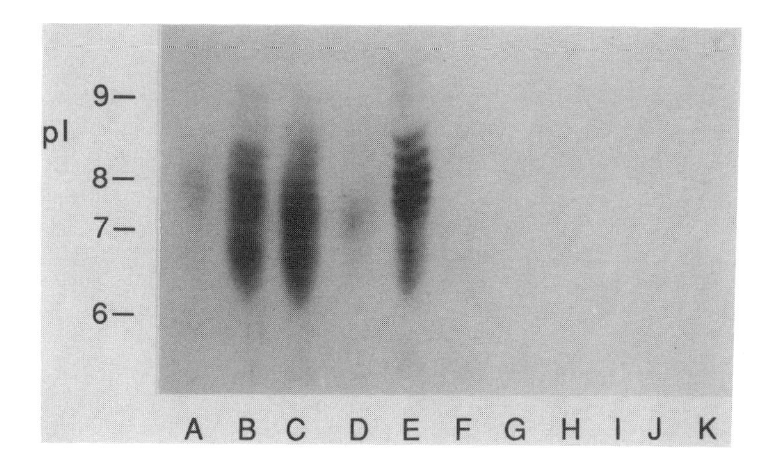

Figure 5. Patient 2: Isoelectric Focusing of Antibodies. Samples A-K span a 100-week time period during which the inhibitor titer reached a peak of 1,000 Bethesda units (E) and was suppressed to undetectable levels. Sample A with a titer of 100 Bethesda units was barely detectable.

period studied. Patient 2 showed a very different pattern of response. During the early period of tolerance induction, only two or three clones of antibody could be detected, a pattern indicative of a clonally restricted response. Spectrotype patterns could not be detected in samples with a Bethesda titer of less than about 200 (see Figure 5).

Neither the spectrotype pattern (detected for about 20 weeks during the course of tolerogenic induction) nor the 73kD fragment specificity varied during therapy. In fact, anti-light chain antibody was detectable even when significant F.VIII coagulant activity could be recovered from the patient. One would be most curious to know the fine pattern of specificity of this antibody during the course of therapy.

The pattern of extreme stability in spectrotype pattern over time is not unusual in patient samples. Dr. Levin has much improved the resolution and sensitivity of the isoelectric focusing technique by using biotinylated F.VIII to probe blots and has examined serial samples in a number of additional patients with lower titers of inhibitor. The modified technique is almost one hundred fold more sensitive than the radioautographic technique used in figures 4 and 5 and appears to have a higher resolution. The first patient studied (LD) had an inhibitor which varied in titer between 10 and 120 Bethesda units over a 26-month time period. The constant pattern of immunoreactive bands is striking (see Figure 6).

The four intense bands in the acidic region of the gel are found in some normals and is likely to represent staining by red cell pseudo-

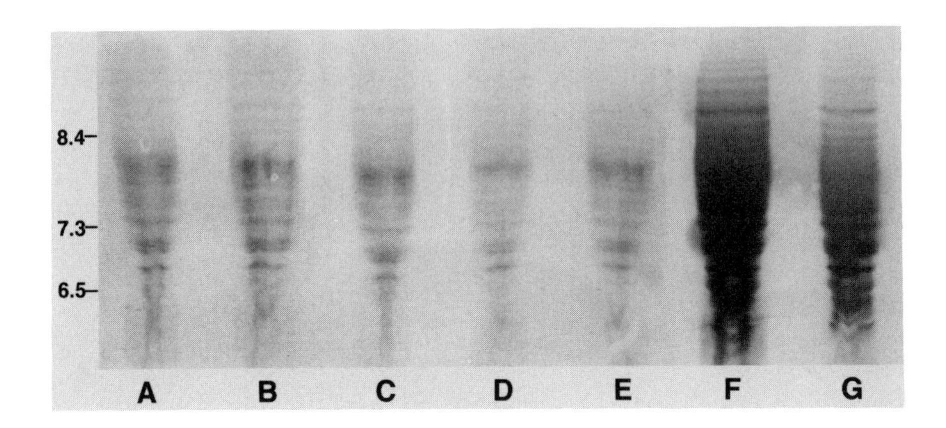

Figure 6. Patient LD: Isoelectric Focusing of Antibodies. Samples A-G span a 26-month period. Samples F & G have inhibitor titers of 120 and 40 Bethesda units. The figure has been computer enhanced to accommodate the wide range of signal.

peroxidase activity in the plasma samples. This will be more clearly seen in Figure 7. The next set of patient samples (DS) spans a 2.5 year period during which the Bethesda titers ranged between 7 and 2 (see Figure 7).

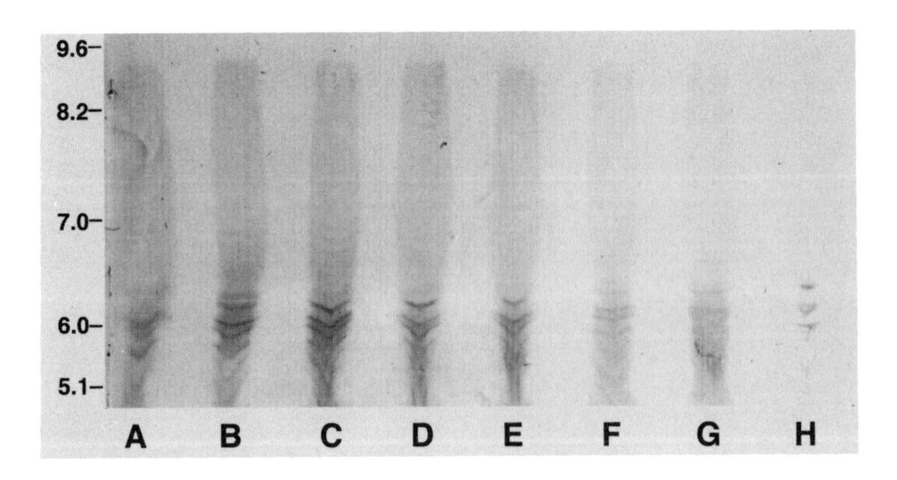

Figure 7. Patient DS: Isoelectric Focusing of Antibodies. Samples A-G span a 30-month period. Sample H is from a single normal individual (see text). The figure has been computer enhanced to demonstrate faint bands of immunoreactivity.

The response is oligoclonal and unvarying in spectrotype pattern. The last sample on this blot is a single normal individual. The dark staining in the acidic region of the gel represents the previously mentioned presumptive psuedo-peroxidase activity, two barely perceptible bands are visible in the more cathodal region. This is reproducible in several samples derived from this individual and possibly represents the anti-F.VIII activity reported to be present in some normal sera by Kazatchkin and others ("V Region-Mediated Interactions Controlling Autoreactivity to Factor VIII," M. D. Kazatchkine and panel discussion, this volume and reference 32). We must emphasize that in our hands this does not appear to be a common finding in normal samples.

We are currently attempting to map the fine epitopic specificity of these and other samples using recombinant molecular techniques. If the conservation of spectrotype pattern is as striking as it appears, we expect little change in specificity with time. Perhaps the immune response in these patients involves predominantly B memory cells that no longer somatically mutate. Certainly we expect the number of epitopes detected by several of the patients we have studied to be limited.

In summary, the factors that control the hemophilic patient's response to F.VIII are likely to be multiple. Changes in the conformation (and potentially the sequence) of therapeutic F.VIII preparations clearly can play a role. More important are genetic factors both at the locus bearing the hemophilia mutation and other loci (notably HLA). Definition of "immune response" loci will require analysis of the antibody response to F.VIII on an epitope by epitope basis as each epitope may well require presentation or processing by different elements. Such epitope by epitope dissection of the immune response to F.VIII by the T- and B-cell repertoire is both feasible and necessary to the definition of the multiple factors which control the inhibitor response.

References

1. Hoyer, L.W. Factor VIII inhibitors: A continuing problem. J.Lab.Clin.Med. 121:385, 1993.
2. Hoyer, L.W. Medical progress: Hemophilia A. N.Engl.J.Med. 330:38, 1994.
3. Nilsson, I.M., Berntorp, E., Zettervall, O. and Dahlbäck, B. Noncoagulation inhibitory factor VIII antibodies after induction of tolerance to factor VIII in hemophilia A patients. Blood 75:378, 1990.
4. Hopp, T.P. Protein surface analysis. Methods for identifying antigenic determinants and other interaction sites. J.Immunol.Meth. 88:1, 1986.
5. Getzoff, E.D., Geysen, H.M., Rodda, S.J., Alexander, H., Tainer, J.A. and Lerner, R.A. Mechanisms of antibody binding to a protein. Science 235:1191, 1987.
6. Scientific Imaging Systems, MacVector, New Haven:Eastman Kodak Company, 1994. Ed. 4th
7. Rothbard, J.B. and Taylor, W.R. A sequence pattern common to T cell epitopes. EMBO Journal 7:93, 1988.
8. Margalit, H., Spouge, J.L., Cornette, J.L., Cease, K.B., DeLisi, C. and Berzofsky, J.A. Prediction of immunodominant helper T cell antigenic sites from the primary sequence. J.Biol.Chem. 138:2213, 1987.
9. Scandella, D., Mattingly, M. and Prescott, R. A recombinant factor VIII A2 domain polypeptide quantitatively neutralizes human inhibitor antibodies that bind to A2. Blood 82:1767, 1993.
10. Shima, M., Scandella, D., Yoshioka, A., et al. A factor VIII neutralizing monoclonal antibody and a human inhibitor alloantibody recognizing epitopes in the C2 domain inhibit factor VIII binding to von Willebrand factor and to phosphatidylserine. Thromb.Haemost. 69:240, 1993.

11. Scandella, D., Timmons, L., Mattingly, M., Trabold, N. and Hoyer, L.W. A soluble recombinant factor VIII fragment containing the A2 domain binds to some human anti-factor VIII antibodies that are not detected by immunoblotting. Thromb.Haemost. 67:665, 1992.
12. Ware, J., MacDonald, M.J., Lo, M., de Graaf, S. and Fulcher, C.A. Epitope mapping of human factor VIII inhibitor antibodies by site-directed mutagenesis of a factor VIII polypeptide. Blood Coag.& Fibrinol. 3:703, 1992.
13. Lubahn, B.C., Ware, J., Stafford, D.W. and Reisner, H.M. Identification of a F.VIII epitope recognized by a human hemophilic inhibitor. Blood 73:497, 1989.
14. Aalberse, R., van der Gaag, R. and van der Leeuwen, J. Serologic aspects of IgG4 antibodies. I. Prolonged immunization results in IgG4 restricted response. J.Immunol. 130:722, 1983.
15. Lusher, J.M., Arkin, S., Abildgaard, C.F., Schwartz, R.S. and Kogenate Prev Untreated Patient Group. Recombinant factor VIII for the treatment of previously untreated patients with hemophilia A. Safety, efficacy, and development of inhibitors. N.Engl.J.Med. 328:453, 1993.
16. Rosendaal, F.R., Nieuwenhuis, H.K., van den Berg, H.M., et al. A sudden increase in factor VIII inhibitor development in multitransfused hemophilia A patients in the Netherlands. Blood 81:2180, 1993.
17. Hironaka, T., Furukawa, K., Esmon, P.C., et al. Structural study of the sugar chains of porcine factor VIII—Tissue- and species-specific glycosylation of factor VIII. Arch.Biochem.Biophys. 307:316, 1993.
18. Frommel, D. and Allain, J.P. Genetic predisposition to develop Factor VIII antibody in classic hemophilia. Clin. Immunol. Immunopathol. 8:34, 1977.
19. Frommel, D., Allain, J.P., Saint-Paul, E., et al. HLA antigens and Factor VIII antibody in classic hemophilia. Thromb.Haemostas 46:687, 1981.
20. Frommel, D., Muller, J.Y., Prou-Wartelle, O. and Allain, J.P. Possible linkage between the major histocompatibility complex and the immune response to Factor VIII in classic haemophilia. Vox Sang 33:270, 1977.
21. Lubahn, B.C. Inhibitor development in hemophilia A: Defining genetic factors that control the human immune response to F.VIII. (Ph. D. Dissertation), Chapel Hill: University of North Carolina at Chapel Hill, 1991.
22. Thompson, A.R. Molecular biology of the hemophilias. Prog.Hemost.Thromb. 10:175, 1991.
23. Millar, D.S., Steinbrecher, R.A., Wieland, K., et al. The molecular genetic analysis of haemophilia A; Characterization of six partial deletions in the factor VIII gene. Hum.Genet. 86:219, 1990.
24. McMillan, C.W., Shapiro, S.S., Whitehurst, D., et al. The natural history of Factor VIII:C inhibitors in patients with hemophilia A: A national cooperative study. II. Observations on the initial development of Factor VIII:C inhibitors. Blood 71:344, 1988.
25. McMillan, C.W. Clinical patterns of hemophilic patients who develop inhibitors. In: Factor VIII Inhibitors, New York: Alan R. Liss, 1984, p. 31-44.
26. Clark, A. Immunogenetics of Inhibitor Response to FVIII:C, (Masters Thesis), 1994.
27. Simonney, N., De Bosch, N., Argueyo, A., Garcia, E. and Layrisse, Z. HLA antigens in hemophiliacs A with or without Factor VIII antibodies in a Venezuelan mestizo population. Tissue Antigens 25:216, 1985.
28. Papasteriades, C., Economidou, J., Pappas, H., et al. Association between HLA antigens and progression of HIV infection in Greek haemophiliacs. Dis.Markers 11:131, 1993.
29. Reisner, H.M., Reisner, E.G., Kostyu, D.D., Lubahn, B.C., McMillan, C. and White, G.C. Possible association of HLA and Gm with the alloimmune response to FVIII. Thromb.Haemostas 58:1222a, 1987.(Abstract)
30. Lubahn, B.C., Reisner, E.G., Dawson, D.V., et al. Genetic susceptibility to inhibitor formation in hemophilia A: Related individuals develop antibodies to similar regions of Factor VIII. In preparation, 1994.
31. Fulcher, C.A., Lechner, K. and De Graaf Mahoney, S. Immunoblot analysis shows changes in factor VIII inhibitor chain specificty in factor VIII inhibitor patients over time. Blood 72:1348, 1988.
32. Gilles, J.G., Gilis, K., Arnout, J., Vermylen, J. and Saint-Remy, J-M. The serum of normal donors contains both anti-FVIII and corresponding anti-idiotype antibodies. Thromb.Haemostas 65:677, 1991.(Abstract)

Factor IX: Molecular structure, epitopes, and mutations associated with inhibitor formation

Katherine A. High

Children's Hospital of Philadelphia
Philadelphia, Pennsylvania

The focus of this overview is the occurrence of inhibitors in hemophilia B. Differences exist between hemophilia A and hemophilia B in terms of disease severity and inhibitor incidence, and these are reviewed. The structure of the Factor IX (F.IX) protein and epitopes of F.IX that have been mapped using monoclonal antibodies are also reviewed. Finally, mutations that have been associated with inhibitor formation in hemophilia B are discussed.

Disease severity and risk of inhibitor formation

Since inhibitors tend to occur in patients with severe disease, as opposed to mild or moderately severe disease, it is germane to point out the difference in disease severity between patients with hemophilia A and hemophilia B. In a pilot study of hemophilia treatment in the United States carried out in the early '70s, hemophilia treaters were surveyed and asked to estimate the total number of patients with severe and moderately severe disease who had been treated with blood components in the previous year.[1] Patients with mild disease were specifically excluded. The results of this survey are shown in Table 1. Of a total of 20,297 hemophilia A patients, 60% had severe disease and 40% had moderately severe disease; of 5,202 patients with hemophilia B, 44% had severe disease and 56% had moderate disease. Thus, it is possible that some of the differences in risk of inhibitors in patients with hemophilia A and hemophilia B may be due to the larger number of patients with hemophilia A who have severe disease. However, as will be seen from incidence and prevalence data presented below, this does not explain all of the differences in risk of inhibitors in these two groups.

A number of studies document the difference in risk of inhibitors between patients with hemophilia A and those with hemophilia B. A

Inhibitors to Coagulation Factors
Edited by Louis M. Aledort *et al.*, Plenum Press, New York, 1995

study of the British hemophilia population by Rosemary Biggs in 1974, documented a 7% prevalence of inhibitors in patients with hemophilia A and a 2.5% prevalence of inhibitors in patients with hemophilia B.[2] A study by Sultan and colleagues in the French hemophilia population in 1992 gave remarkably similar results, with a prevalence of inhibitors in hemophilia A patients of 7% and in hemophilia B patients of 2%.[3] In the French population, among those with severe disease, the prevalence of inhibitors in hemophilia A was 12.8% and in hemophilia B was 4%. Thus, the difference in prevalence of inhibitors between hemophilia A and hemophilia B is not entirely due to the difference in disease severity, since even among patients with severe disease, more patients with hemophilia A are affected.

Table 1. Disease Severity Among Patients with Hemophilia

	Total	Severe	Moderately Severe
Hemophilia A	20,297	12,117 (60%)	8,180 (40%)
Hemophilia B	5,202	2,304 (44%)	2,898 (56%)

Ehrenforth et al., in a study published in *Lancet* in 1992, described the incidence of inhibitor development in a German population of patients born after 1970.[4] Among patients with hemophilia A, 16 of 63 developed inhibitors, whereas among patients with hemophilia B, zero of 17 developed inhibitors. Somewhat different results were reported by I.M. Nilsson and colleagues in *Lancet* 1992.[5] This study followed 117 patients with severe or moderate hemophilia A or B who had been enrolled at treatment centers in Sweden between 1970 and 1990. In this group inhibitors developed in 16 of 77 patients with severe hemophilia A and in 4 of 12 patients with severe hemophilia B. These numbers are clearly different from those reported in other studies. The authors point out, as will be discussed below, that if certain mutations have a high prevalence in the population, there will be a high prevalence of inhibitors also.

The structure of the F.IX protein

Like other vitamin K-dependent clotting factors, the F.IX protein has a pre-pro leader sequence consisting of a signal sequence and a propeptide. These are required for secretion and for carboxylation of the amino terminal glutamic acid residues, but are not part of the mature protein in the circulation, and thus will not concern us further here. The N-terminus of the mature protein contains the γ- carboxyglutamic acid residues that are the hallmark of the vitamin-K dependent clotting

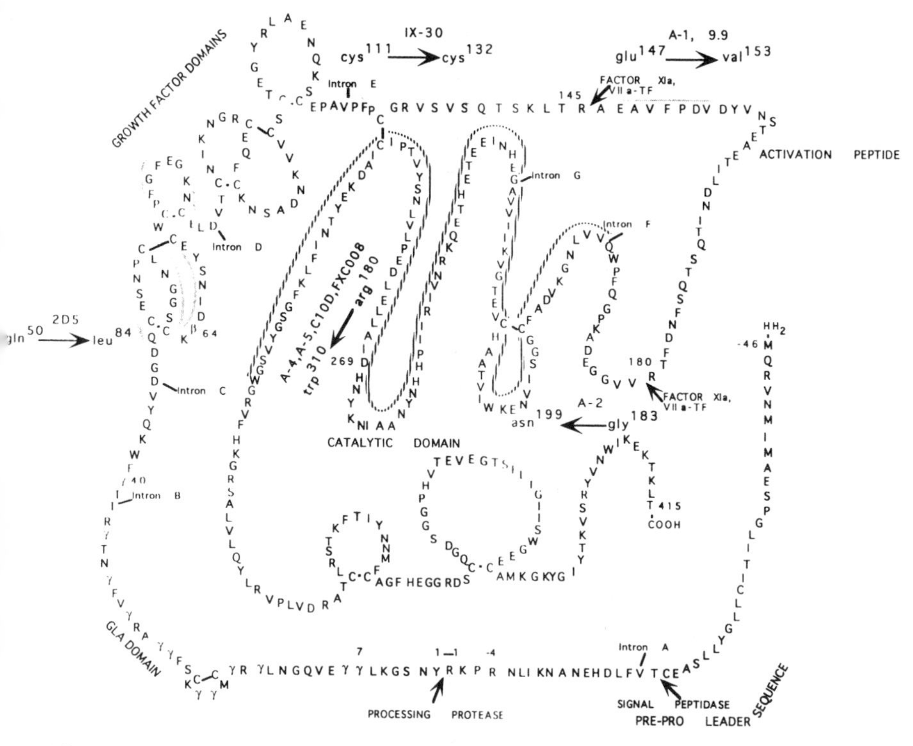

Figure 1. Schematic diagram of F.IX showing binding regions for 9 monoclonal antibodies. Adapted from Frazier et al. *Blood* 74:971,1989.

factors (see Figure 1). The chemical shorthand for γ- carboxyglutamic acid is Gla, and this region is termed the Gla domain. Twelve of the glutamic acid residues in the Gla domain undergo addition of a carboxyl group at the γ- carbon to generate a paired negative charge (two carboxyl groups) at the γ- carbon. This paired negative charge is required for calcium binding. A conformational change takes place in the presence of calcium, and this conformational change is required in order for the molecule to have activity in coagulation. The Gla domain is followed by two epidermal growth factor-like domains, so-called because of the homology to epidermal growth factor. The function of these domains is not known with certainty, but they may be involved in the interaction of with its cofactor, Factor VIII and its substrate Factor X.[6] Following the epidermal growth factor-like domains is the activation peptide. This 35 amino acid stretch is cleaved out when F.IX is activated by either Factor XIa or by VIIa-tissue factor. The carboxy terminus of the protein contains the catalytic domain, a typical trypsin-like serine protease domain. The complete amino acid sequence for F.IX is available for several species including human, canine, bovine, and mouse.[7-10] Comparison of amino acid sequence among these species demonstrates that the highest degree of conservation occurs in the N-terminus of the Gla domain and in the carboxy terminus of the catalytic domain.

F.IX epitopes

There is little known about the ability of F.IX to induce an immune response. If we define an epitope as the part of a molecule specifically recognized by the binding site of an immunoglobulin molecule, then we can describe some epitopes in F.IX, based on work done mapping regions of F.IX that react with mouse monoclonal antibodies. In this experimental approach, fragments of human F.IX cDNA are cloned into λgt11 and expressed as –13–ß-galactosidase fusion proteins. Filter lifts are probed with specific monoclonal antibodies. Positive clones are picked and sequenced; defining regions of overlap among the sequenced clones allows one to identify amino acids that contribute to an epitope. Using this technique, Stafford and colleagues[11] have mapped several mouse monoclonal antibodies to human F.IX, and the epitopes defined by these are shown in Figure 1. The antibody 2D5 defines an epitope extending from glutamine 50 to leucine 84. IX-30 defines an epitope extending from cysteine 111 to cysteine 132. A7 defines an epitope from phenylalanine 32 to Gla 40. A1 and 9.9 recognize the same epitope from glutamic acid 147 to valine 153. The reactivity of these antibodies depends upon the presence of a threonine rather than an alanine at the polymorphic site at residue 148. The antibody A-2 recognizes a region from glycine 183 to asparagine 199. The antibodies A4, A5, C10D, and FXC008 all recognize an epitope extending from arginine 180 to tryptophan 310. Efforts to define this epitope more closely by deleting the 5' or 3' ends of the λgt11 clones resulted in complete loss of reactivity. Thus, it is likely that these antibodies recognize a discontinuous epitope that includes the regions at either end of the sequence.

There is no information in the literature mapping antibodies from inhibitor patients. Thus, whether the same epitopes are the target of the immune response in hemophilia B patients with inhibitors is as yet unknown.

F.IX mutations associated with inhibitor formation

The human F.IX cDNA was isolated in 1982[7] and the F.IX gene in 1985.[12] Since that time, and particularly following the advent of PCR, mutations have been delineated in a large number of patients with hemophilia B. Since 1990, molecular defects causing hemophilia B have been collected in a database and published annually in *Nucleic Acids Research*.[13] The 1993 database contains reports of mutations in 806 patients with hemophilia B. Three hundred seventy-eight (378) of these are unique. For two population groups, the British and the Swedish populations, the mutation is known for a very large percentage of the patients, so that extensive databases are available for these two groups.[14] Based on data in the worldwide database and in the analysis of the British and Swedish population, it is clear that the molecular defect in a high percentage of hemophilia B patients with inhibitors is a gene deletion. Overall, deletions account for only 1 to 3% of mutations in patients with hemophilia B, but 50% of mutations in hemophilia B patients with inhibitors. The size of these deletions varies from several

Table 2. Point Mutations Associated with Inhibitors

Patient	Clotting (norm=100%)	Antigen (norm=100%)	Nucleotide Position & Mutation	Amino Acid Change	Comments
UK 12	<1		6,392, Δ1	6	Frameshift
Madrid 3	<1	<1	6,401-10, Δ10	9	Frameshift
Bonn 2	<1	<1	6,402-6, Δ**5**	9	Frameshift de novo in mother(22)
Chongqing	<1	<1	6,460,C→T	29,R→Stop	*16 cases including 2 inhibitors*
HB 7, Japan	<1	<1	6,680-1,Δ2	39	Frameshift, terminates at aa46, >100 Bethesda units
HB5, Japan	<1	<1	20,551, C→T	191,Q→stop	>100 Bethesda units
Malmo 5	<1	<0.1	20,561,G→A	194,W→Stop	
HB6, Japan	<1	<1	30,821, G→A	Destroys AG splice acceptor site	>100 Bethesda units
UK 140	<1	<1	30,863, C→T	248, R→Stop	*15 cases including 1 inhibitor*
Malmo 1	<1	<0.1	30,950-7, Δ8	277	Frameshift
Varel 1	<1	86	31,213-14, TA→CG	365, S→G	Active site, silent mutation at aa364, 8.1 Bethesda

nucleotides to more than 35 kilobases, encompassing the entire F.IX gene. In a report by Bloom and colleagues in 1987,[15] several deletions associated with inhibitors were mapped, and this report also assembled data from previous reports of deletions associated with inhibitor formation. Of interest was the fact that for several of the deletions, the breakpoint occurred at the site of a repetitive element, either Alu I or Kpn I repeats. It is possible that these repetitive elements serve as a site for a direct intra-chromosomal recombination event, resulting in loss of genetic material. This mechanism of mutation has been demonstrated for other genetic diseases.

In addition to large deletions, there have been a number of point mutations associated with inhibitor formation. These can be extracted from the hemophilia B database, but it should be noted that contributors to the database are not required to note whether a patient has had a history of an inhibitor, so the database must be regarded as incomplete on this point. Table 2 lists point mutations from the 1993 database that have been associated with inhibitor formation. Note that the overwhelming majority of these patients, like individuals with gene deletions, have less than 1% F.IX antigen. The single exception to this is $F.IX_{varel}$; this patient has a normal antigen level of 86%. Also of note is the fact that all of the mutations except one are either frameshift mutations, stop codons, or splice acceptor site mutations. Missense mutations, which cause a substitution at a specific amino acid, are not seen in these patients with the single exception of $F.IX_{varel}$.[16] The mutation in $F.IX_{varel}$ is a double point mutation. A T→C change occurs at the third nucleotide of the codon for amino acid 364 and is silent at the amino acid level. An A→G change at the next nucleotide changes the serine at 365 to a glycine. This patient had a maximum inhibitor titer of 8.1 BU. Note that the nature of the mutation is not the sole determining factor in inhibitor formation since some mutations were reported frequently (see italicized entries), but were only occasionally associated with inhibitors. However, when one knows the nature of the mutation, at least in the case of F.IX, one can use this information to make much more precise predictions about the risk of inhibitor formation. Thus, Giannelli and colleagues, using data from the Swedish population, demonstrated a 3% risk of inhibitor in hemophila B if the defect is unknown, but a near zero risk for patients with single amino acid substitutions. In contrast, for patients with mutations leading to loss of coding information, such as frameshift, stop codons, or splice site mutations, the risk of inhibitor was 20%.[14]

In summary, it appears that the nature of the IX mutation is an important factor in predisposing to inhibitor formation, with mutations leading to loss of coding information much more frequently associated with inhibitor formation then missense mutations that cause a single amino acid substitution. Large deletions account for only 1-3% of all hemophilia patients but represent 50% of mutations in inhibitor patients. Other factors besides the nature of the mutation itself are important, since not all patients with a particular mutation develop an inhibitor. Family studies in which only one of two or three affected patients developed an inhibitor further support this notion. It is likely that differences in immune response or perhaps in treatment regimen may determine which patients develop inhibitors.

It is not completely clear why patients with hemophilia B have a lower incidence of inhibitors than patients with hemophilia A. One factor is that a lower proportion of patients with hemophilia B have severe disease, but this does not explain the entire effect. Another possible factor is the much higher nonplasma level of F.IX (5 μg/ml versus 100 ng/ml for F.VIII). Patients with undetectable levels of IX antigen may nonetheless synthesize small amounts that are adequate for induction of immune tolerance. A third factor may be found in the structure of the other vitamin K-dependent clotting factors. Since there is considerable conservation of amino acid sequence among the vitamin K-dependent clotting factors, the presence of factors II, VII, and X may confer some immune tolerance to F.IX.

References

1. Department of Health and Human Services: National Heart and Lung Institute Blood Resource Studies. Vol 3: Pilot Study of Hemophilia Treatment in the U.S. U.S. Government Printing Office, Washington, 1972.
2. Biggs R. Jaundice and antibodies directed against Factors VIII and IX in patients treated for haemophilia or Christmas disease in the United Kingdom. *British J Haematology* 26:313, 1974.
3. Sultan Y and French Hemophilia Study Group. Prevalence of inhibitors in a population of 3,435 hemophilia patients in France. *Thrombosis and Haemostasis* 67:600, 1992.
4. Ehrenforth S, Kreuz W, Scharrer I, Linde R, Funk M, Gungor T, Krackhardt B, and Kornhuber B. Incidence of development of Factor VIII and Factor IX inhibitors in haemophiliacs. *Lancet* 339:594, 1992.
5. Ljung R, Petrini P, Lindgren AC, Tengborn, and Nilsson IM. Factor VIII and Factor IX inhibitors in haemophiliacs. *Lancet* 339:1550, 1992.
6. Nishimura H, Takeya H, Miyata T, Suehiro K, Okamura T, Niho Y, and Iwanaga S. Factor IX Fukuoka. *JBC* 268:24041, 1993.
7. Kurachi K and Davie EW. Isolation and characterization of a cDNA coding for human factor IX. *PNAS* 79:6461, 1982.
8. Evans JP, Watzke HH, Ware JL, Stafford DW, and High KA. Molecular cloning of a cDNA encoding canine factor IX. *Blood* 74:207, 1989.
9. Katayama K, Ericsson LH, Enfield DL, Walsh KA, Neurath H, Davie EW, and Titani K. Comparison of amino acid sequence of bovine coagulation Factor IX (Christmas Factor) with that of other vitamin K-dependent plasma proteins. *PNAS* 76:4990, 1979.
10. Wu SM, Stafford DW, and Ware J. Deduced amino acid sequence of mouse blood-coagulation factor IX. *Gene* 86:275, 1990.
11. Frazier D, Smith KJ, Cheung WF, Ware J, Lin SW, Thompson AR, Reisner H, Bajaj SP, and Stafford DW. Mapping of monoclonal antibodies to human Factor IX. *Blood* 74:971, 1989.
12. Yoshitake S, Schach BG, Foster DC, Davie EW, and Kurachi K. Nucleotide sequence of the gene for human Factor IX (antihemophilic factor B). *Biochemistry* 24:3736, 1985.

13. Giannelli F, Green PM, High KA, Sommer S, Poon MC, Ludwig M, Schwaab R, Reitsma PH, Goossens M, Yoshioka A, and Brownlee GG. Haemophilia B: database of point mutations and short additions and deletions–fourth edition, 1993. *Nucleic Acids Research* 21:3075, 1993.
14. Green PM, Montandon AJ, Bentley DR, and Giannelli F. Genetics and molecular biology of haemophilias A and B. *Blood Coagulation and Fibrinolysis* 2:539, 1991.
15. Matthews RJ, Anson DS, Peake IR, and Bloom AL. Heterogeneity of the Factor IX locus in nine hemophilia B inhibitor patients. *J Clin Invest* 79:746, 1987.
16. Ludwig M, Schwaab R, Olek K, Brackmann HH, and Egli H. Haemophilia B^+ with inhibitor. *Thrombosis and Haemostasis* 59:340, 1988.

Antibodies to von Willebrand factor in von Willebrand disease

P.M. Mannucci and A. B. Federici
Angelo Bianchi Bonomi Hemophilia and Thrombosis Center
Institute of Internal Medicine and IRCCS Maggiore Hospital
University of Milano, Italy

Introduction

The occurrence of an alloantibody directed against von Willebrand factor in a multitransfused patient with severe (type III) von Willebrand disease was first reported in 2 consecutive studies by Sarji et al. (1974) and Stratton et al. (1975). After this, 14 additional cases of alloantibodies were described and reviewed by Mannucci and Mari (1984). A survey carried out on behalf of the World Federation of Hemophilia identified 6 additional cases that, added to the 15 cases previously reported, bring to 21 the total number of patients reported so far with anti-von Willebrand factor alloantibodies (Mannucci and Cattaneo, 1991). In the frame of the survey, Hanna (1989) reported an antibody directed towards factor VIII (F.VIII) in a girl with severe von Willebrand disease. Since this antibody did not apparently react with von Willebrand factor, it will not be reviewed here.

Frequency of the antibodies

Alloantibodies directed against von Willebrand factor are a rare complication of replacement therapy in multitransfused patients with severe von Willebrand disease. Antibodies are usually reported either during the systematic screening of small populations of patients or because they affect the response to replacement therapy. Hence, selection bias makes it difficult to evaluate their true prevalence and incidence. The only available cross-sectional study was carried out during a survey of severe von Willebrand disease in Western Europe and Israel and was based on the centralized measurement of anti-von Willebrand factor antibodies (Mannucci et al., 1984). With a sensitive assay, antibodies were looked for in the plasma of 106 patients from 21 different countries whose diagnosis of severe von Willebrand disease was confirmed by centralized measurements of von Willebrand factor antigen. Eight patients, all previously reported (Bloom et al., 1979; Egberg and Blomback, 1976; Lenk et al., 1978; Maragall et al., 1979; Ruggeri et al., 1979) were found to have alloantibodies, a prevalence of 7.5%. This

Inhibitors to Coagulation Factors
Edited by Louis M. Aledort *et al.*, Plenum Press, New York, 1995

prevalence of antibodies is very similar to that found in the survey of the World Federation of Hemophilia (9.5%), indicating that antibody development is a rare complication of a rare disease.

Pathogenesis

Ruggeri et al. (1979) found antibodies in 3 of 4 affected siblings, Miller et al. (1983) and Mancuso et al. (1994) in 2 siblings each. The 2 patients reported by Maragall et al. (1979) and Mancuso et al. (1994) are second-degree cousins, suggesting that there is a familial predisposition to develop antibody. In the 2 siblings diagnosed by de Bosch (and subsequently characterized by Lopez-Fernandez et al., 1988; and Mancuso et al., 1994), the sibling who received the most transfusions developed the antibody at the age of 16. Her sister, who received fewer transfusions, developed the antibody at the age of 26 (personal communication). Hence, there is some evidence indicating that antibodies appear after a critical number of exposures to replacement therapy.

Since not all multitransfused patients with severe von Willebrand disease develop antibodies, factors other than exposure to allogeneic von Willebrand factor must influence their development. We hypothesized that antibodies develop only in patients with homozygous deletions of the von Willebrand factor gene (Mannucci and Mari, 1984). So far, all patients with deletions affecting both von Willebrand factor alleles also have antibodies to von Willebrand factor (Shelton-Inloes et al., 1987; Ngo et al., 1988; Peake et al., 1990; Mancuso et al., 1994). Of 46 patients with severe von Willebrand disease and no alloantibodies, none had von Willebrand factor gene deletion (Ginsberg and Sadler, 1993). Therefore, it appears that there is an association between homozygous deletions of the von Willebrand factor gene and development of alloantibodies. The most notable exception, from the report of Zhang et al. (1992), showed that a patient with an anti-von Willebrand factor antibody, previously described by Egberg and Blomback (1976), had a nonsense mutation characterized by a stop codon in exon 28.

Immunological and kinetic properties

In general, the presence of antibodies was demonstrated *in vitro* by the capacity of patient plasma to inhibit, in a time-independent manner, ristocetin-induced aggregation in normal platelet-rich plasma (see Table 1). When antibodies are present in very low levels, this simple screening test may not be sensitive enough and give negative or dubious results. Testing patient's semipurified immunoglobulins instead of plasma may help in these instances. More quantitative antibody assays of greater sensitivity have been described (Ruggeri et al., 1979).

In all the cases in which immunological typing was carried out, antibodies were of the IgG class. IgG contained both types of light chains in the 6 cases characterized by Ruggeri et al. (1979) and Lopez-

Table 1. Demography and characteristics of alloantibodies in 21 patients with severe von Willebrand's disease

Reported by	Patient ID code	Sex and age (in years) at diagnosis	Inhibitory activity against Willebrand factor	Inhibitory activity against F.VIII	Precipitin formation	Gene deletions
Sarji, Stratton	vWF-I(FM)	M/38	yes	no	n.t.	n.t.
Egberg, Zhang	AHP	F/30	yes	no	n.t.	stop codon
	AM	F/13	yes	no	n.t.	n.t.
Mannucci (1976), Ruggeri, Shelton-Inloes, Ngo	GS	M/9	yes	no	yes	yes
	GE	M/6	yes	no	yes	yes
	GT	F/4	yes	no	yes	yes
	SG	M/10	yes	no	yes	yes
Shoai	n. 19	F/25	yes	no	yes	n.t.
Lenk	DW	M/1	yes	no	yes	n.t.
Bloom, Peake	S1(IV5)	M/28	yes	no	no	yes
Maragall, Lopez-Fernandez, Mancuso	AR(HU2)	F/21	yes	no	yes	yes
Miller	EE	-	yes	no	no	n.t.
	AA	-	yes	no	no	n.t.
	KA	-	yes	no	no	n.t.
Mannucci (1984), Shelton-Inloes	CK	F/8	yes	no	yes	yes
Mancuso (related to the Maragall's patient)	JF	F/34	yes	no	yes	yes
Lopez-Fernandez, Mancuso	HU1(GF)	F/16	yes	no	yes	yes
	BM	F/26	yes	no	n.t.	yes
G. White II (personal communication)	ER	F/62	yes	no	n.t.	n.t.
	HC+	M/71	yes	no	n.t.	n.t.
	LZ	F/7	yes	no	n.t.	n.t.

-When more than one first author is indicated in the column "Reported by" it means that the patient(s) was reported on more than one publication.
-When more than one identification code is given (between parentheses), it means that patients were identified differently in different publications.
n.t.=not tested

Fernandez et al. (1988). Stratton et al. (1975), however, found only Kappa light chains in their case (19). It seems, therefore, that most anti-von Willebrand factor alloantibodies are of polyclonal origin. Another striking property of some alloantibodies is their capacity to precipitate von Willebrand factor in normal plasma (Mannucci et al., 1976). This property was found in 10 cases, but not in 4 additional cases (see Table 1). These varied results might be due to different antigenic specificities of the antibodies, but it is more likely that lattice formation depends on antibody potency. In the Italian patients, for instance, Ruggeri et al. (1979) found that only high-potency antibodies would precipitate in complex with von Willebrand factor whereas lattices did not form when antibody levels were lower. Following the early studies of Sarji et al. (1974) and Stratton et al. (1975), it seems well established that anti-von Willebrand factor antibodies do not inactivate purified F.VIII (Ruggeri et al., 1979) and that their capacity to interact with plasma F.VIII is due to steric hindrance of the F.VIII molecule bound to von Willebrand factor. The domains involved in the platelet glycoprotein Ib-ristocetin interaction and in F.VIII binding are relatively close together on the von Willebrand factor molecule, so that an antibody interacting with the platelet glycoprotein-Ib binding site can also affect the F.VIII binding site of von Willebrand factor and interfere with F.VIII coagulant activity.

Clinical consequences

In most reported cases, antibody development was heralded by poor clinical response to replacement therapy accompanied by lower than expected recovery of von Willebrand factor in plasma, lack of correction of the prolonged bleeding time and absence of the delayed and sustained rise of F.VIII coagulant activity. When the inhibitor titer is relatively low, it is usually not difficult to treat soft-tissue bleeding and to prevent bleeding at surgery in these patients. Mannucci et al. (1989) have shown that measurable plasma levels of F.VIII can be attained with cryoprecipitate or F.VIII-containing concentrates. It is much more difficult to shorten the bleeding time. Lack of bleeding time correction is usually associated with poor control of mucosal hemorrhages (menorrhagia, gastrointestinal) as shown by Mannucci et al. (1989). In some patients with particularly high antibody levels, replacement therapy is not only ineffective in controlling mucosal bleeding, it may also trigger severe or life-threatening anaphylactic reactions, associated with marked activation of the classical pathway of the complement system (Mannucci et al., 1987; 1989). An anamnestic rise in antibody level is usually seen 5 to 10 days after replacement therapy with von Willebrand factor, with features typical of a secondary response to a foreign antigen (Mannucci et al., 1989).

Recombinant F.VIII is a useful, potentially life-saving weapon in the management of bleeding episodes in these patients. Unlike available plasma products, this product contains F.VIII but no von Willebrand factor. The rationale for its use is that anti-von Willebrand factor antibodies do not inactivate F.VIII per se, but only when this protein is bound to von Willebrand factor as in plasma, cryoprecipitate, or concentrates. Accordingly, infusion of recombinant F.VIII in a girl with intra-abdominal bleeding who had developed life-endangering anaphylactic reactions and had no increase in plasma F.VIII after infusion with plasma-derived concentrate, was accompanied by a marked rise of plasma F.VIII, even though there was no change in von Willebrand factor plasma levels (manuscript in preparation). The plasma half-life of infused F.VIII was short (approximately 2 hours), because it is not stabilized by von Willebrand factor. Yet, using large doses given by continuous intravenous infusion, it was possible to maintain F.VIII plasma levels at levels of 50 U/dL or higher for many days and to carry out successfully emergency abdominal surgery, even though von Willebrand factor remained unmeasurable and the bleeding time very prolonged. Recombinant F.VIII has the additional advantage of not eliciting anaphylactic reactions, because it contains no von Willebrand factor which would form complexes with the alloantibodies and generate complement anaphylatoxins.

Conclusions

The characteristics of alloantibodies developing in patients with von Willebrand disease are similar to those of polyclonal antibodies that develop in animals immunized with von Willebrand factor preparations. There is evidence that antibodies are more likely to develop in patients with a deleted von Willebrand factor gene. In patients with low levels of antibodies it is possible to achieve measurable levels of plasma F.VIII with cryoprecipitate or concentrates. This approach is usually effective in controlling soft-tissue and surgical bleeding. It is much more difficult to achieve measurable von Willebrand factor levels in plasma, to shorten the bleeding time and hence to stop mucosal bleeding, particularly when the antibody titers are high. Replacement therapy can be complicated by severe anaphylactic reactions triggered by von Willebrand factor-containing immune complexes. Recombinant F.VIII is a novel form of treatment for those bleeding episodes that respond by raising F.VIII levels (soft-tissue and post-operative bleeding), even though the von Willebrand factor deficiency and hence the primary hemostasis defect are not affected by this treatment.

References

Bloom AL, Peake IR, Furlong RA and Davies BL. High potency factor VIII concentrate: more effective than cryoprecipitate in a patient with von Willebrand's disease and inhibitor. Thromb Res 16: 847 (1979).

Egberg N and Blomback M. On the characterization of acquired inhibitors to ristocetin-induced platelet aggregation found in patients with von Willebrand's disease. Thromb

Res 9: 527 (1976).
Ginsburg D and Sadler JE. von Willebrand disease: a database of point mutations, insertions and deletions. Thromb Haemostas 69: 177 (1993).
Hanna WT. Replacement therapy for surgery in type III von Willebrand patient with inhibitors to factor VIII procoagulant. Thromb Haemostas 62: 224 (1989) (abstract).
Lenk H, Weissbach G and Donula M. Ein Hemmkorper bei von Willebrand Jurgens-Syndrome and seine wirkung aur die Eigenschaften des Factor VIII molekuls. Folia Haematol 105: 826 (1978).
Lopez-Fernandez MF, Martin R, Lopez-Berges C, Ramos F, de Bosch N and Battle J. Further specific characterization of von Willebrand factor inhibitors developed in two patients with severe von Willebrand disease. Blood 72: 116 (1988).
Mancuso DJ, Tuley EA, Castillo R, de Bosch N, Mannucci PM and Sadler JE. Partial gene deletions in type III von Willebrand disease with alloantibody inhibitors. Thromb Haemostas, in press (1994).
Mannucci PM, Meyer D, Ruggeri ZM, Koutts J, Ciavarella N and Lavergne JM. Precipitating antibodies in von Willebrand's disease. Nature 262: 141 (1976).
Mannucci PM, Ruggeri ZM, Ciavarella N, Kazatchkine MD and Mowbray JF. Precipitating antibodies to factor VIII-von Willebrand factor in von Willebrand's disease: effects on replacement therapy. Blood 57: 25 (1981).
Mannucci PM and Mari D. Antibodies to factor VIII/von Willebrand factor in congenital and acquired von Willebrand's disease. In: LW Hoyer ed. Factor VIIII inhibitors. New York: Alan R. Liss, 109 (1984).
Mannucci PM, Bloom A, Larrieu MJ, Nilsson IM and West RR. Atherosclerosis and von Willebrand factor. Prevalence of severe von Willebrand disease in Western Europe and Israel. Br J Haematol 57: 166 (1984).
Mannucci PM, Tamaro G, Nerchi G, Candotti G, Federici A, Altieri D and Tedesco F. Life-threatening reaction to factor VIII concentrate in a patient with severe von Willebrand disease and alloantibodies to von Willebrand factor. Eur J Hematol 39: 467 (1987).
Mannucci PM and Cattaneo M. Alloantibodies in congenital von Willebrand disease. Res Clin Lab 21: 119 (1991).
Maragall S, Castillo R, Ordinas A, Liendo F and Rodriquez M. Inhibition of Willebrand factor in von Willebrand's disease. Thromb Res 14: 495 (1979).
Miller CH, Bussel JB and Hilgartner MW. Characteristics of inhibitors in severe von Willebrand's disease. Thromb Haemost 50: 34 (1983) (abstract).
Ngo K, Glotz VT, Koziol JA, et al. Homozygous and heterozygous deletions of the von Willebrand factor gene in patients and carriers of severe von Willebrand's disease. Proc Natl Acad Sci USA 85: 2753 (1988).
Peake IR, Lidde CB, Moodie P, et al. Severe type III von Willebrand's disease caused by deletion of exon 42 of the von Willebrand factor gene: family studies that identify carriers of the condition and a compound heterozygous individual. Blood 72: 6544 (1990).
Ruggeri ZM, Ciavarella N, Mannucci PM, Molinari A, Dammacco F, Lavergne JM and Meyer D. Familial incidence of precipitating antibodies in von Willebrand's disease: a study of four cases. J Lab.Clin Med 94: 60 (1979).
Sarji KE, Stratton RD, Wagner RH and Brinkhous KM. Nature of von Willebrand factor. A new assay and a specific inhibitor. Proc Natl Acad Sci USA 71: 2937 (1974).
Shelton-Inloes BB, Chebab FF, Mannucci PM, Federici AB and Sadler JE. Gene deletions correlate with the development of antibodies in von Willebrand disease. J Clin Invest 79: 1459 (1987).
Shoa'i I, Lavergne JM, Ardaillou N, Obert B, Ala F and Meyer D. Heterogeneity of von Willebrand's disease; study of 40 Iranian cases. Br J Haematol 37: 67 (1977).
Stratton RD, Wagner RH, Webster WP and Brinkhous KM. Antibody nature of circulating inhibitor of plasma von Willebrand factor. Proc Natl Acad Sci USA 72: 4167 (1975).
Zhang ZP, Lindstedt M, Falk G, Blomback M, Egberg N and Anvret M. Nonsense mutations of the von Willebrand factor gene in patients with von Willebrand disease type III and I. Am J Hum Genet 51: 850 (1992).

II. The molecular basis of the human immune response

The two papers in this section describe the complex protein surface epitopes with which antibodies react and the short peptide structures with which T-cell receptors react. It is important to recognize the very different characteristics of these two kinds of determinants. In the case of Factor VIII (F.VIII), considerable data are now available about the protein epitopes with which the antibodies interact. To date, however, there is no information about which segments of F.VIII are presented to T-cell receptors by class II HLA glycoproteins.

Benjamin's summary of epitope definitions provides a useful framework for evaluating the extensive, complex literature on this subject. As there is no three-dimensional structure information for F.VIII, the emphasis in the near future will certainly be on its functional epitopes, those critical surfaces of the F.VIII molecule that bind inhibitor antibodies. It is likely that studies using site-directed mutagenesis will further define which amino acids are essential for this reactivity.

The many distinct F.VIII epitopes that have been characterized to date make it unlikely that therapies can be based on F.VIII fragment infusion. However, a better understanding of the peptides that bind to T-cell receptors has a better likelihood of identifying new approaches to the treatment of inhibitor patients. Alexander and colleagues describe encouraging *in vivo* studies using T-cell receptor antagonists in well-defined autoimmune diseases. As F.VIII peptides are identified that bind to T-cell receptors and thereby support T-helper cell functions, it will be possible to test modified peptides for their ability to block F.VIII antibody formation. A systematic evaluation of T-cell reactivity to F.VIII is very likely to provide essential new information.

B-cell epitopes: Fact and fiction

David C. Benjamin

Department of Microbiology
and the Beirne B. Carter Center for Immunology Research
University of Virginia Health Sciences Center
Charlottesville, Virginia 22908

Antibodies are present as integral membrane receptors on cells of the B lymphocyte lineage and as proteins secreted by members of this cell family. As such they represent one of several macromolecular complexes used by the immune system in the specific recognition of antigenic molecules. In addition to antibody, these include T-cell receptors and the class I and class II molecules encoded by genes within the major histocompatibility complex (MHC). In humans, antibodies are produced in response to foreign particles such as infectious organisms, transplantation antigens and, in genetically deficient individuals, to human or animal proteins provided as therapeutic agents in the treatment of certain diseases. Of particular interest at this symposium are antibodies produced to coagulation factors and the nature of the epitopes on these factors. It has become increasingly apparent that a significant fraction of patients when provided with the factor, isolated either from natural sources or from recombinant material, respond by producing antibody that inhibits the function of that factor. In order to fully understand the mechanisms of the production of these antibody inhibitors to coagulation factors and to have a hope of successfully treating these individuals, one must have a better understanding of the nature of the antigenic sites on the factors to which the immune response is directed.

This presentation is not meant to be an all inclusive review of the current knowledge of the molecular nature of B-cell epitopes. A number of excellent reviews serve that purpose. [1-6] Rather, it is intended to provide the reader of this volume with sufficient background information on the state of the art so that he or she may read and understand the relevant literature. More importantly, it is intended to help clarify the terms and concepts used in discussing the nature of the structures on protein antigens recognized by antibody.

General properties of B-cell epitopes

All antibodies are conformationally specific, i.e., they recognize a

Inhibitors to Coagulation Factors
Edited by Louis M. Aledort *et al.*, Plenum Press, New York, 1995

three-dimensional structure composed of protein, DNA, RNA, carbohydrate, lipid, small molecular weight molecules (haptens), or combinations thereof. The same site on the surface of a protein antigen may be seen differently by different antibodies. Thus, in one sense, a B-cell epitope can only be defined with reference to a particular antibody and, perhaps, with reference to the particular method used to study the interaction of the antigen and antibody. However, there are a number of factors that are common to all B-cell epitopes on protein antigens. First, on native protein antigens all B-cell epitopes are three-dimensional structures on the surface of the protein (see Figure 1). The epitope can be visualized as a complex surface with multiple protrusions and

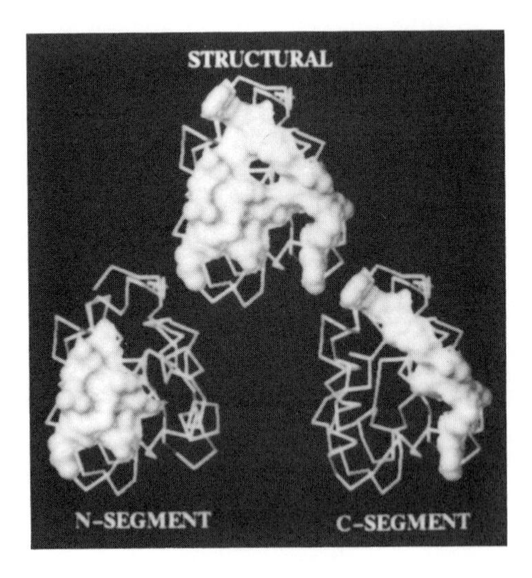

Figure 1. The D1.3 B-cell epitope of hen egg-white lysozyme. The structural epitope was determined by X-ray diffraction and includes all amino acids in contact with the D1.3 antibody.[7] This complex, three-dimensional epitope is composed of amino acids from two distant regions of the lysozyme primary structure (see text). Shown is the Cα backbone of the lysozyme molecule with the D1.3 structural epitope portrayed as a solid water accessible surface.

indentations. This complex surface is complementary to the surface in the antibody combining site with which it interacts. Second, the number of contacts between amino acids in the epitope and those in the antibody combining site is roughly proportional to the size of the epitope (see Table 1). In Table 1, the first two lines represent hapten-antibody complexes, the next two lines peptide-antibody complexes, and the final four lines protein-antibody complexes. Third, the bonds formed between the epitope and the antibody are the same as those formed between any other interacting protein pair, i.e., van der waals (VDW) interactions, hydrogen bonds, and salt bridges, although the number and type may vary (see Table 1). Fourth, the surface area on the protein antigen that is buried on interaction with antibody is approximately the same (about 700 Å^2 to 900 Å^2) regardless of the size of the protein. Fifth, most if not all B-cell epitopes on protein antigens are discontinuous (vide infra). Sixth, there is no chemical or physical property of any portion of the surface of a protein that makes it inherently more antigenic or immunogenic.[1,5]

Thus, a protein B-cell epitope is a large three-dimensional structure composed of amino acids which are brought together on the surface of the protein during folding.

Table 1. Properties of Some B-Cell Epitopes

		Contacts[3]				Area[4]	
Antigen[1]	AA[2]	VDW	HB	Salt	Total	(Å^2)	Ref
FITC (4FAB)	n.a.	19	0	0	22	257	8
Digoxin (1IGJ)	n.a.	27	0	0	27	279	9
Mhr peptide (2IGF)	7(7)	42	12	3	57	438	10
Flu peptide (1HIN)	7(8)	53	12	1	66	422	11
HyHEL-5 (2HFL)	14(23)	60	9	4	73	718	12
HyHEL-10 (3HFM)	15(27)	88	21	1	110	774	13
D1.3 (1FDL)	16(27)	61	18	0	79	692	7
NC41 (1NCA)	16(34)	65	13	0	78	840	14

[1] FITC, fluorescein isothiocyanate; Mhr, peptide comprised of residues 69-87 of myohem-erythrin; Flu, peptide comprised of residues of influenza hemagglutinin protein; HyHEL-5, HyHEL-10 and D1.3 antigenic epitopes of hen egg-white lysozyme; NC41, influenza neuraminidase. Letters in parenthesis are the Brookhaven PDB code for the atomic coordinates of the antigen-antibody complexes used to determine the contacts and surface areas.

[2] Number of contact amino acids within the structural epitope. Numbers in parentheses are the total number of contact and non-contact buried residues.

[3] Atomic contacts between antigen and antibody calculated using CONTACSYM.[15]

[4] Area on the antigen buried by antibody. Calculated by the program Molecular Surface (MS).[16]

Types of B-cell epitopes

There are five different definitions of epitope used in the immunological literature. Unfortunately, the misuse of the terms involved and/or a lack of knowledge of how and why these terms were coined has led to considerable confusion. These definitions are listed in Table 2 and are discussed below.

Continuous epitope

A continuous epitope is comprised of amino acids from a small segment of the polypeptide chain and is usually about 8-10 amino acids in length. This term was originally coined[1] following experiments using synthetic peptides to either inhibit the reaction of anti-protein antibody or to induce the formation of antibody that would react with intact protein. There is no good and consistent evidence for the presence of continuous epitopes on native, intact protein antigens. The arguments for this statement are reviewed in much greater detail elsewhere[1-5,11-14,17] and will not be discussed further here.

Discontinuous epitope

A discontinuous epitope (see Figure 1) is comprised of amino acids which are far apart in the primary structure of the protein but which are found together on the surface of the folded protein. If one accepts the thesis that most, if not all, protein epitopes are discontinuous, then if anti-protein antibody reacts with a small peptide, that peptide most probably represents a "portion" of a more complex epitope. In addition, the failure of a peptide to react with anti-protein antibody does not mean the peptide sequences are not part of the epitope. Rather, it may mean that the peptide does not react with sufficient affinity to be detected with the assay used. Interestingly, of the antigen-antibody complexes for which the crystal structures are known, all are discontinuous. Furthermore, when tested, the formation of these complexes are not inhibited by a peptide representing only a portion of the epitope. Figure 1 portrays the D1.3 epitope of hen egg-white lysozyme (HEL). This epitope is structurally defined (see below and reference 7) and is composed of amino acids from two very distinct and distant regions of the primary structure of HEL. The N-segment is derived entirely from residues within the range of amino acids 18-27 whereas the C-segment is comprised entirely from residues within the range of amino acids 116-129. These two regions are brought together on the surface of HEL following synthesis and folding of the polypeptide chain into the native HEL enzyme. Thus, the D1.3 epitope is clearly discontinuous.

Table 2. Definitions of B-cell epitopes

Continuous epitope: a structure composed of amino acids from a relatively short, continuous segment of the polypeptide chain.

Discontinuous epitope: a structure composed of two or more regions which are separate from each other in the primary sequence of the polypeptide chain but which are brought together on the surface of the protein during folding.

Structural epitope: a three-dimensional structure defined by X-ray diffraction analysis of an antigen-antibody complex. It includes all antigen atoms which are in "contact" with antibody atoms using predefined interatomic distances.

Functional epitope: a structure defined by a functional analysis, i.e., by difference analysis of the reactivity of antibody with two antigens that differ in one or more amino acids within the epitope.

Energetic epitope: those amino acids within the structural or functional epitope which contribute the majority of the binding energy. The remainder of the interacting surface is essentially a passive surface.

Given the large size of structural epitopes on proteins and of the antibody combining site, one might expect that few if any epitope on native globular protein antigens are continuous. Indeed, Barlow et al.[17] have clearly demonstrated that the probability is very low that that all of the amino acids contacted by antibody come from a single continuous segment of the polypeptide chain.

Structural epitope

A structural epitope is a three-dimensional structure defined by X-ray diffraction analysis of an antigen-antibody complex (see figures 1 and 2). It includes all antigen atoms which are in "contact" with antibody atoms using predefined interatomic distances. It is this structure which is used to calculate the properties of B-cell epitopes listed in Table 1. It is also this structure which is used as the starting point for calculating the energy of binding between antigen and antibody which is the basis for the concept of the energetic epitope (vide infra). It is also the structure upon which many experiments were designed to determine the role of individual amino acids in antigenicity. The crystal structure of a number of antigen-antibody complexes have been determined. The number of amino acids within the epitope ranges from 14 to 21 and the number of segments of the polypeptide chain from which these amino acids are derived ranges from 2 to 5. Thus, using X-ray diffraction analysis as the sole criterion, ALL protein epitopes are discontinuous.

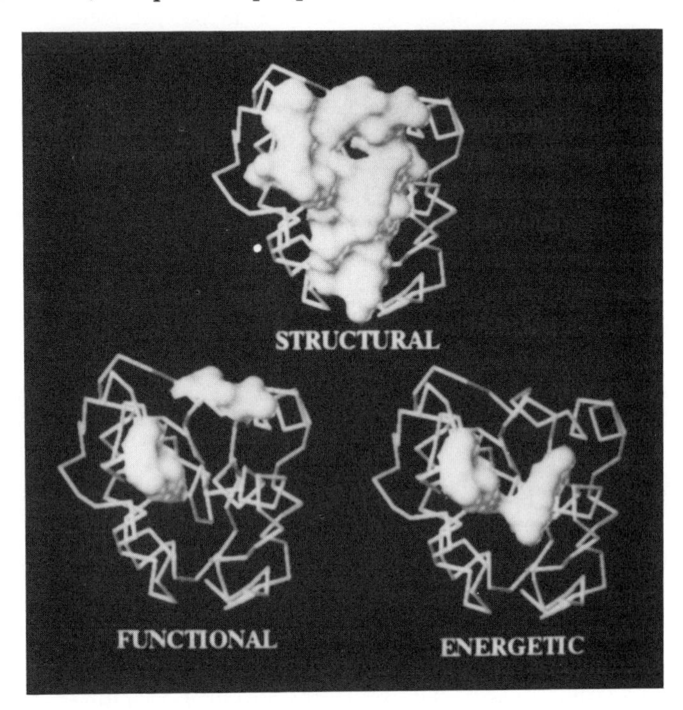

Figure 2. The HyHEL-5 B-cell epitopes on of hen egg-white lysozyme. The structural, functional and energetic epitopes are composed of 14 (ref. 12, Table 1), 3 (ref. 18), and 2 (ref. 19) amino acids, respectively. Shown are the Cα backbone of lysozyme with the epitopes shown in solid shaded surfaces.

Functional epitope

A functional epitope is defined by functional analyses, i.e., an amino acid is assumed to be within the epitope if it can be demonstrated by a nonstructural analysis to be involved in interaction with antibody (see Table 2). For example, two evolutionarily related proteins, X and Y, differ in a single surface amino acid. An antibody to X fails to react with Y in an ELISA assay (or any other functional assay). Therefore, the single surface amino acid that is different between X and Y is said to be part of the epitope on X to which the antibody was made.

Most B-cell epitopes are functional epitopes. For those epitopes that are defined both in structural and functional terms, all functional epitopes are smaller than structural epitopes. For example, Figure 2 portrays the structural, functional and energetic epitopes on hen egg-white lysozyme (HEL) that interact with the HyHEL-5 monoclonal antibody. The functional epitope was defined by the ability of a series of avian lysozymes to inhibit the reaction between the HyHEL-5 antibody and HEL.[18] The HyHEL-5 structural epitope defined by crystallography[11] contains atoms from 14 amino acids that contact the HyHEL-5 antibody.

In contrast, the HyHEL-5 functional epitope contains atoms from only three amino acids. In addition, although there are atoms from 16 amino acids within the D1.3 structural epitope (see Figure 1), the functional epitope (not shown) is defined by only a single amino acid. [7] In both of these cases, and in many others,[1,2] the demonstration that amino acids reside within functional epitopes depends entirely on the availability of related antigens (such as avian egg-white lysozymes) that differ at this site. More recently, site directed mutagenesis of the DNA encoding the antigen molecule has been used to study antigen structure.[2,20-22] Although the size of the functional epitopes in some mutagenesis studies approaches that of the structural epitope, in the single case where the structural epitope is known, the functional epitope is still smaller.[22]

Functional analyses, regardless how carefully or extensively done, can provide only limited information. They cannot tell us which atoms in a given amino acid are in contact with antibody. Furthermore, a change in an amino acid side chain may not result in a loss of reactivity, if that particular amino acid contributes only main chain atoms to the interaction.

Energetic epitope

The term energetic epitope was coined by Jiri Novotny and his colleagues.[19,23] He calculated the contribution of individual amino acids, on the antigen and within the antibody binding site, to the energy of binding of HyHEL-5, HyHEL-10, and D1.3 antibodies to lysozyme. These calculations suggested that although the contact areas between lysozyme and each antibody were large (see Table 1) productive binding appears to be mediated by a smaller number of residues in both the antibody and the lysozyme antigen. He concluded that the energetically most important residues constitute an "energetic epitope." Thus an energetic epitope is defined as those amino acids, within the structural or functional epitope, which contribute the majority of the binding energy. The remainder of the interacting surface is considered to be essentially a passive surface. However, since we do not know which atoms of functional epitopes are involved in binding or what type interactions they contribute, we must restrict discussion of energetic epitopes to those which are also defined in structural terms. The HyHEL-5 energetic epitope is compared to the HyHEL-5 structural and functional epitopes in Figure 2. The HyHEL-5, HyHEL-10 and D1.3 energetic epitopes are comprised of only two or three residues each although the number of contact residues within the structural epitope is much larger (see Table 1).

Mutagenesis to study B-cell epitope structure

Site-directed mutagenesis has been used to study antigenic and antibody structure[2,20-22] Using such methods, the investigator is able to circumvent the major limitations of using variant antigens to study functional epitopes. With site-directed mutagenesis, the researcher is able to systematically alter the coding sequences for the antigen of choice and to produce essentially an unlimited panel of variant antigen molecules with one or more amino acid differences. Although a number of such studies have been and are being carried out, one study[21] is particularly interesting because it addresses many of the issues discussed above and because of the number of variants produced in a single system.

In this study the cDNA encoding human growth hormone (HGH) was subjected to extensive site-directed mutagenesis producing a large panel of single amino acid variants of HGH. The variant HGH molecules were then assayed for their ability to react with 21 different monoclonal antibodies. The results were as follows: 1) all of the 21 epitopes were functionally discontinuous, i.e., they were composed of amino acids from noncontiguous segments of the primary sequence; 2) all these functional epitopes were located in highly accessible regions of the

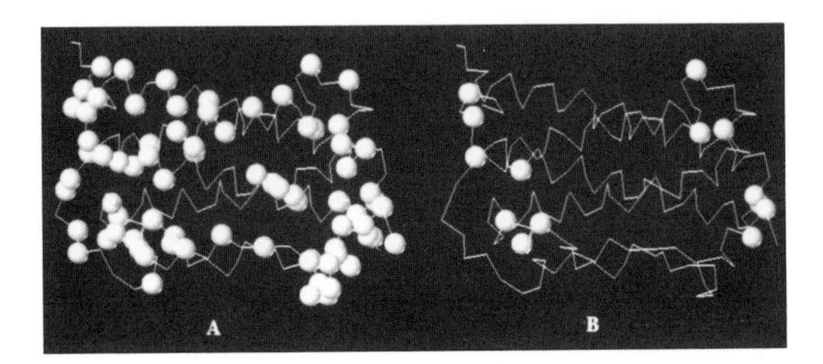

Figure 3. The antigenic structure of human growth hormone (HGH). A. The Cα backbone structure of the HGH molecule with amino acids known to be part of an epitope shown as spheres. B. Two independent epitopes, each composed of amino acids from segments of HGH that are quite distant from each in the primary sequence but which are brought together in the folded molecule. Data taken from ref. 21.

protein surface; 3) the functional epitopes were dominated by a small number of amino acid side chains within each epitope suggesting that certain residues were critical to binding; perhaps these "critical" residues correspond to the energetic epitopes discussed above; 4) there was no correlation between affinity of interaction with antibody and the composition or location of the epitope, i.e., no intrinsic property that dictates antigenicity; 5) the discontinuous functional epitopes approached the size of the structurally defined epitopes reported for other protein antigen molecules; and 6) essentially the entire surface of the HGH molecule was immunogenic in the mouse. Figure 3A shows the Cα backbone of the HGH molecule with the amino acids that were demonstrated to be within one of the epitopes depicted as spheres. This clearly shows that most surface amino acids are part of at least one epitope. Figure 3B shows the amino acids which were demonstrated to reside within two of the 21 epitopes (one at each end). This clearly shows these two epitopes are discontinuous (as were all the others).

Prediction algorithms

A number of methods have been reported for predicting the position of antigenic regions within the primary sequence of protein antigens. These include hydrophilicity;[24] solvent accessibility;[25] secondary structure, i.e., ß-turns;[26] and segmental flexibility.[27] These, however, are merely predictors of properties of protein surfaces. At best, such methods are limited to predicting continuous segments of a more complex discontinuous epitope. Given that many antigens cannot be isolated in sufficient quantities to prepare antibodies or determine antigenicity, the use of one or more of these predictive algorithms may allow the synthesis of peptide immunogens. Whether these anti-peptide antibodies do react with intact, native protein antigens depends on a number of factors, as discussed below. In addition, whether such anti-peptide antibodies have any utility depends entirely on the needs of the individual investigator. Nevertheless, one must be very careful in interpreting any results forthcoming from the use of these reagents or these predictive algorithms.

Peptides as antigens and as immunogens

It should be quite clear that any antibody made to native globular protein antigen is most probably directed to a discontinuous epitope. Should a peptide contain a large proportion of the amino acids contributing to the binding energy of the epitope, then that peptide should be able to inhibit the binding of the antibody to the native protein antigen. However, in order to do so the conformation of the peptide in solution must be the same as, or very similar to, the conformation of that part of the native protein from which it is derived. That is, the peptide must be able to energetically and conformationally mimic a large portion of the native epitope.

Antibodies made in response to peptide immunogens present an entirely different set of problems and such reagents must be used with care. First, there is no guarantee that the peptide alone or coupled to a carrier protein is capable of assuming a conformation even similar to that of the region of the native antigen from which it is derived. If it is unable to do so, any antibody produced will not recognize the "native" structure of the protein and its usefulness is restricted to detecting denatured forms of the antigen. Second, even if an anti-peptide antibody does react with the intact protein antigen, it may not react with the native conformation. A recent study[28] directly addressed the ability of anti-peptide antibodies to interact with intact protein antigen in solution and in solid phase assays. It was quite clear that although the anti-peptide antibodies bind "immobilized" antigen in a concentration dependent manner, they fail to react with native antigen in solution. These results were interpreted as showing that: 1) immunobilized antigen undergoes a conformational change as a result of absorption to the solid surface; and 2) binding of anti-peptide antibodies to peptides, either immobilized or free, cannot be used to infer that the peptide has assumed a conformation corresponding to that of the same sequence in the native protein.

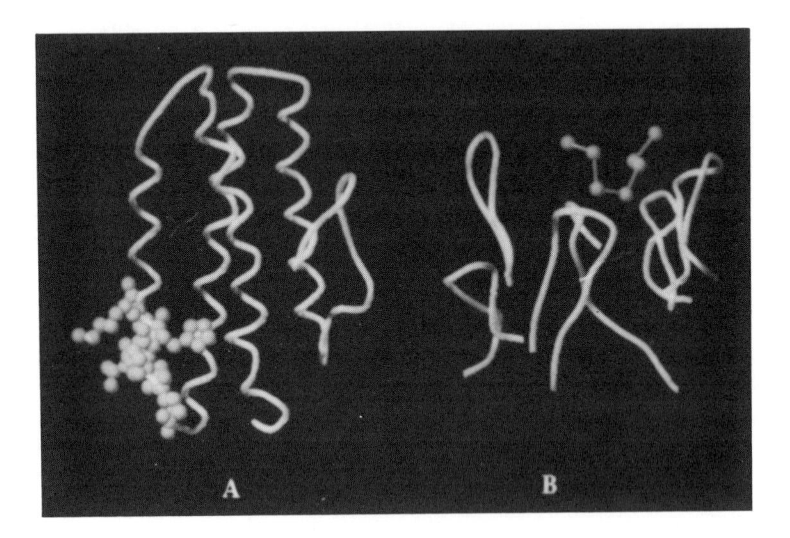

Figure 4. The antigenic peptide from myohemerythrin is shown as part of the native myohemerythrin protein structure (A) where it is part of an α- helix, and when bound to antibody (B) where it is a β-turn.

A graphic example for one of the anti-peptide antibodies used in this study is shown in Figure 4. A peptide representing amino acids 69-87 of the protein myohemerythrin was used as immunogen.[29] The B13I2 anti-peptide antibody derived from this immunization was shown to react with intact myohemerythrin in solid phase assays[28,29] but not in solution.[28] The molecular basis for this discrepancy becomes quite clear when one examines the conformation of these sequences in the protein myohemerythrin and of the peptide when bound by antibody. This region of the intact protein is an α-helix. However, when bound to antibody it has a ß-turn configuration (see Figure 4, ref. 29). In fact only seven of the 18 amino acids (residues 69-75) of the antibody-bound peptide can be seen by X-ray diffraction analyses indicating that the conformation of the remaining 11 residues is essentially random. Therefore, it is reasonable to conclude, as did Spangler, [28] that the immobilized myohemerythrin has unfolded and that the region corresponding to amino acids 69-75 has assumed a nonnative ß-turn configuration.

Anti-peptide antibodies can be highly useful reagents but one must be very careful in their use and in the interpretation of data from experiments in which they are used.

Overlapping epitopes

There are three possible interpretations of data showing competition between two antibodies: 1) the two antibodies see the same site although they may do so in different ways; 2) they see two sites that share some, but not all, amino acids, i.e., overlapping epitopes; and 3) react with two distinct antigenic sites but are prevented from binding at the same time due to steric hindrance between noncontact regions of the antibodies. Until recently, there was no information which would permit the investigator to distinguish between these alternatives. It is possible that detailed and extensive site-directed mutagenesis studies will enable us to distinguish between the alternatives for some antibodies but not for others.

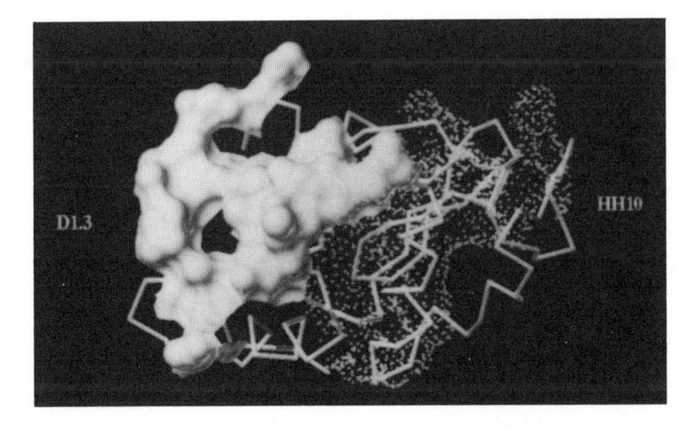

Figure 5. The D1.3 and HyHEL-10 structural B-cell epitopes on hen egg-white lysozyme. These two epitopes share several amino acids, yet the two monoclonal antibodies that define them can bind simultaneously.[30] The D1.3 epitope is shown as a solid shaded surface and the HyHEL-10 epitope is shown as a dotted surface.

A recent study by Smith-Gill and colleagues[30] has shed a surprising light on this issue. Figure 5 shows the D1.3 and the HyHEL-10 structural epitopes of hen egg-white lysozyme. The crystallographic data clearly shows that these two epitopes share at least two amino acids and thus should be an example of overlapping epitopes (alternative 2 above). However, Smith-Gill and her colleagues have shown that the two antibodies can bind lysozyme at the same time suggesting that they see two independent nonoverlapping sites (alternative 1 above). The discrepancy can be explained if crystallographic data is examined closely. Although these two antibodies contact some of the same amino acids, they do so by contacting "different atoms" on these amino acids. Indeed, the orientation of these shared amino acids is such that the two antibodies can bind at the same time. Interestingly, as close as the two antibodies must come to each other during binding to the same lysozyme molecule, there was no evidence of steric hindrance.[30] Whether steric hindrance, in the true sense, occurs in antigen-antibody interaction is now open to question.

Summary

It is quite clear that B-cell epitopes on intact, native protein antigens in solution are of the discontinuous type whether defined structurally or functionally. It is also clear that although B-cell epitopes share a number of common features, they can only be defined in detail in terms of the individual antibody with which they react and in terms of the method used to describe them. Therefore, in order to ensure open communication and eliminate misunderstanding between individual investigators, it is wise to clearly state the conditions under which the epitope is defined. Given these conditions, one can view protein antigenicity in terms of the multideterminant, regulatory hypothesis presented some years ago.[1] This hypothesis states that "The surface of a protein consists of a complex array of overlapping potential antigenic determinants; in aggregate these approach a continuum. Most determinants depend upon the conformational integrity of the native molcule. Those to which an individual responds are dictated by the structural differences between the antigen and the host's self-proteins and by host regulatory mechanisms, and are not necessarily an inherent property of the protein molecule reflecting restricted antigenicity or limited antigenic sites."

Acknowledgments

This work was supported in part by NIH grant AI20745. The figure graphics were prepared on a Silicon Graphics 4D computer using the molecular modeling package SYBYL from Tripos Associates, St. Louis, Missouri.

References

1. D.C. Benjamin, J.A. Berzofsky, I.J. East, F.R.N. Gurd, C. Hannum, S.J. Leach, E. Margoliash, J.G. Michael, A. Miller, E.M. Prager, M. Reichlin, E.E. Sercarz, S.J. Smith-Gill, P.E. Todd, and A.C. Wilson, The Antigenic Structure of Proteins: A Reappraisal. Ann. Rev. Immunol. 2:67 (1984).
2. D.C. Benjamin, Molecular Approaches to the Study of B-cell Epitopes. Intern. Rev. Immunol. 7:149 (1991).
3. D.C. Benjamin and S.S. Perdue, Techniques for Determining Epitopes for Antibodies and T-Cell Receptors. Ann. Rev. Medicinal Chem. 27:189 (1993).
4. E.D. Getzoff, J.A. Tainer, and R.A. Lerner, The Chemistry and Mechanism of Antibody Binding to Protein Antigens. Adv. Immunol. 43:1 (1988).
5. J.A. Berzofsky, Intrinsic and Extrinsic Factors in Protein Antigenic Structure. Science. 229:932 (1985).
6. D.R. Davies, S. Sheriff, and E.A. Padlan. Antibody-Antigen Complexes, J. Biol. Chem. 263:10541 (1988).
7. A.G. Amit, R.A. Mariuzza, S.E.V. Phillips, and R.J. Poljak, Three-Dimensional Structure of an Antigen-Antibody Complex at 2.8 Å Resolution. Science. 233:747 (1986).
8. J.N. Herron, X.M. He, M.L. Mason, E.W. Voss, Jr., and A.B. Edmunson, Three-Dimensional Structure of a Fluorescein-Fab Complex Crystallized in 2-Methyl-2,4-Pentanediol. Proteins 5:271 (1989).
9. P.D. Jeffrey, R.K. Strong, L.C. Sieker, C.Y.Y. Chang, R.L. Campbell, G.A. Petsko, E. Haber, M.N. Margolies, and S. Sheriff, 26-10 Fab-Digoxin Complex: Affinity and Specificity Due to Surface Complementarity. Proc. Nat. Acad. Sci. USA 90:10310 (1993).
10. R.L. Stanfield, T.M. Fieser, R.A. Lerner, and I.A. Wilson, Crystal Structures of an Antibody to a Peptide and Its Complex with Peptide Antigen at 2.8 Angstroms. Science 248:712 (1992).
11. J.M. Rini, U. Schulze-Gahmen, and I.A. Wilson, Structural Evidence for Induced Fit as a Mechanism for Antigen-Antibody Recognition. Science 255:959 (1992).
12. S. Sheriff, E.W. Silverton, E.A. Padlan, G.H. Cohen, S.J. Smith-Gill, B.C. Finzel, and D.R. Davies. Three-Dimensional Structure of an Antibody-Antigen Complex. Proc. Nat. Acad. Sci. USA. 84:8075 (1987).
13. E.A. Padlan, E.W. Silverton, S. Sheriff, G.H. Cohen, S.J. Smith-Gill, and D.R. Davies, Structure of an Antibody-Antigen Complex: Crystal Structure of the HyHEL-10 Fab-Lysozyme Complex. Proc. Nat. Acad. Sci. USA. 86:5938 (1989).
14. P.M. Colman, W.G. Laver, J.N. Varghese, A.T. Baker, P.A. Tulloch, G.M. Air, and R.G. Webster, Three-Dimensional Structure of a Complex of Antibody with Influenza Virus Neuraminidase. Nature 326:358 (1987).
15. S. Sheriff, Some Methods for Examining the Interactions between Two Molecules. Immunomethods 3:191 (1993).
16. M.L. Connolly, Analytical Molecular Surface Calculation. J. Appl. Crystallography. 16:548 (1983).
17. D.J. Barlow, M.S. Edwards, and J.M. Thornton, Continuous and Discontinuous Protein Antigenic Determinants. Nature 322:747 (1986).
18. S.J. Smith-Gill, T.B. Lavoie, and C.R. Mainhart, Antigenic Regions Defined by Monoclonal Antibodies Correspond to Structural Domains of Avian Lysozyme. J. Immunol. 133:384 (1984).
19. J. Novotny, R.E. Bruccoleri, and F.A. Saul., On the Attribution of Binding Energy in Antigen-Antibody Complexes McPC 603, D1.3 and HyHEL-5. Biochemistry 28:4735 (1989).
20. A.M. Smith and D.C. Benjamin, The Antigenic Surface of Staphylococcal Nuclease. II. Analysis of the N-1 Epitope by Site-Directed Mutagenesis. J. Immunol. 146:1259 (1991).
21. L. Jin, B.M. Fendly, and J.A. Wells, High Resolution Functional Analysis of Antibody-Antigen Interactions. J. Mol. Biol. 226:851 (1992).
22. L. Prasad, S. Sharma, M. Fandonselaar, J.W. Quail, J.S. Lee, E.B. Waygood, K. Wilson, Z. Dauter, and L.T.J. Delbaere, Mutagenesis for Epitope Mapping: Structure of an Antibody-Protein Antigen Complex. J. Biol. Chem. 268:10705 (1993).
23. J. Novotny, Protein Antigenicity: A Thermodynamic Approach. Mol. Immunol. 28:201 (1991).

24. T.P. Hopp and K.R. Woods, Prediction of Protein Antigenic Determinants from Amino Acid Sequences. Proc. Nat. Acad. Sci. USA. 78:3824 (1981).
25. J. Novotny, M. Handschumacher, E. Haber, R.E. Bruccoleri, W.B. Carlson, E.W. Fanning, J.A. Smith, and G.D. Rose, Antigenic Determinants in Proteins Coincide with Surface Regions Accessible to Large Probes (Antibody Domains). Proc. Nat. Acad. Sci. USA. 83:226 (1986).
26. E. Westhof, Correlation of Segmental Mobility and the Location of Antigenic Determinants in Proteins. Nature 311:123 (1984).
27. M.H.V. Van Regenmortel, J.P. Briand, Z. Al Moudallal, D. Altschuh, and E. Westhof, in: "Current Communications in Molecular Biology: Immune Recognition of Protein Antigens," W.G. Laver and G.M. Air, Eds., Cold Spring Harbor Press, Cold Spring Harbor, N.Y. (1985).
28. B.D. Spangler, Binding To Native Proteins by Antipeptide Monoclonal Antibodies. J. Immunol. 146:1591 (1991).
29. T.M. Fieser, J.A. Tainer, H.M. Geysen, R.A. Houghton, and R.A. Lerner, Influence of Protein Flexibility and Peptide Conformation on Reactivity of Monoclonal Anti-Peptide A antibodies with a Protein α- helix. Proc. Nat. Acad. Sci. USA. 84:8568 (1987).
30. S.J. Smith-Gill, Protein Epitopes: Functional and Structural Differences. Res. in Immunology. In press (1994).

Antigen analogs as therapeutic agents

Jeff Alexander,[1] Jörg Ruppert,[1] Dawne M. Page,[2] Stephen M. Hedrick,[2] Alessandra Franco,[1] Glenn Y. Ishioka,[1] Howard M. Grey,[1] and Alessandro Sette[1]

[1]Cytel, 3525 John Hopkins Court, San Diego, California 92121
[2]Department of Biology and the Cancer Center, University of California, San Diego, La Jolla, California 92093-0063

The immune system's function is to protect us against invading pathogens such as viruses, bacteria, parasites, and even the runaway growth of cancer cells. T lymphocytes play a major role in these immune responses, and in most cases the outcome is beneficial, such as killing infected cells or helping B-cells produce antibodies. In many cases, medical science has been capable of augmenting or enhancing the beneficial immune responses. Successes are numerous. Vaccines are now available against viral diseases such as influenza, hepatitis, and polio, and against bacterial infections such as diptheria, whooping cough, and tetanus. Clearly, however, much more progress is required in developing vaccines efficacious against diseases such as cancer and AIDS.

In contrast to beneficial immune responses discussed above, unwanted T-cell responses to self antigens or benign antigens can also occur, resulting in severe pathological states such as autoimmune diseases and allergic reactions. Familiar examples include multiple sclerosis, insulin-dependent diabetes mellitus, rheumatoid arthritis, and common pollen and pet allergies.

A complex series of molecular events is involved in T-cell activation. Class I and class II major histocompatibility complex (MHC) glycoproteins[1] bind a vast collection of peptides that are processed intracellularly from larger polypeptides. The peptide/MHC complexes, upon formation, are transported to the surface of specialized antigen processing cells (APC). The T-cell recognizes this ligand through its T-cell receptor (TCR). Formation of this trimolecular complex and delivery of appropriate co-stimulatory signals[2] leads to a cascade of biochemical events such as inositol phosphate (IP) turnover and CA^{2+} influx, ultimately leading to T-cell activation, proliferation, and differentiation. In recent years, several potential therapeutic strategies for regulating these unwanted immune responses were suggested, based on our current understanding of the chain of events leading to T-cell activation.

Inhibitors to Coagulation Factors
Edited by Louis M. Aledort *et al.*, Plenum Press, New York, 1995

Different therapeutic approaches to interfering with APC/T-cell interaction

There are several target areas for selective immunointervention.[3,4] A partial list includes CD4 blockade, MHC blockade, tolerance induction, and TCR antagonism. The specific disease situation would dictate the most likely effective means of immunotherapy. For example, when multiple MHC types are suspected to be involved in the disease, broad effector agents such as anti-MHC or anti-CD4 monoclonal antibodies could be the agents of choice. Unfortunately, these broadly selective agents may also be considerably toxic and lead to an immunocompromised host. If a single or a few MHC types are suspected to be involved, such as in rheumatoid arthritis or multiple sclerosis, then it is conceivable that a therapeutic approach based on specific MHC blockers may be used. Finally, when a single antigen is recognized, selective approaches such as induction of specific tolerance, TCR vaccination, or TCR antagonism may be the therapeutic agent of choice.

In the following paragraphs, we will describe in more detail the novel TCR antagonism approach.

The phenomenon of TCR antagonism

During a series of experiments aimed at developing MHC-specific blockers, we examined a large set of peptides of various origin in parallel for their ability to either bind MHC or inhibit antigen presentation to T-cells. Using a DR1 restricted T-cell clone specific for influenza hemagglutin (HA) 307-319 in addition to a DR1 restricted T-cell clone specific for tetanus toxoid (TT) 830-843, it was observed that nonantigenic antigen analogs were superior in their capacity to inhibit antigen presentation relative to unrelated peptides that were equivalent MHC binders.[5] This suggested that some other mechanism in addition to competition for the MHC was in operation.

To examine this phenomenon more directly, an antigen prepulse assay was developed. MHC competition relies on saturation of the MHC by peptide blockers to prevent antigen from binding and subsequently activating the T-cell. In the antigen prepulse assay, a suboptimal antigen dose was allowed to bind MHC on the surface of APC before the antigen analogs were added to the assay, thus making it impossible for MHC competition to take place. The APC, which carried both antigen and antigen analogs bound to these MHC, then presented these antigens to the T-cells, which were followed in a three-day proliferation assay as a measure of T-cell activation. The HA 307-319 analogs, but not TT 830-843 analogs or the unrelated binder, myelin basic protein (MBP) 74-98, were able to inhibit antigen presentation and prevent proliferation for the HA 307-319 specific T-cell clone. Conversely, only analogs of TT 830-843 were able to inhibit antigen presentation and subsequent proliferation of the TT 830-843 specific T-cell clone. Therefore, it was concluded that this phenomenon of preferential inhibition by antigen analogs was distinct from MHC competition, and it was thus termed "TCR antagonism."

A cartoon contrasting interaction of either antigen, antigen analog, or unrelated MHC binder with the TCR is given in Figure 1. By definition, the antigen makes appropriate contact with TCR, which subsequently leads to T-cell activation, proliferation, and differentiation (see Figure 1a). An antigen analog would have less than the optimal contacts with the TCR. This weaker interaction would not be sufficient to activate the cascade of biochemical events necessary for T-cell activation (see Figure 1b). Finally, a completely unrelated MHC binder antigen, although occupying the MHC groove, would not have the correct amino acid side chains neccessary for interaction with the TCR (see Figure 1c). We therefore postulated that TCR antagonists function by preventing proper signaling to the T-cell by the MHC/antigen/TCR trimolecular complex.

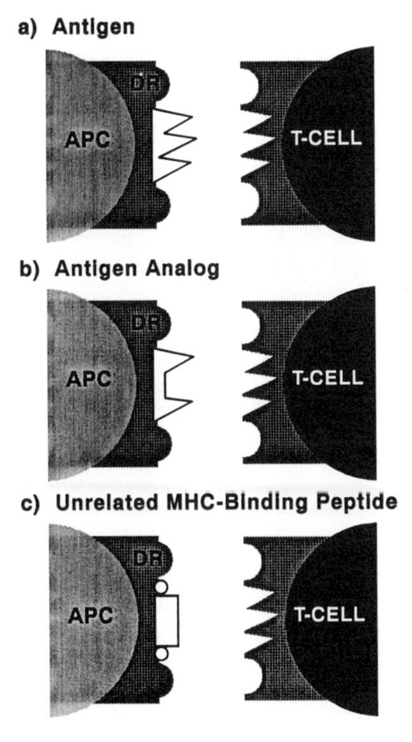

Figure 1. Hypothetical model illustrating the mechanism of action of a TCR antagonist.

Since these initial observations, TCR antagonism has also been demonstrated for several other human and murine CD4+ T-cells restricted to class II molecules. TCR antagonism has been reported in the DR4 restricted/HA 307-319 specific human system,[6] the IE^k restricted/ cytochrome C 88-104 specific murine system, and the IA^s restricted/

proteolipid protein 139-151 specific murine system. In addition, TCR antagonism has also been demonstrated for class I restricted T-cells.[7]

The rules dictating the generation of TCR antagonists

In order to characterize the molecular mechanisms involved in the phenomenon of TCR antagonism, a systematic analysis of the structure/ function relationship was undertaken, in which the effect of single amino acid substitutions in the antigen molecule on TCR antagonism was examined.[8] Although no general rules were determined, several trends were observed. First, a TCR antagonist could be generated by modifying any of the five major TCR contact residues. Second, conservative amino acid changes resulted in either poor agonists or efficient antagonists. Radical substitutions resulted in either poor TCR antagonists or analogs that functioned as neither an agonist nor an antagonist ("null" peptides). These results suggested that the affinity of TCR for its ligand determined whether the T-cell would be activated or antagonized.

The next series of experiments further supported the affinity argument.[8] A selected set of analogs was generated with a polyalanine backbone (the MHC amino acid contact residues) and increasing additions of TCR contact residues. One analog with a single TCR contact residue out of a possible five demonstrated no antigenic or antagonistic capacity. Analogs with two TCR contact residues were poor antagonists, while those with three contact residues were good antagonists. Analogs with four TCR contact residues were either good antagonists or poor agonists. Finally, the analog with all five TCR contact residues contained within the polyalanine backbone was a good agonist. In conclusion, as the number of TCR contact residues was increased (and presumably affinity also increased), the antagonism capacity increased up to a point where the analogs became antigenic.

These results suggest that an affinity threshold of a TCR for its ligand may have to be reached before T-cell activation can occur. An affinity model also may help explain studies done by other groups which identified partial agonists of T-cell signaling. If the affinity of the MHC/ peptide analog for TCR is below the crucial affinity threshold, then some, but not all, of the signals required for T-cell activation may be delivered. In fact, it was shown that an antigen analog of hemoglobin 64-76, in context with MHC, could elicit IL-4 production but not induce proliferation by a Th2 T-cell clone.[9] This same group also demonstrated that antigen analogs could induce Th1 cytolysis, but not proliferation and lymphokine production.[10]

A summary of affinity and T-cell effector functions is given in Figure 2. A null peptide/MHC complex would have little or no affinity for the TCR, resulting in no T-cell activation. As the affinity increases, the analog/MHC complex would function as a TCR antagonist. In that way, it has the capacity to competitively inhibit antigen-induced signaling. A partial agonist/MHC complex would have still higher affinity, resulting in inducing some, but not all effector functions. Finally, for affinities above the crucial threshold, T-cell activation, proliferation, and differentiation occurs.

Affinity: Low⟶Intermediate⟶High

TCR/MHC/Peptide
Interaction: Null⟶TCR Antagonist⟶Partial Agonist⟶Agonist

Figure 2. Affinity interaction of TCR with its ligand dictates the type of T-cell response.

A molecular model of TCR antagonists

The data described above suggest an affinity model to explain the phenomenon of TCR antagonism, but do not offer any insight as to the molecular mechanism by which the TCR antagonists may operate. As previously stated, TCR antagonists are able to function in the presence of suboptimal antigen doses. Therefore, a simplistic receptor saturation model seems unlikely, since only a small number of MHC/antagonist complexes, roughly corresponding to a few hundred to a thousand, are apparently sufficient to demonstrate antagonism.[5] This number corresponds to only a few percent of the number of TCR available on the surface of T-cells. Therefore, it seems unlikely that a small excess of antagonist would saturate the TCR. In addition, the identification of partial agonists/antagonists is indicative of more complex mechanisms of receptor conformational changes or interference with its interaction with coreceptors.[11]

We have examined the effects of TCR antagonists on the early events of T-cell activation in order to determine which biochemical step may be blocked by TCR antagonists.[5,12] No early T-cell activation events such as IP turnover or Ca^{2+} influx were detected in the presence of TCR antagonists alone, or in combination with suboptimal antigen doses. Moreover, TCR antagonists were, neither alone[5] nor in conjunction with suboptimal antigen doses (conditions necessary to demonstrate TCR antagonism[13]), capable of inducing T-cell anergy.

Further investigation of membrane-associated events demonstrated that MHC/antagonist complexes did not inhibit the formation of antigen-induced APC/T-cell conjugates.[12] Therefore, these data suggest that engagement of the TCR by MHC/antigen complexes in itself (conjugate formation) is necessary, but not sufficient, for the activation of early biochemical events leading to T-cell activation. Engagement of TCR by MHC/antigen complexes results in TCR clustering and subsequent T-cell activation. However, when the APC carries both MHC/antigen and MHC/antagonist complexes, then "mixed" TCR clusters may form, and the signals required for early biochemical activation events are consequently inhibited. Of note is that the 3-D structure of DR1 has suggested that the formation of complexes of TCR dimers engaging MHC/antigen dimers may be involved in signal transduction leading to T-cell activation.[14] This observation further supports the notion that TCR antagonists may act by interfering with proper TCR clustering.

In vivo ***immunomodulation by TCR antagonists***

A question that remains to be answered is whether TCR antagonists will be useful as a clinical approach to control allergies and autoimmune diseases. As a first step to address this question, we sought to examine the efficacy of TCR antagonists in an *in vivo* disease model.

Therefore, we have examined whether TCR antagonists would be effective *in vivo* to protect SJL (H-2^s) mice from development of experimental autoimmune encephalomyelitis (EAE). This disease is induced in SJL mice by a well-defined T-cell peptide epitope, murine proteolipid protein (PLP) 139-151. This model is attractive, in that the pathogenesis of the disease involves a heterogenous T-cell response (a more stringent test for inhibition of disease by a TCR antagonist). In addition, a previous study in our laboratory demonstrated the ability of unrelated antigen binders to inhibit EAE by the MHC blockade approach.[15]

Independently, previous studies by McDevitt's group[16,17] in a different mouse strain demonstrated that pre- or post-treatment of animals with an encephalitogenic analog prevented EAE when animals were immunized with the encephalitogenic antigen. It is tempting to speculate that either TCR antagonism or partial agonism may be involved in the inhibition of EAE observed in these studies.

In our case, we first sought to determine the MHC and TCR amino acid contact residues of PLP 139-151. Single amino acid substitutions at each position were subsequently tested for their binding capacity to MHC molecules. Those analogs that retained good binding capacity were also tested for their antigenic potential in a T-cell proliferation assay. It was determined that residues L_{145} and P_{148} were important for MHC binding, while residues G_{142}, K_{143}, W_{144}, G_{146}, and H_{147} were important for T-cell activation. The most critical residue was W_{144}, because most of the substitutions introduced at this position led to loss of antigenicity for a panel of seven different representative PLP 139-151 specific T-cell clones.

Nonantigenic peptides were next tested for antagonism in the antigen prepulse assay. Several antagonistic peptides were thus identified. However, while several antagonists inhibiting multiple clonal specificities were identified, not one peptide was able to inhibit proliferation of all seven PLP 139-151 T-cell clones. These results were reminiscent of earlier studies by Snoke et al.[18] which described two different TCR antagonists that were able to inhibit five of six different DR4 restricted,

HA 307-319 specific T-cell lines, but could not identify a "pan-inhibiting" TCR antagonist.

Since we were unable to obtain an antagonist capable of inhibiting all seven PLP 139-151 specific T-cell clones, the two most effective antagonists were pooled. In particular the two antagonists selected had the potential to inhibit six of seven T-cell clones tested. When tested *in vivo,* the antagonist pool was remarkably effective. However, in accordance with their proposed mechanism of action, when tested separately, these same two antagonists were much less effective in inhibiting EAE. The antagonist pool was also compared to our most effective unrelated MHC blocking peptide, in order to examine the effectiveness of TCR antagonism versus the MHC blockade therapeutic approach. As shown in Table 1, when either the antagonists pool or the MHC blocker were administered together with the encephalitogenic PLP 139-151, inhibition of EAE was observed. However, the TCR antagonist pool was approximately tenfold more effective in inhibiting EAE than the MHC blocker. Further studies also demonstrated that the TCR antagonist pool was moderately effective when administered before the antigenic challenge. These studies suggest that TCR antagonists may be useful for controlling unwanted T-cell responses *in vivo.*

Table 1. Relative *in vivo* efficacy of IA^s blockers versus TCR antagonist peptides in inhibiting EAE induction[1]

Molar Excess[2]	ROIV Peptide	No. Expts.	TCR Antagonist Pool	No. Expts.
10x	44.7± 36.1[3]	3[4]	92.0 ± 1.2	5
3.3x	27.0± 38.1	2	55.0 ± 2.8	3
1.1x	11.6± 20.6	2	49.6 ± 13.1	3
0.1x	ND[5]		2.7 ± 11.3	2

1 Data from several experiments are summarized above, comparing the efficacy of an unrelated IA^s blocking peptide, ROIV, with a pool of L_{144} and Y_{144} TCR antagonist analogs in preventing EAE induction. Mice were treated by co-immunizing with the inhibitor peptides and with the encephalitogenic PLP 139-151 peptide.

2 Molar excess of inhibitor peptide co-injected into mice relative to the PLP 139-151 peptide, which was administered at a dose of 42.6 nmole (25 μg) per animal.

3 Mean relative % inhibition of EAE ± SD. The relative % inhibition of EAE was calculated by plotting the mean clinical score of each group versus time from day 8 through day 35 post-induction of disease. The area under the curve for each group treated with inhibitor and for the untreated control group was then calculated using the formula: % inhibition = 100 - [(Area of test peptide + Area of untreated control) x 100]

4 Number of independent experiments.

5 Not done.

TCR antagonists and negative selection

As discussed previously, the data available so far indicate that TCR antagonists may function by interacting with TCR with a lower affinity than antigen. This suggests that different signals may be delivered to the T-cell, depending on the affinity of the interaction between MHC/ peptide and TCR. It has also been suggested[19,20,21] that thymic differentiation events may be dependent on affinity or avidity differences. We have recently studied a system, in which TCR transgenic mice were generated, where the T-cell receptor transgene is restricted by IE^k and is specific for Pigeon Cytochrome C (PCC) 88-104. Various peptide analogs were screened for binding to IE^k and subsequently examined for their antagonism capacity in the antigen prepulse assay. Furthermore, it was found that the TCR antagonist peptides were able to induce negative selection, as assayed in an *in vitro* thymocyte death assay.[22] Thus, although these PCC 88-104 TCR antagonists were unable to activate a mature T-cell clone, they nonetheless were able to induce negative selection.

These studies suggest that a lower affinity threshold is required for thymocyte deletion than mature T-cell activation. Interestingly, the most effective TCR antagonists were also able to upregulate the IL-2 receptor. Previous studies by Sloan-Lancaster et al.[23] identified partial agonists that also upregulated the IL-2 receptor, but in addition caused T-cell anergy. Thus, it is possible that similar signals may be involved in inducing negative selection on thymocytes and anergy in mature T-cell clones.

Conclusions

This presentation has illustrated how the concept of TCR antagonism was developed and how it can be applied to the rational design of therapeutic agents. It is not known at this time whether TCR antagonism or partial activation of T-cells will turn out to be a viable therapeutic strategy. However, it is encouraging to note that TCR antagonism is more effective than MHC blockade in inhibition of the experimental autoimmune disease, EAE. Tantamount to the use of TCR antagonists or partial agonists is the precise knowledge of the disease-inducing antigen. Recent developments[24] in sequencing of minute amounts of naturally processed peptides should help identify pathogenic antigens. Also, it is interesting to note that MBP and $PLP^{139-151}$ have been suggested as the autoantigens involved in multiple sclerosis in humans. In the near future the evaluation of TCR antagonism and partial T-cell activation as therapeutic agents could be initiated.

Acknowledgment

We wish to thank Joyce Joseph for her assistance in preparation of the manuscript.

References

1. A. Sette and H.M. Grey, Chemistry of peptide interactions with MHC proteins, *Curr. Opin. Immunol.* 4:79 (1992).
2. D.L. Mueller, M.K. Jenkins, and R.H. Schwartz, Clonal expansion versus functional clonal inactivation: A costimulatory signalling pathway determines the outcome of T-cell antigen receptor occupancy, *Ann. Rev. Immunol.* 7:445 (1989).
3. L. Adorini, J-C. Guéry, G. Rodriguez-Tarduchy, and S. Trembleau, Selective immunosuppression, *Immunol. Today* 14:285 (1993).
4. A. Lanzavecchia, Identifying strategies for immune intervention, *Science* 260:937 (1993).
5. M.T. De Magistris, J. Alexander, M. Coggeshall, A. Altman, F.C.A. Gaeta, H.M. Grey, and A. Sette, Antigen analog-major histocompatibility complexes act as antagonists of the T-cell receptor, *Cell* 68:625 (1992).
6. K. Snoke, J. Alexander, A. Franco, L. Smith, J.V. Brawley, P. Concannon, H.M. Grey, A. Sette, and P. Wentworth, The inhibition of different T-cell lines specific for the same antigen with TCR antagonist peptides, *J. Immunol.*, in press (1993).
7. S.C. Jameson, F.R. Carbone, and M.J. Bevan, Clone-specific T-cell receptor antagonists of major histocompatibility complex class I-restricted cytotoxic T-cells, *J. Exp. Med.* 177:1541 (1993).
8. J. Alexander, K. Snoke, J. Ruppert, J. Sidney, M. Wall, S. Southwood, C. Oseroff, T. Arrhenius, F.C.A. Gaeta, S.M. Colón, H.M. Grey, and A. Sette, Functional consequences of engagement of the T-cell receptor by low affinity ligands, *J. Immunol.* 150:1 (1993).
9. B.D. Evavold and P.M. Allen, Separation of IL-4 production from Th-cell proliferation by an altered T-cell receptor ligand, *Science* 252:1308 (1991).
10. B.D. Evavold, J. Sloan-Lancaster, B.L. Hsu, and P.M. Allen, Separation of T helper 1 clone cytolysis from proliferation and lymphokine production using analog peptides, *J. Immunol.* 150:3131 (1993).
11. T. Tenakin, Agonist, partial agonists, antagonists, inverse agonists and agonist/antagonists? *Trends Pharmacol. Sci.* 8:423 (1987).
12. J. Ruppert, J. Alexander, K. Snoke, M. Coggeshall, E. Herbert, D. McKenzie, H.M. Grey, and A. Sette, Effect of T-cell receptor antagonism on interaction between T-cells and antigen-presenting cells and on T-cell signaling events, *Proc. Natl. Acad. Sci. USA* 90:2675 (1993).
13. J. Alexander, J.Ruppert, K. Snoke, and A. Sette, TCR antagonism and T-cell tolerance can be independently induced in a DR restricted, HA specific T-cell clone, *Int. Immunol.*, in press (1993).
14. J.H. Brown, T.S. Jardetzky, J.C. Gorga, L.J. Stern, R.G. Urban, J.L. Strominger, and D.C. Wiley, Three-dimensional structure of the human class II histompatibility antigen HLA-DR1, *Nature* 364:33 (1993).
15. A.G. Lamont, A. Sette, R. Fujinami, S.M. Colón, C. Miles, and H.M. Grey, Inhibition of experimental autoimmune encephalomyelitis induction in SJL/J mice by using a peptide with high affinity for IA^s molecules, *J. Immunol.* 145:1687 (1990).
16. D.C. Wraith, D.E. Smilek, D.J. Mitchell, L. Steinman, and H.O. McDevitt, Antigen recognition in autoimmune encephalomyelitis and the potential for peptide-mediated immunotherapy, *Cell* 59:247 (1989).
17. D.E. Smilek, D.C. Wraith, S. Hodgkinson, S. Dwivedy, L. Steinman, and H.O. McDevitt, A single amino acid change in a myelin basic protein peptide confers the capacity to prevent rather than induce experimental autoimmune encephalomyelitis, *Proc. Natl. Acad. Sci. USA* 88:9633 (1991).
18. K. Snoke, J. Alexander, A. Franco, L. Smith, J.V. Brawley, P.O. Concannon, H.M. Grey, A. Sette, and P. Wentworth, The inhibition of different T-cell lines specific for the same antigen with TCR antagonist peptides, *J. Immunol.* 151:1 (1993).
19. J. Sprent, D. Lo, E.-K. Gao, and Y. Ron, T-cell selection in the thymus, *Immunol. Rev.* 101:173 (1988).
20. P. Kisielow, H.S. Teh, H. Bluthmann, and H. von Boehmer, Positive selection of antigen-specific T-cells in thymus by restricting MHC molecules, *Nature* 335:730 (1988).
21. H. Pircher, U.H. Rohrer, D. Moskophidis, R.M. Zinkernagel, and H. Hengartner, Lower receptor avidity required for thymic clonal deletion than for effector T-cell function, *Nature* 351:482 (1991).

22. N.J. Vasquez, J. Kaye, and S.M. Hedrick, In vivo and in vitro clonal deletion of double-positive thymocytes, *J. Exp. Med.* 175:1307 (1992).
23. J. Sloan-Lancaster, B.D. Evavold, and P.M. Allen, Induction of T-cell anergy by altered T-cell-receptor ligand on live antigen-presenting cells, *Nature* 363:156 (1993).
24. D.F. Hunt, H. Michel, T.A. Dickinson, J. Shabanowitz, A.L. Cox, K. Sakaguchi, E. Appella, H.M. Grey, and A. Sette, Peptides presented to the immune system by the murine class II major histocompatibility complex molecule I-A^d, *Science* 256:1817 (1992).

III: Genetic variation and the immune response

"Is there any other point to which you wish to draw my attention?"

"To the curious incident of the dog in the night-time"

"The dog did nothing in the night-time"

"That was the curious incident," remarked Sherlock Holmes

From: *Silver Blaze* by Sir Arthur Conan Doyle

If the basis of the immune response is the recognition of self (safe) from non-self (worrisome), then why don't all hemophiliacs produce antibodies or immune T cells to infused coagulation protein? Our favorite polymath wonders why the dog doesn't bark. Section I looked at this question from the point of view of the proteins (F.VIII, F.IX and von Willebrand Factor) which are, perhaps, foreign to the treated patient. In section III, two other likely sources of genetic variation in the hemophiliac are considered as possible players in the immune response to infused coagulation factor. T cell receptor genes and immunoglobulin heavy chain variable region genes, the subjects of this section, are often given short shrift by the immunologist enamored by the major histocompatibility complex (MHC) and the genetics of the immune response. But the antibody molecule is the critical end effector that does the inhibiting, and that molecule is certainly made of variable regions. The T-cell receptor (TCR) must recognize co-presented peptide and class II MHC antigens for a T-cell dependent response to occur.

As Dr. Robinson points out in her paper "T-cell receptors in immune responses" several sources of variation are important in generating TCR diversity. However, in contradistinction to Ig V region diversity, somatic mutation does not play a role. In an overly simplified view, the TCR could be looked at as being closer to the germline genes, which are its ultimate origin. The presence of allelic variation in TCR B V region genes may well relate to disease susceptibility and perhaps, by extension, to the inhibitor response.

Drs. Pascual and Capra consider the effector molecule in the inhibitor response in their review "Immunoglobulin heavy chain variable region gene usage in human autoimmune diseases." Although autoimmune and heteroimmune responses to some alloantigens (most notably red blood cell antigens) tend to be preferentially encoded by some V_H gene segments, most autoimmune antibody responses are not restricted. The fact that at least some of these antibodies are identical with germ line sequences (i.e., not somatically mutated) suggests that genetic variation in germ lines V_H, D or J_H Ig gene segments have the *potential* to play a role in inherited variation in the response to F.VIII.

T-cell receptors in immune responses

M. A. Robinson

Laboratory of Immunogenetics
National Institute of Allergy and Infectious Diseases
National Institutes of Health
Rockville, Maryland 20852

T lymphocytes play a central role in the generation of both humoral and cell-mediated immune responses. Although T-cells mediate diverse effector functions, most recognize antigenic peptide presented by molecules encoded within the major histocompatibility complex (MHC). The fine specificity of a T-cell is determined by a heterodimeric receptor for antigen (TCR) displayed on the cell surface. The TCR present on the majority of T-cells is composed of an alpha and a beta chain which are similar to immunoglobulins (Ig) in both sequence and gene organization. Genes for the alpha and beta chains of the TCR are encoded in germline DNA as discontinuous gene segments that rearrange specifically in T-cells during development. The variable regions of each of these chains are responsible for antigen recognition and are encoded by juxtaposed variable (V), diversity (D) (for beta chains), and joining (J) gene segments (reviewed in Toyonaga and Mak, 1987; Wilson et al., 1988; Davis and Bjorkman, 1988). A critical difference between TCR and Ig is that somatic mutation does not appear to play a role in the generation of TCR diversity (Ikuta et al., 1985), which emphasizes the importance of germline polymorphism of TCR gene segments. Knowledge of both the extent and variability of the germline TCR repertoire is critical to understanding the diversity of potential T-cell specificities.

Genetic variations in immune responses

The ability to recognize a vast array of antigens with a high degree of specificity is a function of both B and T lymphocytes. Although B lymphocytes produce antibodies that directly recognize antigens, T lymphocytes are required to provide "help" to the B lymphocytes for the production of large quantities of high-affinity antibodies. The CD4+ subset of T lymphocytes provides "help" to B cells and recognizes processed peptide antigen in the context of self MHC Class II molecules. This cellular interaction, central to immune response, is depicted schematically in Figure 1.

Inhibitors to Coagulation Factors
Edited by Louis M. Aledort *et al.*, Plenum Press, New York, 1995

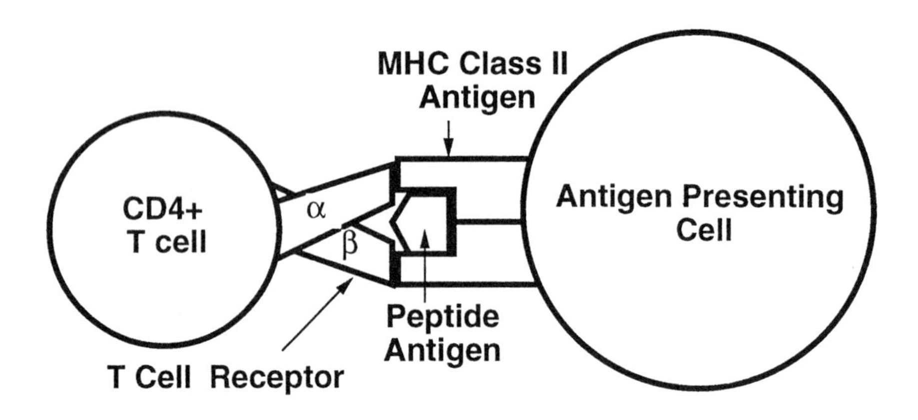

Figure 1. Helper T lymphocytes are CD4+ and recognize peptide antigens in the context of self MHC Class II antigens.

It has long been known that variation in MHC genes has an important impact on immune responses. The products of MHC genes are highly polymorphic; there are multiple alleles of each Class I and Class II antigen. A known function of MHC antigens is to present antigen to T cells. Each MHC antigen is capable of presenting many different peptides, but a given MHC antigen cannot present all peptides. The failure to produce a response to a specific peptide may be due to the inability of the individual's MHC antigens to present the peptide (see Figure 2A). The differential ability of MHC antigens to bind peptide antigens has been referred to as "epitope selection" (Schaeffer et al., 1989).

There is a good correlation between responses and binding of peptides to MHC antigens; however, the correlation is not absolute. Certain peptides have been found to bind an MHC product, even with a high affinity, and yet not elicit a T-cell response in certain individuals. Thus, epitope selection is not the only mechanism by which T-cell responses are determined and polymorphism of MHC products is not the only factor critical to the regulation of an immune response. One theory to explain the failure of a MHC-peptide combination to elicit a T-cell response in certain individuals has been referred to as a "hole in the T cell repertoire" (Schaeffer et al., 1989) and assumes that the defect lies in the T cell. This theory is depicted in Figure 2B. Possible reasons for limitations in the T-cell repertoire include the lack of relevant functional genes in germline TCR repertoire, the depletion of reactive cells from the peripheral repertoire, or a state of nonresponsiveness caused by suppression or anergy. The present review will focus on the germline TCR repertoire, the mechanisms used to generate diversity in TCR and a description of the extent and polymorphism of TCR genes in the human population.

Generation of diversity in TCR

Diversity in TCR is generated using a number of mechanisms that are similar to those employed in the generation of immunoglobulin (Ig) diversity. Both products, T-cell receptor and Ig molecules, are composed of two chains; TCR has alpha and beta chains and Ig has H and L chains. Each chain is encoded by discontinuous gene segments that rearrange to form a mature product. Diversity arises from the multiple combinations of these gene segments that are possible. TCR alpha and beta chains and Ig H and L chains are composed of variable (V), joining (J), and constant gene segments and TCR ß and Ig H chains have diversity (D) gene segments as well.

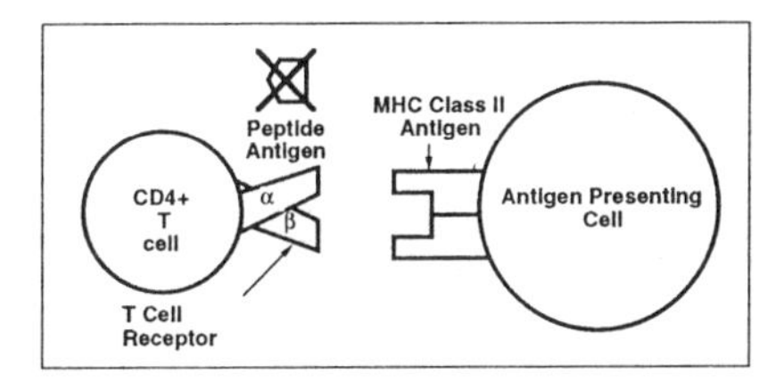

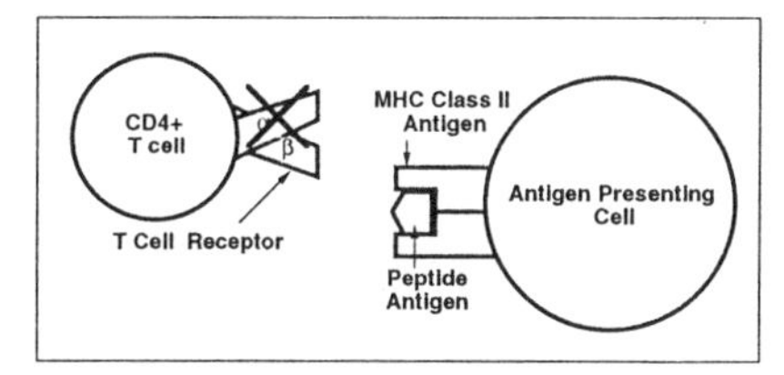

Figure 2. Determinant selection (panel A) and "Holes in the T-cell repertoire" (panel B) are two theories to describe mechanisms of MHC control of immune responses. A) Determinant selection predicts that control of an immune response is at the level of MHC binding and polymorphism in Class II molecules determines the peptides that will be recognized by T cells. B) The "Hole in the T-cell repertoire" theory predicts that the control of immune responses lies with the T cell. The inability to produce a particular immune response is due to the lack of T cells that recognize a specific MHC-peptide combination.

Two mechanisms are responsible for diversity at the junctions of V, D, and J gene segments. First, there is considerable flexibility in the number of nucleotides incorporated into the final rearranged product that derives from the D gene segments and from the 5' end of the J segment. The human TCR-ß-D1 gene can be translated in all three reading frames, and TCR-ß-D2 can be translated in two different reading frames allowing for variation. The second mechanism for diversity at the junctions of V, D, and J gene segments is the random addition of nucleotides. These additions, termed N regions, may be variable in length. In order for the rearranged product to be functional, an open reading frame must be maintained by the cumulative number of additions (N regions) and subtractions (use of portions of D and J gene segment) of nucleotides at junctions.

Additional diversity in TCR derives from combinations of different α and ß chains. The rearrangement process occurs independently at the loci encoding both chains of the TCR (and Ig) molecules. If a rearrangement process produces a nonfunctional product, the gene encoded on the other chromosome can undergo rearrangement. However, once the gene for a given chain has undergone a productive, in-frame rearrangement, there is no further rearrangement activity at that locus. This phenomenon is referred to as allelic exclusion. Thus, each T cell has one productively rearranged α chain gene and one productively rearranged ß chain gene. However, recent evidence has reported that allelic exclusion of α chains is not stringent. More than one α chain was found to be expressed by individual T cells (Padovan et al., 1993). This may provide an additional mechanism for diversity of TCR.

There is one very important distinction between Ig and TCR in the mechanisms used to generate diversity. Although somatic mutation is an important mechanism in the maturation of an antibody response, somatic mutation does not appear to play a role in the generation of TCR diversity (Ikuta et al., 1985). Sequences of the V gene segments in mature TCR transcripts are identical to the sequences of germline genes. This observation emphasizes the importance of the extent and polymorphism of the germline TCR repertoire.

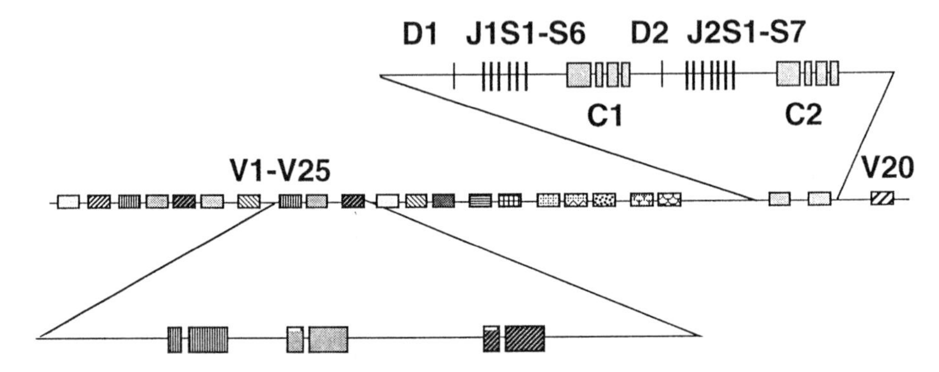

Figure 3. A schematic representation of the human TCR ß gene complex showing the relative locations of variable (V), diversity (D), joining (J), and constant (C) gene segments. There are 62 TCR ßV gene segments grouped into 25 TCR ßV families (labeled V1-V25). More than half of the families contain a single V gene segment and others may contain as many as 9 V gene segments (family members share >75% sequence identity). A total of 52 of the TCR ßV gene segments encode functional genes. The TCR ßV20 gene is located 3′ of the constant region. Individual D and J gene segments and the intron/exon structure of C genes are shown above and the intron/exon structure of V genes is shown below.

Organization of the human TCRß gene complex

The human TCRß gene complex spans approximately 600kb (Robinson et al., 1993) and is encoded at chromosome 7q35 (Isobe et al., 1985). A schematic representation of the TCRß gene complex is shown in Figure 3. It contains 2 C gene segments, each associated with one D gene segment and 6-7 J gene segments that are located immediately 5′ of the C genes. The majority of the TCRßV gene segments are located 5′ of the constant region and are dispersed over a distance of approximately 500kb.

The first step toward determining the extent of the TCRß V repertoire was to determine how many TCRß V genes were present in the gene complex. One approach involved counting novel cDNA sequences and comparison of this number with the number of bands observed on Southern blots detected using TCRß V probes (Concannon et al., 1986; Concannon et al., 1987; Robinson, 1991). These methods have the potential for either over- or under-estimations of TCRß V gene numbers. For example, gene segments expressed at low levels may not be detected by cDNA cloning (Robinson, 1992). In contrast, the number of distinct cDNA sequences that have been reported for some TCRß V families vastly exceeds the number of bands observed on Southern blots. Counting V gene segments by comparison of sequences can lead to over-estimation of the number of TCRß V gene segments due to germline polymorphism or to errors in reported sequences. The presence of errors is of particular concern, since many of the reported sequences have been obtained by cloning PCR amplified products. Taq polymerase, used in PCR, has been shown to introduce sequence errors in amplified DNA. In contrast, detection of TCRß V gene segments by Southern blotting may fail to detect V gene segments clustered on common fragments, or located on distinct fragments that migrate at the same position. Finally, Southern blotting does not allow discrimination between functional genes and pseudogenes and will detect only members of families for which probes have been previously isolated.

Another approach that has proven useful toward establishing the extent of the human TCRß germline repertoire is genomic mapping. Genomic mapping allows assignment of distinct TCRß V sequences to specific sites within the TCRß locus making it possible to unambiguously discriminate between the products of alleles of the same gene segment as opposed to different gene segments. Genomic cloning studies confirmed the existence of 63 unique human TCRß V gene segments grouped into 25 families.

A TCRß V family is a group of sequences that share >75% identity at the level of nucleotide sequence. This is a useful definition since genes sharing >75% identity will generally cross hybridize on Southern blots. Although the total number of gene segments is 63, not all gene segments may contribute to the expressed TCRß V repertoire. Genomic mapping and sequence analysis has facilitated identification of functional genes and pseudogenes. One cluster of TCRß V genes was found to be the result of a translocation event and even though some of these genes appeared to be functional by sequence analysis, all genes within the translocated cluster were characterized as pseudogenes.

TCR orphon genes

A cluster of duplicated V gene segments maps outside of the TCRß gene complex encoded on chromosome 7 and is located on chromosome

9p21 (Robinson et al., 1993; Charmley et al., 1993b). These V genes on chromosome 9 have been designated "orphon genes" using the nomenclature established for immunoglobulin V genes encoded outside of the Ig gene complexes (Huber et al., 1990; Borden et al., 1990; Matsuda et al., 1990). TCRß V orphon genes include TCRß V2(O), TCRß V4(O), TCRß V10(O), TCRß V11(O), TCRß V15(O) and TCRß V19(O). The orphons are presumably nonfunctional even though some of them have sequences containing no apparent defects that might preclude translation. In addition, although the orphon genes have heptamer-nonamer rearrangement sequences similar to those 3′ of functional TCRßV genes, there is no evidence for rearrangement of orphon V genes with the constant gene segments encoded on chromosome 7. The rearrangement signals 3′ of the orphon TCRß V genes may, however, be involved in aberrant rearrangements in certain neoplasia. The 9p21-22 region has been reported to be nonrandomly involved in deletions and translocations in leukemia patients (Diaz et al.,1990).

The cluster of orphon TCR genes does not represent a precise duplication of TCRß V genes encoded within the TCRß gene complex. Two of the TCRß V genes present within the cluster of functional TCRß genes encoded on chromosome 7 do not have orphon counterparts. It appears that these two genes were deleted during the duplication or translocation process. Although the orphon genes have no known function, they may have evolutionary significance. Such translocation events have the possibility of generating a novel locus. Gene duplication has been postulated as a major force in the evolutionary process whereby new gene families have developed and diversified.

Extent of the expressed TCRß V repertoire

Based upon sequence analysis and detection of the gene segments in cDNA clones, a total of 52 of the 63 TCRß V gene segments are thought to be functional. A list of the 25 TCRß V families (sequences sharing >75% sequence identity), the number of genes in each family, and the number of these genes that appear to be functional is shown in Table 1. In addition to the orphon genes, five gene segments within the TCRß gene complex are pseudogenes. Two TCRß V6, two TCRß V8 and one of the TCRß V9 gene segments contain in frame stop codons within the coding region and therefore do not encode a functional molecule. Almost half of the human TCRß V families contain a single gene segment. The TCRß V gene families that contain more than one gene most likely derived from duplication events. In some cases groups of genes were duplicated as a unit and appear as repeating clusters. For example, TCRß V5, TCRß V6, and TCRß V13 are the largest families and are found together in the gene complex as a tandem series of duplicated genes. Likewise genes in the TCRß V8, TCRß V12, and TCRß V21 families are present together at three positions within the TCRß gene complex. The

TCRß V genes in these clusters have diverged sufficiently to have distinctive sequences but are related to family members in both coding region and flanking sequences.

Table 1. Estimated size of human germline TCRβ V repertoire

TCRß V Family	Bands on Southern blots	Gene number	Number of functional genes
BV1	1	1	1
BV2	2	2	1
BV3	1	1	1
BV4	2	2	1
BV5	5-8	6	6
BV6	8-10	9	7[a]
BV7	3	3	3
BV8	5	5	3
BV9	2	2	1
BV10	2	2	1
BV11	2	2	1
BV12	3	3	3
BV13	6-9	9	9
BV14	1	1	1
BV15	2	2	1
BV16	1	1	1
BV17	1	1	1
BV18	1	1	1
BV19	2	2	1
BV20	1	1	1[a]
BV21	3	3	3
BV22	1	1	1
BV23	1	1	1
BV24	1	1	1
BV25	1	1	1
Total	58-66	63	52

[a] Common null alleles exist for a TCRß V gene in these families (Barron and Robinson, in press and Charmley et al., 1993a)

Allelic variation in TCRß V genes

Additional variability resulting in either expansion or contraction of the basic TCRß V repertoire derives from polymorphism of germline TCRß V genes. Armed with the knowledge of the number and location of the TCRß V gene segments, it is possible to establish whether closely related but distinct sequences correspond to allelic or alternative isotypic genes. Nucleotide substitutions in coding regions have been reported for a number of TCRß V genes; those resulting in replacement substitutions within the TCRß V coding regions are listed in Figure 4. The locations of the substitutions are shown with respect to regions thought to correspond to functional regions of the molecule.

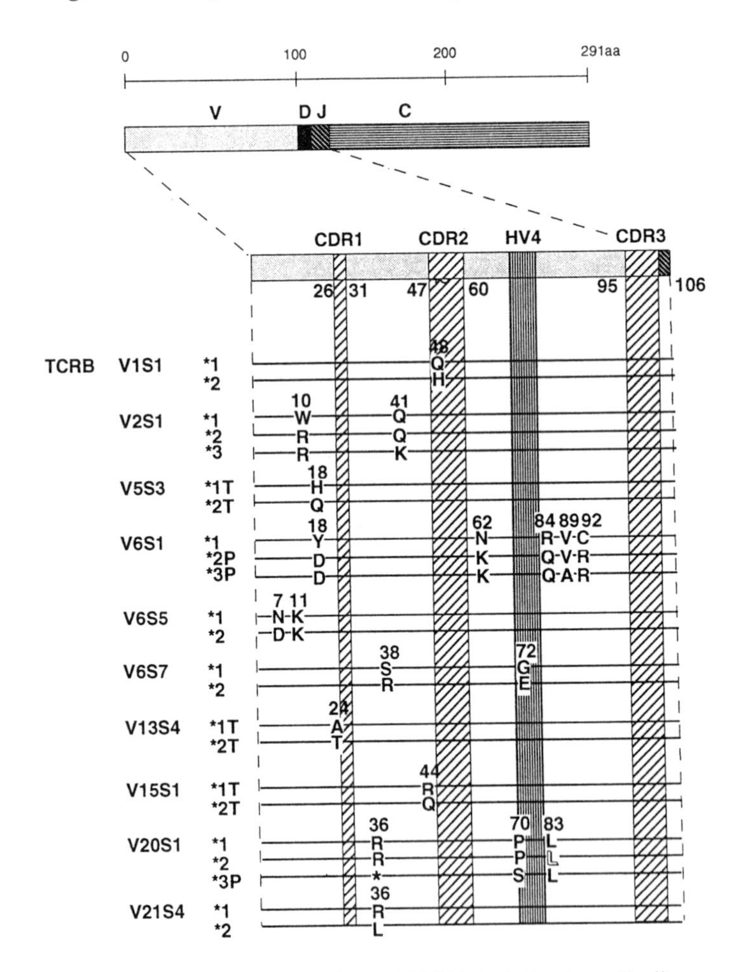

Figure 4. Listing of TCRß V alleles. A TCRß V chain is schematically represented showing V, D, J, and C regions. Regions of hypervariability are highlighted and residue numbers are given for CDR1, CDR2, and CDR3. Alleles of TCRß V genes are listed showing the locations of polymorphic residues. An asterisk (*) represents a stop codon and outline type (L) represents a silent substitution. Sources for sequences are as follows: TCRß V1S1 (Robinson, 1989); TCRß V2S1 (Cornelis et al., 1993); TCRß V5S3 (Plaza et al., 1991); TCRß V6S1 (Barron and Robinson, in press); TCRß V6S5 (Hansen et al., 1992); TCRß V6S7 (Li et al., 1990); TCRß V13S4 (Plaza et al., 1991); TCRß V15S1 (Baccala et al., 1991); TCRß V20S1 (Charmley et al., 1993a); and TCRß V21S4 (Hansen et al., 1992). Substitutions reported for TCRß V7S2, TCRb V8S2, TCRß V12S2, and TCRß V16 (Plaza et al., 1991, Day et al., 1992) are silent and thus do not result in protein sequence variation for allelic forms of these genes. Certain sequences apparently correspond to alleles, but since segregation in families has not been reported these are given tentative allelic designations and marked (T).

Structural models of TCR based upon sequence homology with Ig and the known crystal structure of Ig molecules suggest that the TCR antigen/MHC binding site localizes to three areas (Chothia et al., 1988). Two of these regions are encoded within the V gene segment and the third is generated by combinatorial and junctional mechanisms during the assembly of V, D, and J gene segments. A fourth region of hypervariability is found in TCRß V chains and is thought to contain residues important for interactions between superantigens and TCR molecules. Replacement substitutions occurring in the putative binding site may alter TCR function by changing the antigen/MHC binding site; substitutions that occur in framework regions may have an impact upon conformation of the molecule and affect expression by changing alpha/beta chain pairing patterns. Identification of alleles makes it possible to examine the impact substitutions may have on specific immune responses and knowledge of the position of the substitutions may reveal functional regions of the molecules.

Comparison of the TCR alpha and beta chain repertoire of inbred strains of mice responding to defined peptide antigens have revealed that even a single replacement substitution within a TCRß V chain can have a significant impact on the TCR gene segments utilized. Different TCR usage between two strains of mice in response to a specific antigen was attributed to an allelic substitution in TCRß V3 located in CDR1. Consistent with the hypothesis that portions of the TCR molecule have discrete functions, responses to stimulation by a superantigen specific for TCRß V3 was the same between the two strains (Gahm et al., 1991). In another study, strains of mice having TCRß V17 molecules that differ for two residues within HV4, showed different response patterns to superantigens (Cazenave et al., 1991).

Perhaps the most dramatic type of allelic variation is where one allele contains substitutions causing it to be nonfunctional. For example, studies of mice have revealed that strains of mice with deletions in the germline TCRß V repertoire or depletions of TCR due to self-tolerance may be nonresponsive to certain antigens (Nanda et al., 1991). The listing of TCRß V alleles in Figure 4 includes two genes, TCRß V6S1 and TCRß V20, that have null alleles or alleles that contain a defect causing them to not be expressed. The null allele of TCRß V20 contains an in-frame stop codon. The two null alleles of TCRß V6S1 contain an arginine replacing the conserved cysteine at position 92 (Luyrink et al., 1993 and Barron, et al., in press). A cysteine at this position is found in all TCRß V chains and all TCRAV chains. It is thought that replacement of the cysteine with an arginine in the null alleles of TCRß V6S1 does not allow the translated products to appropriately associate with an TCRA chain and therefore preclude its expression on the surface of a T cell.

The human TCRβ gene complex has two regions of insertion or deletion (Seboun et al., 1989b) and one of these regions contains TCRß V gene segments. These polymorphisms, which have been termed insertion/deletion related polymorphisms (IDRP), each have two alternate forms that are present in the population at approximately equal frequencies. The IDRP (~30kb) located in the TCRß V region was found to contain three TCRß V gene segments; however, the TCRß C region IDRP (~20kb) was found to contain no TCRß V genes. One of the TCRß V genes in the V region IDRP has a unique sequence: one is an identical copy of a TCRß V gene present in all haplotypes and one is a pseudogene. Thus, the genomic TCR repertoire is expanded in individuals with inserted TCRß V haplotypes (MAR, unpublished results).

Conclusion

Knowledge of the extent of the germline TCR repertoire derived from the development of an extended map of the human TCRß gene complex. The TCRß complex encoded at 7q35 (Isobe et al., 1985) spans a distance of approximately 600 kb. There are a total of 63 TCRßV genes segments; however, approximately 10% of the known TCRß V gene segments are located outside of the TCRß gene complex. Orphon TCRß V genes encoded on chromosome 9 are not involved in productive rearrangement events with other gene segments within the TCRß gene complex and thus do not contribute to TCR diversity. When the orphon status of TCRß V genes and the number of TCRß V genes within the TCRß gene complex that are pseudogenes is considered, the number of functional TCRß V gene segments can be reduced to 52.

The involvement of T lymphocytes in the pathogenesis of a variety of autoimmune diseases has prompted numerous studies to determine the role of TCR genes in disease (Nepom and Concannon, 1992; Robinson and Kindt, 1992; Posnett et al., 1988). Population based association studies (Beall et al. 1989) and linkage studies in families (Seboun et al., 1989a) suggest that germline polymorphism of TCRß V gene segments is functionally relevant to disease susceptibility. Knowledge of the genomic organization and polymorphism of human TCRß V gene segments will be necessary in order to further localize TCRß V gene segments that have an impact upon disease susceptibility.

References

Baccala, R., Kono, D.H., Walker, S., Balderas, R.S., and Theofilopoulos, A.N.: Genomically imposed and somatically modified human thymocyte Vß gene repertoires. *Proc Natl Acad Sci USA 88*: 2908-2912, 1991

Barron, K.S. and Robinson, M.A.: The human T-cell receptor variable gene segment, TCRß V6S1 has two null alleles. *Human Immunol.*: in press

Beall, S.S., Concannon, P., Charmley, P., McFarland, H.F., Gatti, R.A., Hood, L.E., McFarlin, D.E. and Biddison, W.E.: The germline repertoire of T-cell receptor beta-chain genes in patients with chronic progressive multiple sclerosis. *J Neuroimmunol 21*: 59-66, 1989

Borden, P, Jaenichen, R., and Zachau, H.G.: Structural features of transposed human VK genes and implications for the mechanism of their transpositions. *Nucleic Acids Res 18*: 2101-2107, 1990

Cazenave, P-A., Marche, P., Jouvin-Marche, E., Voegtle, D., Bonhomme, F., Bandeira, A. and Coutinho, A.: Vß17-gene polymorphism in wild-derived mouse strains: Two amino acid substitutions in the Vß17 region alter drastically T-cell receptor specificity. *Cell 63*: 717

Charmley, P., Wang, K., Hood, L., and Nickerson, D.A.: Identification and physical mapping of a polymorphic human T-cell receptor V beta gene with a frequent null allele. *J Exp Med 177*:135-143, 1993a

Charmley, P., Wei, S., and Concannon, P.: Polymorphisms in the Tcrß-V2 gene segments localize the Tcrb orphon genes to human chromosome 9p21. *Immunogenetics 38*:283-286, 1993b

Chothia, C., Boswell, D.R. and Lesk, R.M.: The outline structure of the T-cell alpha beta receptor. *EMBO J*: 3745-3755, 1988

Concannon, P., Gatti, R.A. and Hood, L.E.: Human T-cell receptor Vß gene polymorphism. *J. Exp. Med. 165*: 1130-1140, 1987

Concannon, P., Pickering, L.A., Kung, P. and Hood, L.: Diversity and structure of human T-cell receptor beta-chain variable region genes. *Proc Natl Acad Sci USA 83*, 6598-6602, 1986

Cornelis, F., Pile, K., Loveridge, J., Moss, P., Harding, R., Julier, C.,and Bell, J.: Systematic study of human alpha beta T-cell receptor V segments shows allelic variations resulting in a large number of distinct T-cell receptor haplotypes. *Eur J Immunol 23*: 1277-1283, 1993

Davis, M.M. and Bjorkman, P.J.: T-cell antigen receptor genes and T-cell recognition, *Nature 334*: 395-402, 1988

Day, C.E., Zhao, T. and Robinson, M.A.: Silent allelic variants of a T-cell receptor Vß12 gene are present in diverse human populations. *Hum Immunol 34*: 196-202, 1992

Diaz, M.O., Ziemin, S., Le Beau, M.M., Pitha, P., Smith, S.D., Chilcote, R.R., Rowley, J.D.: Homozygous deletion of the alpha- and beta 1-interferon genes in human leukemia and derived cell lines. *Proc Natl Acad Sci USA 85*: 5259-5263, 1988

Gahm, S-J., Fowlkes, B.J., Jameson, S.D., Gascoigne, N.R.J., Cotterman, M.M., Kanagawa, O., Schwartz, R.H., and Matis, L.A.: Profound alteration in an αß T-cell antigen receptor repertoire due to polymorphism in the first complementarity-determining region of the ß chain. *Proc Natl Acad Sci.USA*, 88: 10267-10271, 1991

Hansen, T., Ron ningen, K.S., Ploski, R., Kimura, A., and Thorsby, E.: Coding region polymorphisms of human T-cell receptor V beta 6.9 and V beta 21.4. *Scand J Immunol 36*: 285-290, 1992

Huber, C., Thiebe, R., Hameister, H., Smola, H., Lötscher, E. and Zachau, H.G.: A human immunoglobulin kappa orphon without sequence defects may be the product of a pericentric inversion. *Nucleic Acids Res 18*: 3475-3478, 1990

Ikuta, K., Ogura, T., Shimizu, A. and Honjo, T.: Low frequency of somatic mutation in beta-chain variable region genes of human T-cell receptors. *Proc Natl Acad Sci USA 82*: 7701-7705, 1985

Isobe, M., Erikson, J., Emanuel, B., Nowell, P., and Croce, C.: Location of gene for beta subunit of human T-cell receptor at band 7q35, a region prone to rearrangements in T cells. *Science 228*: 580-583, 1985

Li, Y., Szabo, P., Robinson, M.A., Dong, B., Posnett, D.N.: Allelic variations in the human T-cell receptor V beta 6.7 gene products. *J Exp Med 171*: 221-230, 1990

Luyrink, L., Gabriel, C.A., Thompson, S.D., Grom, A.A., Maksymowych, W.P., Choi, E., and

Glass, D.: Reduced expression of a human Vß6.1 T-cell receptor allele. *Proc Natl Acad Sci USA 90*: 4369-4373, 1993

Matsuda, F., Shin, E.K., Hirabayashi, Y., Nagaoka, H., Yoshida, M.C., Zong, S.Q. and Honjo, T.: Organization of variable region segments of the human immunoglobulin heavy chain: Duplication of the D5 cluster within the locus and interchromosomal translocation of variable region segments. *EMBO J 9*: 2501-2506, 1990

Nanda, N.K., Apple, R., and Sercarz, E.: Limitations in plasticity of the T-cell receptor repertoire, *Proc Natl Acad Sci USA 88:* 9503-9507, 1991

Nepom, G.T. and Concannon, P.: Molecular genetics of autoimmunity. In: Rose, N.R., and Mackay I.R. (eds). The Autoimmune Diseases. Academic Press, Inc., N.Y. pp. 127-152, 1992

Padovan, E. , Casorati, G., Dellabona, P., Meyer, S., Brockhaus, M., and Lanzavecchia, A., Expression of two T-cell receptor χ chains: Dual receptor T-cells. *Science 262*: 422-424, 1993

Plaza, A., Kono, D.H. and Theofilopoulos, A.N.: New human Vß genes and polymorphic variants. *J Immunol 147*: 4360-4365, 1991

Posnett, D.N., Gottlieb, A., Bussel, J.B., Friedman, S.M., Chiorazzi, N., Li, Y., Szabo, P., Farid, N.R., Robinson, M.A.: T-cell antigen receptors in autoimmunity. *J Immunol 141*: 1963-1969, 1988

Robinson, M.A.: Allelic sequence variations in the hypervariable region of a T-cell receptor ß chain: Correlation with restriction fragment length polymorphism in human families and populations. *Proc Natl Acad Sci USA 86*: 9422-9426, 1989

Robinson, M.A.: The human T-cell receptor beta-chain gene complex contains at least 57 variable gene segments: Identification of six Vbeta genes in four new families. *J Immunol 146*, 4392-439,1991

Robinson, M.A.: Usage of human T-cell receptor Vß, Jß, Cß, and Vα gene segments is not proportional to gene number. *Hum Immunol. 35*: 60-67, 1992

Robinson, M.A., and Kindt, T.J.: Linkage between T-cell receptor genes and susceptibility to multiple sclerosis: A complex issue. *Regional Immunol 4*: 274-283, 1992

Robinson, M.A., Mitchell, M.P., Wei, S., Day, C.E., Zhao, T.M., Concannon, P.: Organization of human T-cell receptor beta-chain genes: Clusters of V beta genes are present on chromosomes 7 and 9. *Proc Natl Acad Sci USA 90*: 2433-2437, 1993

Schaeffer, E.B., Sette, A., Johnson, D.L., Bekoff, M.C., Smith, J.A., Grey, H.M., and Buus, S. Relative contribution of "determinant selection" and "holes in the T-cell repertoire" to T-cell responses. *Proc Natl Acad Sci USA 86:* 4649-4653, 1989

Seboun, E., Robinson, M.A., Doolittle, T.H., Ciulla, T.A., Kindt, T.J., and Hauser, S. L.: A susceptibility locus for multiple sclerosis is linked to the T-cell receptor beta chain complex. *Cell 57*: 1095-1100, 1989a

Seboun, E., Robinson, M.A., Kindt, T.J. and Hauser, S.L.: Insertion/deletion-related polymorphisms in the human T-cell receptor beta gene complex. *J Exp Med 170*: 1263-1270, 1989b

Toyonaga, B. and Mak, T.W.: Genes of the T-cell antigen receptor in normal and malignant T cells. *Annu Rev Immunol 5*: 585-620, 1987

Wilson, R.K., Lai, E., Concannon, P., Barth, R.K. and Hood, L.E.: Structure, organization and polymorphism of murine and human T-cell receptor alpha and beta chain gene families. *Immunol Rev 101*: 149-204. 1988

Immunoglobulin heavy chain variable region gene usage in human autoimmune diseases

Virginia Pascual and J. Donald Capra

Department of Microbiology
UT Southwestern Medical Center
Dallas, Texas 75235

The variable regions of immunoglobulin heavy and light chains are assembled in pre-B cells by the somatic joining of genetic elements which, in the chromosome, are separated by thousands of base pairs. The repertoire of germline variable region genes represents the substrate upon which diversity is generated in the antibody system, since the recombination of one among different available V_H, D, and J_H, as well as V_L and J_L gene segments, provides each B lymphocyte with a unique immunoglobulin receptor molecule. Diversity is enhanced several orders of magnitude by the generation of junctional amino acids during the process of rearrangement and the combinatorial association of heavy and light chains. Somatic mutation and gene conversion are additional mechanisms which, working upon the already assembled antibody molecule, lead to a practically unlimited number of antibody specificities (reviewed in Reference 1).

The genes encoding the variable region of the human immunoglobulin heavy chain are located in chromosome 14, band 14q32.33. Current estimations suggest the presence of around 100 V_H, 30 D, and 6 J_H gene segments. Based on nucleotide sequence homology V_H genes are classified into 6-7 families ranging in size from one single member to more than 25 (see Table 1). Two important characteristics of the human V_H locus are the high content in pseudogenes (30 to 50% of the total number of genes), and the high level of interdigitation among members of different families.[1,2]

During murine fetal development, the V_H and D gene segments located more 3' in the chromosome are preferentially rearranged. Programmed rearrangement is maintained in adult bone marrow, suggesting that immunoglobulin gene expression during ontogeny is the result of intrinsic mechanisms (position in the chromosome, accessibility to the recombinase machinery) independent of antigen and/or environ-

Inhibitors to Coagulation Factors
Edited by Louis M. Aledort *et al.*, Plenum Press, New York, 1995

mental influences.[3,4] A restricted repertoire of V_H gene segments has also been described during human fetal development.[5,6] However, the most

Table 1. Complexity of the human V_H families

	Number of genes
V_H1	25
V_H2	4
V_H3	50
V_H4	12
V_H5	3
V_H6	1

commonly expressed fetal liver V_H genes do not map 3' in the V_H locus, suggesting that factors other than chromosomal position influence the repertoire even at early stages of development.[7,8]

V_H utilization in autoimmune situations

A large body of information regarding the usage of immunoglobulin genes in autoimmune diseases has accumulated in the last three to four years. Interestingly, most of the V_H genes expressed in human autoantibodies are also found in the fetus. A striking difference regarding the frequency of expression of some human V_H families is observed, however, when both repertoires are compared. To date, we have sequenced up to 100 human autoantibodies from patients with autoimmune diseases and found that members of one of the intermediate size families, V_H4, represent over 1/3 of our sample. Interestingly, up to 50% of these V_H4-expressing antibodies use a single member of the family, the V_H4-21 gene segment.[9]

Most autoantibody responses are not restricted in the usage of V_H, D, or J_H gene segments

Structural data on the V_H regions of human autoantibodies became available at the protein level through the study of human monoclonal paraproteins with rheumatoid factor (RF) activity. These antibodies were found to express motifs recognized by anti-idiotypic cross-reactive antibodies, suggesting a common germline origin.[10] Since autoimmune manifestations are not usually present in patients with monoclonal gammopathies, the question of how these monoclonal paraproteins with RF activity compared to the autoantibodies from patients with Rheumatoid arthritis (RA) was raised. In an attempt to answer this question, polyclonal and monoclonal reagents recognizing cross-reactive idiotypes (CRI) expressed on monoclonal RF heavy and light chains were used to test their expression in the polyclonal population of RF autoantibodies present in the serum of patients with RA. These studies showed that the CRI characteristic of monoclonal RF paraproteins represented a minor proportion of the IgM RF present in patients with classical RA, suggesting that RFs in RA derived from a different subset of germline genes or that they had been the subject of extensive somatic mutation. In the past few years we have analyzed the nucleotide sequences of 17 monoreactive and polyreactive antibodies with RF activity derived from the synovial tissue of two patients with classical RA and one patient with the juvenile form of the disease. Our results led us to conclude that RF produced locally in the synovium derive from a diverse array of V_H and V_L gene segments, some of them displaying evidence of having undergone somatic mutation in the context of an affinity maturation process (reviewed in reference 11). These observations apply to most autoimmune responses; anti-acetylcholine receptor and anti-striated muscle antibodies, the hallmark of patients with myasthenia gravis, are also encoded by a wide array of heavy and light chain gene segments. Sequence analysis reveals that some of these gene segments display 100% identity with the germline, although examples of somatically mutated molecules are easily found as well.[12,13]

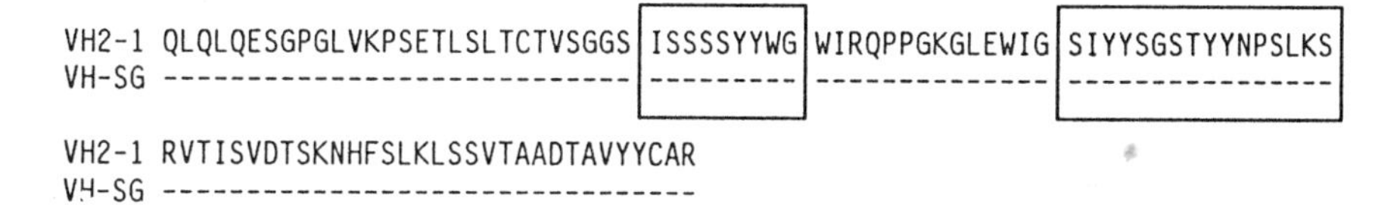

Figure 1. The deduced amino acid sequence of the heavy chain variable region of an anti-Topoisomerase I antibody (VH SG)[14] compared to its germline counterpart (VH2-1). Dashes represent amino acid identity. Boxes denote the two hypervariable regions.

```
GL-SJ  QVQLVESGGGVVQPGRSLSCAASGFTFS SYAMH WVRQAPGKGLEWVA VISYDGSNKYYADSVKG
SJ-2   ---------------------------- ----- -------------- -----------------
SJ-1   -------------------------S-- --G-- ----C--------- ---D-------------

GL-SJ  RFTISRDNSKNTLYLQMNSLRAEDTAVYYCAR
SJ-2   -------------S------------------
SJ-1   ----------K-----S----D---------K
```

Figure 2. The deduced amino acid sequence of the heavy chain variable regions of two clonally related RF (SJ-2 and SJ-1) isolated from the synovial membrane of a RA patient[15]. GL-SJ represents their closest germline V_H gene segment. Dashes represent amino acid identity. Boxes denote the two hypervariable regions.

Is autoimmunity imprinted in the germline?

Examples of autoantibodies fully encoded within the germline are abundant in the literature. Even though some of the more thoroughly studied autoantibody responses are not exclusively found in patients with autoimmune disorders (RF arise, for example, after certain infections and even immunizations), we and others have evidence to support that autospecificities arising in patients with disease can be germline encoded (see Figure 1). Examples of highly mutated, high affinity autoantibodies are also profuse. We have described, for example, the nucleotide sequences of two clonally related RF isolated from the synovial tissue of one patient with RA. While one of the antibodies (RFSJ2) was almost germline encoded, the second (RFSJ1) had accumulated extensive somatic mutations and displayed 100-fold higher affinity toward the antigen.[15] In the same context, we recently had the opportunity to analyze a series of monoclonal and recombinant antibodies derived from patients with primary biliary cirrhosis (PBC). These autoantibodies react with one of the components of the pyruvate kinase dehydrogenase complex (PDC-E2), and are selectively found in patients with PBC.[16] Interestingly, while we found that some of these antibodies were encoded by heavy and light chain genes identical to the germline, analysis of a combinatorial IgG library obtained from the mesenteric lymph node of a PBC patient undergoing liver transplantation disclosed several examples of anti-PDC-E2 Fabs encoded by clonally related, highly mutated heavy and light chain immunoglobulin genes which very likely arose in the context of an *in vivo* germinal center reaction.[17] These examples highlight the concept that autoimmune responses behave in many ways like secondary responses against foreign antigens.

Some V_H gene segments seem to be particularly relevant to the autoimmune repertoire

Although most autoimmune responses are not restricted in the usage of heavy or light immunoglobulin genes, there are a few exceptions to this rule. Cold agglutinins, for example, are antibodies that bind to the surface of erythrocytes and cause agglutination at 4°C. We and others have found that the majority of molecules with cold agglutinin activity recognizing the I/i carbohydrates on human red blood cells are encoded by a single member of the human V_H4 family, the V_H4-21 gene segment, in combination with different D, J_H, and light chain gene segments.[18-20] The same V_H gene segment has been described encoding a wide array of auto and allospecificities (see Table 2), and represents the most common V_H gene segment in the sample of autoantibodies that we have analyzed.[9,21]

The overexpression of the V_H4-21 gene segment in the autoimmune repertoire should, however, be cautiously interpreted. Several lines of evidence suggest that this gene segment might possess structural advantages for being preferentially transcribed and/or rearranged; it has been described as sterile transcript in T-cell leukemias,[22] and rearranged in the scarce B-cell repertoire of one patient with X-linked agammaglobulinemia[23]. From a functional perspective it seems likely that the product of this V_H gene is particularly relevant to antibody responses against the red blood cell. For example, the V_H4-21 gene segment is not only almost universally involved in recognition of the I/i carbohydrates characteristic of the cold agglutinin response, but the IgM allogeneic response of individuals immunized against different blood group antigens, with no structural similarity with the I carbohydrate, is also

Table 2. Examples of autospecificities encoded by the V_H4-21 gene segment[1]

Disease	Specificity
Rheumatoid arthritis	Monospecific RF
	Polyreactive RF
Systemic lupus erythematosus	Anti-DNA
Myastenia gravis	Anti-striated muscle
Wiscott-Aldrich	Anti-lymphocyte
Cold agglutination disease	Cold agglutinin
Normal donors	Anti-blood group antigens

[1] Reviewed in reference 9.

preferentially encoded by this V_H gene segment in combination with different D, J_H, and light chains.[21] These observations led us to hypothesize about the existence of a "non conventional" binding site within the region encoded by the V_H4-21 gene segment.[24]

The V_H4-21 gene segment, on the other hand, is hardly found during human ontogeny in the fetal liver (Pascual and Capra, unpublished observations), suggesting that the balance between tolerance and expansion of autoreactive clones is inclined toward tolerance during ontogeny, especially in those organs where hemo- and lymphopoiesis take place, and where the presence of the putative ligand for this human V_H gene segment is overwhelming. We believe that under normal circumstances an active regulatory network controls the expansion of autoreactive clones with a strong emphasis on the product of the V_H gene segment. Immune disregulation in autoimmune situations very likely allow the V_H4-21 expressing cell to undergo differentiation and expand the pathogenic potential of this self-reactive gene segment.

References

1 V. Pascual, and J. D. Capra, Human immunoglobulin heavy chain variable region genes: Organization, polymorphism, and expression, Adv. Immunol. 49:1 (1991).
2. F. Matsuda, E. K. Shin, H. Nagaoka, R. Matsumura, M. Haino, Y. Fukita, S. Taka-ishi, T. Imai, J. H. Riley, R. Anand, E. Soeda, and T. Honjo, Structure and physical map of 64 variable segments in the 3' 0.8 megabase region of the human immunoglobulin heavy chain locus, Nature Genet. 3:88 (1993).
3. G. D. Yancopoulos, S. V. Desiderio, M. Paskind, J. F. Kearney, D. Baltimore, and F. W. Alt, Preferential utilization of the most J_H proximal V_H segments in pre-B cell lines, Nature 311:727 (1984).
4. R. M. Perlmutter, J. F. Kearney, S. P. Chang, and L. H. Hood, Developmentally controlled expression of immunoglobulin V_H genes, Science 227:1597 (1985).
5. H. W. Schroeder, J. L. Hillson, and R. M. Perlmutter, Early restriction in the human antibody repertoire, Science 238:791 (1987).
6. H. W. Schroeder, and J. Y. Wang, Preferential utilization of conserved immunoglobulin heavy chain variable gene segments during human fetal life, Proc. Natl. Acad. Sci. U.S.A. 87:6146 (1990).
7. K. W. Van Dijk, L. A. Milner, E. H. Sasso, and E. C. B. Milner, Chromosomal organization of the heavy chain variable region gene segments comprising the human fetal antibody repertoire, Proc. Natl. Acad. Sci. U.S.A. 89:10430 (1992).
8. V. Pascual, L. Verkruyse, M. L. Casey, and J. D. Capra, Analysis of Ig H chain gene segment utilization in human fetal liver, J. Immunol. 151:4164 (1993).
9. V. Pascual, and J. D. Capra, V_H4-21, a human V_H gene segment overrepresented in the autoimmune repertoire, Arthritis Rheum. 35:11 (1992).
10. H. G. Kunkel, V. Agnello, F. G. Joslin, R. J. Winchester, and J. D. Capra, Cross-idiotypic specificity among monoclonal IgM proteins with anti-gamma globulin activity, J. Exp. Med. 137:331 (1973).
11. I. Randen, K. M. Thompson, V. Pascual, K. Victor, D. Beale, J. Coadwell, O. Forre, J. D. Capra, and J. B. Natvig, Rheumatoid factor V genes from patients with rheumatoid arthritis are diverse and show evidence of an antigen-driven response, Immunol. Rev. 128:49 (1992).
12. K. D. Victor,V. Pascual, A. K. Levfert, and J. D. Capra, Human anti-acetylcholine antibodies use variable gene segments analogous to those used in autoantibodies of various specificities, Mol. Immunol. 29:1501 (1992).
13. K. D. Victor, V. Pascual, C. L. Williams, V. A. Lennon, and J. D. Capra, Human monoclonal striational autoantibodies isolated from thymic B lymphocytes of patients with myasthenia gravis use V_H and V_L gene segments associated with the autoimmune repertoire, Eur. J. Immunol. 22:2231 (1992).
14. D. Vazquez-Abad, V. Pascual, M. Zanetti, and N. F. Rothfield, Analysis of human anti-topoisomerase-I idiotypes, J. Clin. Invest. 92:1302 (1993).
15. I. Randen, D. Brown, K. M. Thompson, N. Hughes-Jones, V. Pascual, K. Victor, J. D. Capra, O. Forre, and J. B. Natvig, Clonally related IgM rheumatoid factors undergo affinity maturation in the rheumatoid synovial tissue, J. Immunol. 148:3296 (1992).
16. M. E. Gershwin, and I. R. Mackay. Primary biliary cirrhosis: Paradigm or paradox for autoimmunity, Gastroenterol. 100:822 (1991).
17. V. Pascual, S. Cha, M. E. Gershwin, J. D. Capra, and P. S. C. Leung, Nucleotide sequence analysis of natural and combinatorial anti-PDC-E2 antibodies in patients with primary biliary cirrhosis: Recapitulating immune selection with molecular biology. J. Immunol. In press.
18. V. Pascual, K. Victor, D. Leslz, M. B. Spellerberg, T. J. Hamblin, K. M. Thompson, I. Randen, J. B. Natvig, J. D. Capra, and F. K. Stevenson, Nucleotide sequence analysis of the V regions of two IgM cold agglutinins: Evidence that the V_H4-21 gene segment is responsible for the major cross-reactive idiotype, J. Immunol. 146:4385 (1991).
19. J. Leoni, J. Ghiso, F. Goni, and B. Frangione, The primary structure of the Fab fragment of protein KAU, a monoclonal immunoglobulin M cold agglutinin, J. Biol. Chem. 266:2836 (1991).
20. L. E. Silberstein, L. C. Jefferies, J. Goldman, D. Friedman, J. S. Moore, P. C. Nowell, D. Roelcke, W. Pruzanski, J. Roudier, and G. J. Silverman, Variable region gene analysis of pathologic human autoantibodies to the related i and I red blood cell antigens, Blood 78:2372 (1991).
21. K. M. Thompson, J. Sutherland, G. Bardem, M. D. Melamed, I. Randen, J. B. Natvig, V.

Pascual, J. D. Capra, and F. K. Stevenson, Human monoclonal antibodies against blood group antigens preferentially express a V_H4-21 variable region gene-associated epitope, Scand. J. Immunol. 34:509 (1991).
22. R. Baer, A. Foster, I. Lavenir, and T. Rabbitts, Immunoglobulin V_H genes are transcribed in T cells in association with a new 5' exon, J. Exp. Med. 167:2011 (1988).
23. E. Timmers, M. Kenter, A. Thompson, A. E. M. Kraakman, J. E. Berman, F. W. Alt, and R. K. B. Schuurman, Diversity of immunoglobulin heavy chain gene segment rearrangement in B lymphoblastoid cell lines from X-linked agammaglobulinemia patients, Eur. J. Immunol. 21:2355 (1991).
24. V. Pascual, and J. D. Capra, B-cell superantigens? Current Biol. 1:315 (1991).

IV: Treatment of inhibitors

The induction of inhibitors in patients with congenital coagulation defects is seen when patients are exposed to exogenous factor replacement. Antibodies to factor VIII (F.VIII) are much more common than those to factor IX. In addition, these F.VIII antibodies can be acquired in the post-partum period, in patients with autoimmune disorders, and in elderly persons without other diseases. Once present they commonly neutralize the antigen so that treatment of the spontaneous hemorrhage seen in these patients is a great clinical challenge.

This section deals with innovative ways to deal with patients who have developed antibodies with and without hemophilia as an underlying disorder. Dr. Hay, a pioneer in the use of porcine F.VIII, details the use of porcine F.VIII in both populations. Historically, porcine F.VIII has been used in England for surgical intervention in VIII-deficient hemophilia patients. Allergic reactions and thrombocytopenia were substantial enough side effects to curtail its use. Using a new fractionation technology (employing polyelectrolytes), a far better, safer product was produced. This new product has few side effects, and its use over the past 10 years has proven it free of transfusion-transmitted diseases.

The major clinical application of porcine F.VIII is with patients who have F.VIII inhibitors. Hemophiliacs with low-titered inhibitors usually do not cross-react with porcine F.VIII. Because of this, measurable F.VIII levels can be achieved. Nonhemophiliac patients, including those who present with high-titered antibody, will have little cross-reactivity with porcine F.VIII. Because of its ability to achieve measurable F.VIII levels, porcine F.VIII is frequently the first line of defense for bleeding episodes in these patients. However, anamnesis may occur and needs to be monitored.

Many patients with high-affinity, low-titered or high-titered inhibitors will not respond to porcine F.VIII. As a result, the armamentarium for the treatment has been expanded by the availability of activated and nonactivated prothrombin complex (APCC) and a new recombinant F.VIIa.

Drs. Moliterno and Bell describe their experience using FEIBA, an activated, steam-treated APCC in nonhemophiliac patients who have acquired inhibitors. Early studies demonstrated that PCCs and APCCs were capable of stopping hemorrhage in a majority of hemophiliac patients. Although many clinicians have extrapolated these findings to those with acquired inhibitions, little work has been published using this material in these patients.

In this study, 26 patients with acquired inhibitions were treated with FEIBA (one APCC) or EACA and supportive care. Almost all the patients had inhibitions to F.VIII, and the rest were against X, XI, and XIII. Hemorrhages were significant in the patients. The majority of bleeds, whether musculoskeletal, gastrointestinal, genitourinary, or oropharyngeal were successfully treated with FEIBA.

An inhibitor is frequently recognized without the luxury of clear-cut laboratory definition of the specific factor or the magnitude of the inhibitor titer. The use of FEIBA, although not capable of raising the factor level, can be a life-saving intervention. Often the choice of product, i.e., APCC versus porcine F.VIII, depends on level of familiarization of product by the clinician, laboratory data, and when possible, prior knowledge of the patients' response to therapy.

The traditional "waterfall" or "cascade" theory of coagulation has and continues to undergo serious reevaluation. The seminal work of Dr. Nemerson on the importance and probable primary role in hemostasis of tissue factor (T.F.), coupled with the work in St. Louis on the tissue factor pathway inhibitor (TFPI), has made possible a new approach to the treatment of inhibitors. A key role of Factor VIIa in combination with T.F. in activating Xa as a therapy for hemophilia with and without inhibitions is now being investigated by NOVO.

In describing a series of elegant experiments, Dr. Nemerson will detail his concept of how coagulation is initiated, the impact of blood flow, and the kinetics of this activated system. Later, data about clinical trials with Factor VIIa is provided.

Porcine factor VIII therapy in patients with factor VIII inhibitors

Charles Hay[1] and Jay N. Lozier[2]
[1]Liverpool University Department of Hematology, Royal Liverpool University Hospital, Prescot St. Liverpool, UK. L73BX.
[2]University of North Carolina School of Medicine, Department of Medicine, Division of Hematology, CB# 7035, Chapel Hill, North Carolina 27599-7035

Introduction

A review is presented of the use of porcine factor VIII (F.VIII) in patients with F.VIII inhibitors and acquired or congenital hemophilia. This review is drawn partly from the published literature and partly from an international survey of the use of porcine F.VIII in patients with congenital hemophilia conducted under the auspices of the ISTH F.VIII and IX scientific standardization subcommittee. The treatment of congenital and acquired hemophilia are compared, since they differ in a number of important respects.

Survey of the use of porcine F.VIII in congenital hemophilia

Clinicians from around the world were circulated with a questionnaire requesting a detailed account of their use of porcine F.VIII for the management of F.VIII inhibitors in patients with congenital hemophilia. Data were reported from 154 patients treated with porcine F.VIII. These patients had been given almost 5000 infusions of porcine F.VIII for 2472 bleeding episodes, including prophylaxis and home therapy. Over 15,000,000 units of porcine VIIIC were infused over more than 7000 treatment days.

Antibody cross-reactivity

Patients were most commonly selected for treatment with porcine F.VIII on the basis of the cross-reactivity of their inhibitor to porcine VIIIC. This is taken as the ratio of the anti-human to anti-porcine inhibitor titer expressed as a percentage. The Bethesda assay is used with either human plasma or porcine F.VIII concentrate reconstituted in human hemophilic plasma. The pretreatment cross-reactivity of 147 patients in whom this data was available is shown in Figure 1.

The mean cross-reactivity was 24%. Similar cross-reactivity has been observed by others.[1-3] Although 25% of these patients had no apparent

Inhibitors to Coagulation Factors
Edited by Louis M. Aledort *et al.*, Plenum Press, New York, 1995

cross-reactivity using the Bethesda assay, some had reduced F.VIII recovery and half-life indicating low-level inhibitor activity undetected by the Bethesda method. Ten percent of these patients had greater than

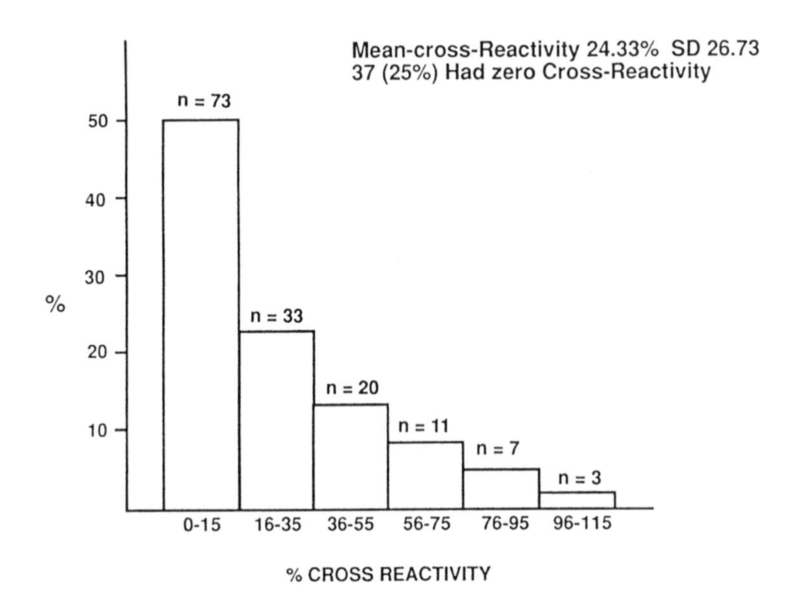

Figure 1. Distribution histogram of pretreatment inhibitor cross-reactivity to porcine VIIIC in 147 patients.

100% pretreatment cross-reactivity even though they had not been treated with porcine VIIIC before. Moderate variations in cross-reactivity occur independent of treatment with porcine VIIIC. Indeed, cross-reactivity to porcine F.VIII is commonly observed to change during the course of successful immune-tolerization using human F.VIII concentrate. If possible, current inhibitor measurements should be used when choosing the most appropriate replacement therapy for this reason.

Antibody cross-reactivity to porcine VIIIC is generally very much lower in acquired than in congenital hemophilia.[1-4] The inhibitor titer measured against porcine VIIIC is usually less than 5% of the inhibitor titer measured with human VIIIC. Therefore, porcine VIIIC is usually the most effective replacement therapy in this situation. The reaction kinetics of acquired F.VIII antibodies are often complex,[5] however, and F.VIII recovery is commonly less than might be predicted from the Bethesda inhibitor titer. The clinical response to treatment of acquired hemophilia with porcine F.VIII, although excellent in 78% of patients in one series, was unrelated to the anti-porcine inhibitor titer, the dose of porcine VIIIC given, or the F.VIII increment achieved.[4] This is in marked contrast with the clinical response in congenital hemophilia, where the pretreatment inhibitor titer and the factor VIII increment are highly predictive of the clinical response.[1]

Efficacy in congenital hemophilia

Four hundred ninety one of the bleeding episodes reported to us were managed in hospital. These included 264 hemarthroses; 93 muscle bleeds; 16 episodes of intracranial bleeding; 57 surgical interventions, including 35 emergency and 22 elective procedures; 13 episodes of trauma; and 48 other bleeds. The use of porcine F.VIII for surgery is reported in greater detail by Lozier.[6] Many clinicians reserved porcine VIIIC only for the treatment of more severe bleeding episodes.

Seven patients used porcine F.VIII for 2 to 13 years as home therapy for routine hemarthroses, or as prophylaxis. Four and a half million units of porcine VIIIC were used in this way. Their pretreatment cross-reactivity varied between zero and 30%. Home therapy or home prophylaxis was initiated only after several months of treatment on-demand, uncomplicated by side effects or anamnesis. Relatively small doses of between 20 and 60 U/Kg were generally used, which may account for the remarkably low incidence of side effects observed in this setting.

High-responding inhibitors disappeared in five of these patients during treatment with porcine F.VIII.[7] Home therapy with human F.VIII was re-established without recurrence of their original inhibitor in three of these, suggesting that they had been "desensitized" by porcine F.VIII. Although three of these patients were HIV-seropositive, none of them had AIDS or AIDS-related complex or a CD4 count <400 at the time of their treatment with porcine VIIIC. All had continued B-cell competence as evinced by protective levels of anti-HBsag antibody and the transient appearance of low titer anti-porcine antibodies 12 months after treatment with porcine VIIIC began. Their inhibitor loss is not, therefore, thought to be attributable to HIV infection.[7]

Twenty-three patients were treated intermittently or regularly on-demand, in hospital, over a period of 2 to 7 years. They were treated for an average of 9 treatment episodes, with a range of 5 to 24. No anamnestic rise in inhibitor titer was observed in about half of this group. An intermediate anamnestic response was observed in the remainder.

Clinicians were asked to grade the clinical response as excellent, good, fair, none, or worse. Four hundred twenty-six in-patient bleeding episodes were graded in this way. 8.7% of responses were excellent, 71% good, and 13.4% fair. No response was observed in 7% of episodes, giving an overall response rate in excess of 90%. Efficacy correlated with the inhibitor titer and the F.VIII increment achieved, as reported by others.[1-3] The response is much less reliable in patients with high-level inhibitors, but some patients did achieve a good or fair response, even in the absence of a measurable F.VIII increment.

Treatment failures generally occurred when patients were treated as an emergency, where porcine F.VIII was chosen empirically or on the basis of earlier inhibitor measurements. The cause of therapeutic failure could be identified in 25/30 cases. It could be attributed to an anti-porcine inhibitor titer greater than 10 BU in 37% of cases, to moderate level inhibitors which might have been quenched by a larger dose of porcine VIIIC in 23%, to an inadequate dose less than 10 U/Kg in 17%, and to permanent irremediable injury unlikely to respond to replacement therapy in 7%.

Dosage calculation

Two approaches have been used to calculate the starting dose of porcine F.VIII required to overcome inhibitor activity and provide an adequate F.VIII increment. The first method involves the administration of a standard dose of 50 to 150 U/Kg, depending on the clinical circumstances and the most recent inhibitor titer.[1] Subsequent doses are based on the current inhibitor level and the F.VIII response. This approach works well in experienced hands, although our data suggest that both under- and over-dosing are common.

This formula has also been used to calculate the initial dose of porcine VIIIC:[2]

Dose (Units) = inhibitor titer (BU) X plasma volume (mls)

The body weight (Kg) and hematocrit are used to calculate the plasma volume and the current rather than the most recent anti-porcine inhibitor titers are used. Our data suggest that this formula may overestimate the dose required to achieve the chosen F.VIII increment, since very large doses, which often gave rise to a F.VIII increment in excess of 200%, were often reported by centers using this formula.

The formula assumes a linear relationship between the inhibitor titer (BU) and response %, an assumption we were able to test. The response % is taken as the % rise in F.VIII per U/Kg infused.

We analyzed 99 of the 137 first infusions for which contemporaneous inhibitor, body weight, and pre- and post-infusion F.VIII measurements were reported. Thirty-eight of these infusions, in which inhibitor titers of <1BU or F.VIII increments greater than 130% were reported, were excluded from the analysis to minimize assay-based inaccuracies at the extremes of the assay range. These cutoffs are clearly arbitrary. A weak inverse correlation was observed between the inhibitor titer and response %, but log-transformation of the inhibitor titer increased the power of this association dramatically from $r = -0.26$ to $r = -0.633$. This indicates a parabolic rather than a linear relationship between the inhibitor titer and response %, and suggests that the dosage formula is based on a false premise.

Linear regression was used to derive the following formula which roughly predicted the initial dose of a course required to achieve a chosen F.VIII increment:

Dose (U/Kg) = 40.7+25.2(Log inhibitor*)+1.39(% VIII required)
(* in Bethesda units)

This formula is simple to use, if one is armed with a pocket calculator, but requires prospective confirmation and may be refined further. A similar approach might be used to predict the dose of human F.VIII required to treat inhibitor patients not completely refractory to human F.VIII concentrate. Prospective studies would be useful to explore various mathematical approaches to a formula that might accurately predict the starting dose. Such an approach is unlikely to be of value in acquired hemophilia, however, because the complex reaction kinetics and marked interpersonal variation found in this group of patients are likely to confound any attempt to predict the response mathematically.

Side-effect profile

Reactions followed 2.3% of hospital-based infusions, a similar incidence to that reported by Brettler et al.[3] but significantly less than the 8% incidence observed by Gringeri et al.[8] Forty-five mild reactions and 24 reactions of moderate severity were reported. One severe reaction, described as "anaphylactoid," was reported. This reaction accompanied an unusually large dose of 630 U/Kg which gave rise to a F.VIII increment of 750%, and was accompanied by a precipitous fall in platelet count. This was the first infusion in a course of treatment. Subsequent, smaller doses of the same batch were uncomplicated by reactions. This reaction is described in greater detail by Gringeri.[8]

Reactions to porcine F.VIII concentrate, like those to human F.VIII concentrate, are not generally batch-related, and may be more common if the product is infused rapidly[1,2,8] and in large doses. Most patients who suffered reactions to porcine VIIIC, including the patient who suffered the most severe reaction reported, had no side effects with subsequent infusions. A minority suffered reactions on more than one occasion and should probably not be treated with porcine F.VIII again. Single reactions to porcine F.VIII are not a contraindication to its continued use.[8] Hydrocortisone and chlorpheniramine was administered prior to infusion of porcine F.VIII concentrate by some, but not all, clinicians. It is not clear whether this reduces the incidence of reactions, but it is unlikely that hydrocortisone will act sufficiently rapidly to influence events.

Only one minor reaction was reported by patients using porcine F.VIII as home therapy. Although transfusion reactions followed even small doses of porcine F.VIII, they were more commonly associated with larger doses. This may account for the very low incidence of moderate severity reactions observed in Brettler's series[3] and the very low incidence of side effects observed among the patients treated on-demand. Larger doses were generally used by Gringeri, and this may account for the generally higher incidence of reactions reported by this group.[2,8]

Porcine F.VIII has long been recognized to cause a reduction in platelet count, although the clinical significance and the frequency with which this occurs has been disputed.[1-3,8] The mean platelet count in our series declined from a pretreatment mean of 247 X 10^9/l to a posttreatment mean of 207 X 10^9/l following 175 infusions in 75 patients. Although the platelet count commonly declined following infusion of porcine F.VIII this reduction was usually modest.

Subnormal platelet counts <150 X 10^9/l, were observed following 46 infusions and <50 X 10^9/l after 5 infusions. A subnormal platelet count commonly resulted from a progressive reduction in platelet count over a period of several days. This was usually observed when very intensive replacement therapy was used for surgery or a major bleed. The association of this side effect with surgery and trauma is well recognized and

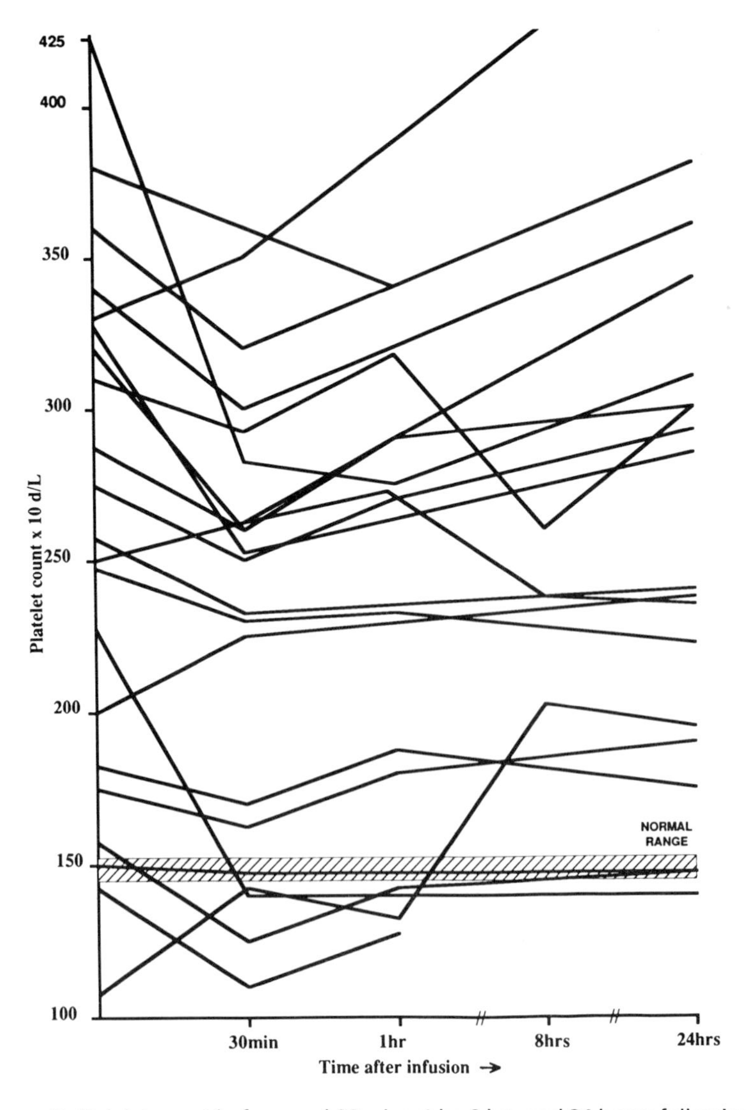

Figure 2. Platelet count before, and 30 mins, 1 hr, 8 hrs, and 24 hours following 21 single porcine F.VIII infusions.

suggests that the reduction in platelet count is not caused by porcine von Willebrand factor alone, but also by pro-coagulants released during trauma.[9,10] A fall in platelet count does not accompany every course of treatment in multiply treated patients and is not batch-related. Severe thrombocytopenia is uncommon and has not been reported to cause bleeding, perhaps because it tends to arise after several days of treatment and is transient.

The change in platelet count over a 24-hour period following 21 single infusions of porcine VIIIC is shown in Figure 2. This data is described in greater detail by Fukui.[11] The platelet count declined following 12 infusions, usually modestly. The nadir of the platelet count was generally observed 30 minutes postinfusion. This effect was usually very transient, and complete recovery to pretreatment values occurred within an hour in all but two individuals.

Immunogenicity

Porcine F.VIII may be less immunogenic than human F.VIII. An anamnestic rise in anti-human inhibitor titer is thought to occur less frequently following the administration of porcine than following the administration of human F.VIII concentrate.[1] An anamnestic rise in anti-human inhibitor titer follows the use of porcine VIIIC in between 20 and 35% of subjects.[1-3,8,11] We were unable to investigate this because human F.VIII replacement therapy frequently preceded treatment with porcine F.VIII in the cases reported to us.

We were able to assess the specific anti-porcine anamnestic response in 65 of our subjects, as evinced by a change in cross-reactivity following treatment with porcine F.VIII. An intermediate specific anamnestic response was arbitrarily defined as less than a 3-fold increase and a brisk response as greater than a 3-fold increase in cross-reactivity. The anamnestic response following the first course of treatment was absent, intermediate, or brisk in 29%, 40%, and 31% of subjects, respectively. A similar total dose was used in each group, suggesting that the magnitude of the initial anamnestic response is unrelated to treatment intensity. Furthermore, the anamnestic response that followed porcine F.VIII replacement was not predicted by the patient's anamnestic response following treatment with human F.VIII.

Fifty-two patients were treated for more than one bleeding episode. The number of treatment episodes managed with porcine F.VIII was strongly influenced by their anamnestic response. Four brisk responders, for example, were treated for a mean of only 2.3 episodes, rapidly becoming refractory to further treatment. In contrast, 20 nonresponders, including all the patients treated at home, were intensively treated, receiving a mean of 89 courses.

Intermediate responders were treated for a mean of 13 episodes. This group demonstrated one of two patterns of immune response. Some showed a progressive increase in cross-reactivity following each course of treatment, becoming relatively or completely refractory to porcine VIIIC after several courses of treatment. Others experienced an anamnestic increase in cross-reactivity only when treated intensively. Several of these patients were treated successfully on-demand for several years without anamnesis, developing a specific anamnestic response only after a period of intensive replacement therapy. There is thus a subgroup of patients, identifiable only by their absent or intermediate, dose-related, anamnestic response who may tolerate treatment with this product on-demand over a period of months or years.

An anamnestic rise in anti-human or anti-porcine inhibitor titer has been reported to be uncommon following porcine F.VIII therapy for acquired hemophilia, occurring in only 10 of 65 patients reported in a recent study.[5] Most of these patients were treated for only a single bleeding episode, however, and anamnesis may be observed more commonly if such patients are treated repeatedly or over a prolonged period.

Current inhibitor policy

So what place does porcine VIIIC occupy in the therapeutic armamentarium in the United Kingdom? Our own hospital policy could be summarized as follows:

We desensitize as many patients as we can, using human F.VIII concentrate. I suspect that this practice is more widespread in Europe than in the United States at the present time. We would expect to successfully desensitize most low and intermediate responders, although this approach is far less successful among high-responders.

Patients with non-cross-reacting antibodies are treated with human or porcine VIIIC. Those with anti-human titers less than 5 BU are treated with human or porcine F.VIII, depending on their cross-reactivity and the severity of the bleed. Those with anti-human titers greater than 5 BU and lower anti-porcine titers are treated with porcine VIIIC, other than for mild bleeds which may be treated with PCCS. High-level anti-porcine inhibitors of >10BU are treated with PCCs, recombinant VIIa, or possibly with porcine (VIIa) F.VIII. We suspect that rVIIa will come to be used increasingly for this category of patient.

Conclusion

Porcine F.VIII is usually effective in appropriately selected patients given adequate doses. Patient selection should be based on current inhibitor measurements.

Reactions are uncommon and are, to a large degree, dose-related. A postinfusion decline in platelet count is usually modest, transient, and clinically insignificant. The platelet count should be monitored during very intensive replacement therapy.

A subgroup of patients, identifiable only by their absent or modest anamnestic response may be treated with modest doses of porcine F.VIII on a regular basis. Whether this treatment should be given in hospital or in a domiciliary setting is a matter of clinical judgement, but since side-effects are to some degree dose related, it would seem wise to give intensive porcine replacement therapy only under direct hospital supervision.

Acknowledgment

We particularly thank Dr. CA Lee, Dr. E Santagostino, Prof. N Ciavarella, Prof. N Fukui, Prof. A Yoshioka, Dr. J Teitel, and professors P.M. Mannucci and C.K. Kasper for their valuable contributions to the survey, and others too numerous to mention individually.

References

1. Kernoff PBA, Thomas ND, Lilley PA, Clinical experience with polyelectrolyte-fractionated porcine FVIII concentrate in the treatment of haemophiliacs with antibodies to factor VIII. Blood 1984, 63; 31-41.
2. Gatti L, Mannucci PM, Use of porcine factor VIII in the management of seventeen patients with FVIII antibodies . Thrombosis and Haemostasis 1984, 51; 379-392.
3. Brettler DB, Forsberg A, Levine PH, et al., The use of porcine factor VIII:C in the treatment of patients with inhibitor antibodies to VIII:C: A multicenter US trial. Arch intern Med 1989, 149; 1381-1385.
4. Morrison AE, Ludlam CA, Kessler C, Use of porcine factor VIII in the treatment of patients with acquired hemophilia. Blood 1993, 81; (6): 1513-1520.
5. Gawryl MS, Hoyer LW. Inactivation of factor VIII coagulant activity by two different types of human antibodies. Blood 1982, 60; 1103-9.
6. Lozier JN, Santagostino E, Kasper CK, Teitel JM, Hay CRM, The use of porcine factor VIII for surgical procedures in haemophilia A patients with inhibitors. Seminars in Hematology, 1993, 30; (2): supp 1; 10-21.
7. Hay CRM, Laurian Y, Verroust F, Preston FE, Kernoff PBA, Induction of immune tolerance in patients with haemophilia A and inhibitors treated with porcine factor VIIIC by home therapy. Blood 1990, 76; 882-886.
8. Gringeri A, Santagostino E, Tradati F, Giangrande PLF, Mannucci PM, Adverse effects of treatment with porcine factor VIII. Thrombosis and Haemostasis 1991, 65; 245-247.
9. Kernoff PBA, The factor VIII-related antigen and antibodies to factor VIII. MD Thesis, University of London 1974.
10. Altieri DC, Capitanio AM, Mannucci PM, Von Willebrand factor contaminating porcine factor VIII concentrate (Hyate:C) causes platelet aggregation. Brit J Haematol 1986, 63;703-711.
11. Fukui H, Yoshioka A, Shima M, Mori K, Yorifuji H, et al. Multicenter clinical trial of a highly purified porcine factor VIII concentrate (MTI-9002 (Hyate:C)) in haemophilia A patients with inhibitors. Japanese J Thrombosis and Haemostasis 1991, 2; (6): 487-50.

Clinical presentation and management of patients with circulating anticoagulants

Alison Moliterno and William R. Bell

Johns Hopkins University School of Medicine
Division of Hematology
600 N. Wolfe Street, Blalock 1002
Baltimore, Maryland 21287-4928

Acquired hemophilia is characterized by the development of antibodies to coagulation proteins. Although these inhibitors are rare, they pose serious threats to health as a result of their ability to produce spontaneous hemorrhage. As these inhibitors occur infrequently, epidemiological analyses of their patterns of occurrence, details of clinical presentations, and practical management of acute hemorrhage are difficult to gather. In an effort to augment the clinical data base, we reviewed the records of 26 patients treated at our institution from 1980 to 1993.

Patient population

Sixteen of the 26 patients were female, 3 of whom were postpartum at diagnosis. The age range of the nonpregnant patients was 46 to 84 years, with a mean of 68 years. The average age of the postpartum group was 29 years.

Comorbid conditions included pulmonary disease (chronic obstructive pulmonary disease or asthma) in 10 patients, diabetes mellitus in 6, autoimmune processes in 6, coronary artery disease in 4, and malignancy in 2. Comorbid conditions were absent in 5 patients (see Table 1).

Clinical presentations

Twenty-two patients were found to have inhibitors to factor VIII (F.VIII), 2 patients to Factor XI, 1 patient to Factor XIII, and 1 to Factor X. The patients with inhibitors to Factor XI did not develop spontaneous hemorrhage; their inhibitors were discovered after evaluation of prolonged activated partial thromboplastin times.

Of the 24 patients afflicted with hemorrhage, 21 initially presented with involvement of either a joint or the soft tissue of an extremity. Two patients presented with gastrointestinal hemorrhage and 1 patient presented with hematuria. The majority of patients were plagued by

Inhibitors to Coagulation Factors
Edited by Louis M. Aledort *et al.*, Plenum Press, New York, 1995

recurrent soft tissue bleeds, most commonly involving the extremities. Spontaneous pharyngeal hemorrhage complicated the course of 4 patients, hematuria the course of 2, and gastrointestinal hemorrhage the course of 1.

Table 1. Comorbid conditions

		Female	Male
Pulmonary disease		5	5
Diabetes mellitus		2	4
Autoimmune			
	Thyroid	3	0
	Polymyositis	1	0
	Rheumatoid arthritis	0	1
	Ankylosing spondylitis	0	1
Coronary artery disease		1	3
Malignancy		2	0
No associated conditions		5	0

Managment of acute hemorrhage

Soft tissue/extremity. Twenty-four patients experienced 43 episodes of soft tissue or joint hemorrhage. Supportive measures alone (ice, elevation of the extremity, epsilon-amino caproic acid) were employed in 18 of the 43 episodes (42 %) successfully. Activated prothrombin complex concentrates were employed when nerve compression developed or hemorrhage persisted. Treatment with FEIBA (Immuno, Vienna, Austria) was successful in each of the 25 episodes.

Pharynx/oral mucosa. Six episodes of pharyngeal or oral mucosa hemorrhage occurred. One was minor and was managed with supportive measures only. Four were severe, producing respiratory compromise or swallowing difficulties and were managed successfully with FEIBA. One patient developed a severe pharyngeal hemorrhage and underwent emergent tracheostomy without replacement measures. Postoperatively she developed persistent hemorrhage which did not respond to F.VIII supplementation (activated complexes were not available at this time). She succumbed to complications of her hemorrhage.

Gastrointestinal tract. One patient developed severe gastrointestinal hemorrhage due to diffuse A-V malformations. She was treated on 6 occasions with FEIBA with successful control of hemorrhage. Ultimately the patient succumbed to complications associated with recurrent GI hemorrhage.

Urogenital tract. Urogenital hemorrhage complicated the course of 3 patients. One patient was found to have a carcinoma in situ of the bladder. One patient developed hematuria from a hemorrhage of a renal calyx; this was managed successfully with FEIBA. Finally, 1 patient developed recurrent hematuria of unknown etiology. This hemorrhage did not respond to multiple courses of FEIBA but did eventually remit spontaneously.

Comment

Acquired hemophilia is a rare disorder that occurs after development of antibodies directed against specific coagulation factors, most commonly to F.VIII. Prior reports support associations with autoimmune processes, pregnancy, malignancy, and drugs.[1,2,3] In addition to these associations, our series reports pulmonary disease as the most frequent comorbid condition.

The majority of clinical bleeding in our series occurred in the joints and soft tissues of the extremities. Far less commonly, bleeding occurred in the gastrointestinal tract and urogenital tract; additionally, bleeding from these sites was often associated with a particular lesion as opposed to spontaneous hemorrhage.

Management of acute hemorrhage requires careful evaluation of the extent of bleeding and prompt institution of supportive measures. When hemorrhage is severe, we have found the use of FEIBA to be a safe and highly effective method to control hemorrhage in patients with spontaneously appearing circulating anticoagulants.

References

1. Green, D., and Lechner, K., A survey of 215 non-hemophilic patients with inhibitors to Factor VIII, Thromb Haemost. 45:200 (1981).
2. Lottenberg, R., Kentro, T.B., and Kitchens, C.S., Acquired hemophilia. A natural history study of 16 patients with Factor VIII inhibitors receiving little or no therapy, Arch Intern Med. 147:1077 (1987).
3. Soriano, R.M., Matthews, J.M., and Guerado-Parra, E., Acquired hemophilia and rheumatoid arthritis, Br J. Rheumatol. 26:381 (1987).

The function of factor VIIa in hemophilia A: An hypothesis

Yale Nemerson
Department of Biochemistry and Division of Molecular Medicine
Department of Medicine
Mount Sinai School of Medicine of the City University of New York
One Gustave L. Levy Place, New York, New York 10029

Since accepting the assignment to talk about this subject, I have reminded myself that I informed my host I hadn't the slightest idea of how Factor VIIa functions in hemophilia! This disclaimer notwithstanding, I do have an hypothesis, albeit one with no experimental validation nor any particular intellectual appeal. It is simply that the activation of Factor VII (F.VII) *in vivo* is the rate-limiting step in hemostasis, at least in hemophiliacs. I will discuss this concept below.

First, however, it is important to appreciate that the dosage of VIIa used clinically is very large, achieving levels in the circulation that are essentially equivalent to the circulating levels of F.VII. This is based on the smallest recommended dose of rVIIa,[1] 65 μg/Kg body weight. This means that a 70 Kg patient would receive ≈ 2500 μg rVIIa. If this were to distribute throughout the plasma volume (≈ 2500 ml) an instantaneous concentration of ≈ 1 μg/ml would result. If the recovery were 100%, this would represent a two-fold excess over the *mass* of circulating F.VII. This very high concentration of VIIa is sufficient to activate Factor X (F.X) in the presence of phospholipids.[2] Indeed, this has recently been proposed as a possible mechanism for the action of F.VIIa inasmuch as monocytes lacking TF have been shown to support F.X activation *via* a purely phospholipid-dependent mechanism. I suspect, however, that this probably is not how VIIa functions *in vivo*, mainly because this reaction is so very slow relative to the TF-dependent reactions. In our original study of this phenomenon, we calculated that at plasma concentrations of (bovine) F.X, the reaction would proceed 3000-fold faster in the presence of TF compared to simply in the presence of phospholipids.[2] This striking disparity in reaction rates suggests that this problem will be refractory to a definitive solution simply because it is so difficult to detect a single TF site amongst 3000 molecules of catalytically active VIIa-phospholipid complexes. Moreover, if TF is very sparse, it might be very difficult to inhibit by antibody as the antibody-antigen reaction is second order.

If rVIIa functions *via* a TF-dependent manner, its high plasma concentration effectively rules out a simple catalytic role for F.VIIa in the sense

Inhibitors to Coagulation Factors
Edited by Louis M. Aledort *et al.*, Plenum Press, New York, 1995

of TF:VIIa complexes converting TF:VII complexes in a cascading manner. The reason for this is that factors VII and VIIa bind to TF with experimentally equivalent dissociation constants.[3] From this it follows that upon the exposure of TF to the circulating blood, the ratio of TF:VII and TF:VIIa complexes will more or less reflect the concentrations of VII and VIIa in the circulation. This formulation is predicated on the rates of formation of the two complexes being the same, a reasonable assumption. Even though it has been clearly shown that TF:VIIa converts TF:VII to TF:VIIa when the TF is inserted in phospholipid vesicles,[4] it is unlikely that this mechanism exists on cell surfaces because of the restricted long-range diffusion of transmembrane proteins in plasma membranes.[5,6] Further, if this mechanism occurred *in vivo*, then one would imagine that much smaller doses of VIIa would be effective in achieving hemostasis in hemophiliacs.

It is also reasonably clear that normal subjects have circulating levels of F.VIIa higher than those found in hemophiliacs.[7] The consequences of lower VIIa levels are certainly not clear. Again, the amount of VIIa required clinically brings the VIIa levels in these patients into a large supranormal range. This argues against the observed lower levels of VIIa in hemophiliacs being a hemostatically significant phenomenon. That is, VIIa infusions probably do not function simply by raising VIIa levels in hemophiliacs to normal levels.

What explanations are left? Unfortunately, not too many! One phenomenon that might bear on this problem is the mobility of tissue factor on cells and inserted into phospholipid vesicles. Rather amazingly, the lateral mobility (two-dimensional diffusion) of transmembrane proteins is very fast, perhaps only two orders of magnitude slower than three-dimensional diffusion through water. On the other hand, transmembrane proteins inserted into the plasma membranes of cells are thought to be extremely slow, at least with respect to long-range diffusion. This means that rapid interactions between transmembrane proteins are unlikely to occur on cell surfaces.

One route to understanding the function of VIIa may reside in the consideration of how blood clots on macroscopic surfaces. The latter is a somewhat imprecise term describing a surface that is very large with respect to protein molecules. Certainly, when coagulation occurs *in vivo* the reactions take place mainly on cell surfaces, either intact or damaged. Because of this, classical (Michaelis-Menten type) kinetics do not apply and indeed, may yield significantly misleading conclusions. The main difference between classical kinetics and macroscopic surfaces lies in the dimensionality of the reactions which, in turn, affects the frequency with which surface-bound enzymes collide with their substrates. This is very important because collision frequency gives the upper limit of any enzymatic reaction; that is, the rate of product formation cannot exceed the rate at which substrate molecules collide with the enzyme.

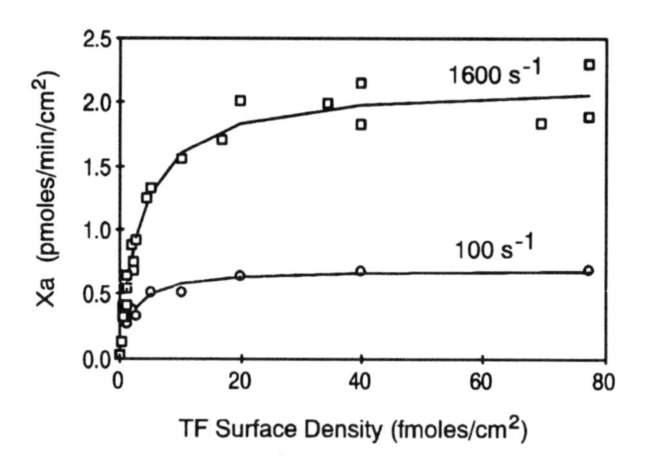

Figure 1. F.X activation in the flow reactor as a function of TF surface density

Now consider a large surface which is in contact with a solution containing substrate molecules. In the absence of stirring, the rate of substrate molecules encountering the surface is a function of the concentration of the molecules and their diffusion coefficients. If the surface is extremely catalytic, then one can readily envision a situation in which every molecule that encounters the surface is converted to product. When this occurs, the reaction is considered to be transport rate-limited. If, on the other hand, the surface has a low catalytic activity, then not every encounter leads to product formation. In this situation, the reaction is considered to be kinetically limited.

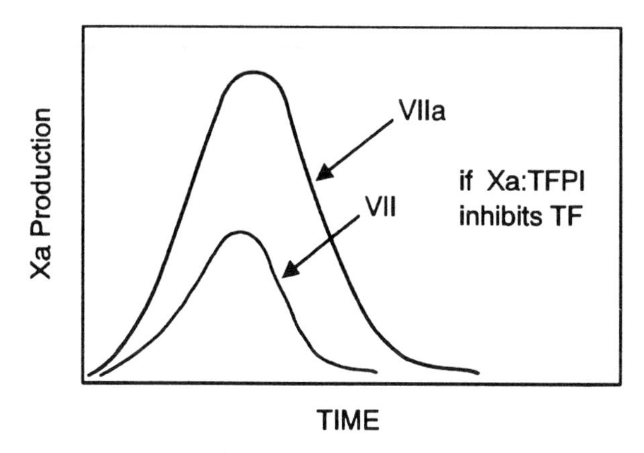

Figure 2. One way F.VIIa may work

We have studied tissue factor imbedded in a glass-supported phospholipid bilayer in order to define the kinetics of F.X activation on a macromolecular surface.[8] At a constant substrate concentration, we found that we reached half-maximal factor activation rates when the surface coverage was only 1.7 and 3.3 fmoles/cm^2, at wall shear rates of 100 sec^{-1} and 1600 sec^{-1}, respectively (these rates correspond to venous and arteriolar conditions). The half-maximal rates are achieved when only 0.03-0.06% of the surface is occupied with tissue factor. Maximum activation, of course, occurs only when the surface is entirely covered with tissue factor, a physically unrealistic situation. While no estimates are available relating to the tissue factor coverage of cells expressing this protein, our impression is that it is lower than the coverages used experimentally.

In a physiological setting, if the tissue factor surface density is, in effect, on the plateau shown in Figure 1, there obviously would be no requirement for the antihemophilic factors, VIII and IX. On the other hand, if the tissue factor surface density is shifted far to the left in Figure 1, not all the F.X (and, presumably, Factor IX) would be directly activated. Accordingly, in the absence of either F.VIII or IX, there could be sub-optimal activation of F.X.

We have previously shown that under certain conditions, the addition of factors VIII and IX to the perfusate used in the flow reactor, results in higher production rates of Factor Xa. Further, we also showed that in the presence of the tissue factor pathway inhibitor, the gap between the direct activation of F.X and that observed in the presence of factors VIII and IX is increased. Now, if we consider, as pointed out above, that F.VII and VIIa compete for tissue factor sites and that the activation of F.VII to VIIa might be the slow step in coagulation, the initial burst of procoagulant activity that occurs on the complexation of tissue factor and F.VII/VIIa would be expected to be higher when the initiating complex contains F.VIIa (see Figure 2).

This gives a theoretical and experimental framework in which to consider the therapeutic role of F.VIIa in hemophilia. In order to extend this beyond mere hypothesis, however, much experimentation must be conducted. We are currently studying the rates of activation of factors IX and X in the flow reactor as a function of tissue factor surface density, wall shear rate, and the participation of F.VIII. Perhaps these experiments will yield some convincing evidence regarding this perplexing problem.

References

1. Macik, B.G., et al., 1993, Safety and initial clinical efficacy of three dose levels of recombinant activated F.VII (rFVIIa): Results of a phase I study, *Blood Coagul Fibrinolysis.* **4:** 521-527.
2. Silverberg, S.A., Nemerson, Y. and Zur, M., 1977, Kinetics of the activation of bovine coagulation factor X by components of the extrinsic pathway. Kinetic behavior of two-chain F.VII in the presence and absence of tissue factor, *J.Biol.Chem.* 252: 8481-8488.
3 Bach, R., Gentry, R. and Nemerson, Y., 1986, F.VII binding to tissue factor in reconstituted phospholipid vesicles: Induction of cooperativity by phosphatidylserine, *Biochem.* **25:** 4007-4020.
4. Neuenschwander, P.F., Fiore, M.M. and Morrissey, J.H., 1993, F.VII autoactivation proceeds via interaction of distinct protease-cofactor and zymogen-cofactor complexes, *J.Biol.Chem.* 268: 21489-21492.
5. Eisinger, J., Flores, J. and Petersen, W.P., 1986, A milling crowd model for local and long-range obstructed lateral diffusion, *Biophys. J.* 49: 987-1001.
6. Sassaroli, M., Vauhkonen,M., Perry, D. and Eisinger J., 1990, Lateral diffusivity of lipid analogue excimeric probes in dimyristoylphosphatidylcholine bilayers, *Biophys. J.* 57: 281-290.
7. Wildgoose, P., et al., 1992, Measurement of basal levels of F.VIIa in hemophilia A and B patients, *Bloodd* 80: 25-28.
8. Andree, H.A.M., Contino, P.B., Repke, D., Gentry, R. and Nemerson, Y., 1994, Transport rate-limited catalysis on macroscopic surfaces: The activation of Factor X in a continuous flow enzyme reactor, *Biochem.* 33: 4368-4374.

Clinical update on the use of recombinant factor VII

Steven Glazer, Ulla Hedner, and Jorn F. Falch

Novo Nordisk A/S
Biopharmaceuticals Division
DK-2820 Gentofte
Denmark

Much effort has converged on finding an agent capable of activating the final common pathway of the coagulation cascade inducing hemostasis independently of factor VIII (F.VIII) or factor IX (F.IX). Such an agent should be not only hemostatically active but also safe with respect to thromboembolic complications. Factor VIIa (F.VIIa) becomes proteolytically active when complexed with tissue factor or other phopholipids minimizing the risk of inducing systemic activation of the coagulation system.

Limited experiences with plasma-derived highly purified F.VIIa indicated that F.VIIa is hemostatically active in dental surgery and joint bleeds[1,2] in a few hemophilia A patients. No adverse events were reported during therapy. However, the plasma concentration of factor VII (F.VII) is only about 0.5 μg/mL making large-scale production unfeasible. In addition, the advent of gene technology eliminates the risk of transmitting blood-borne human viruses.

Recombinant F.VIIa was cloned and expressed in mammalian cells[3] and found to be very similar to plasma-derived F.VIIa with regard to protein characteristics[4] and function.[5,6] Recombinant F.VIIa was reported to be hemostatically active in hemophilia A and B dogs.[7]

The safety and clinical efficacy of recombinant F.VIIa has been investigated in clinical trials since 1988 in hemophilia patients with antibodies to F.VIII or F.IX and in nonhemophilacs with antibodies against F.VIII (acquired hemophilia). Recent preliminary results indicate that recombinant F.VIIa may shorten the bleeding time in patients with thrombocytopenia and may stop or reduce thrombocytopenic bleeding that cannot be controlled by conventional treatment.

F.VIIa as a bypassing agent

F.VIIa in complex with tissue factor (TF) or other phospholipids exposed at the site of injury activates F.X into F.Xa inducing local hemostasis.[8] This process is independent of the presence of factors VIII or IX and is not influenced by inhibitors to these coagulation factors.

Inhibitors to Coagulation Factors
Edited by Louis M. Aledort *et al.*, Plenum Press, New York, 1995

Infused F.VIIa is active only when complexed with TF, minimizing the risk of inducing generalized activation of the coagulation cascade and disseminated intravascular coagulation (DIC). Furthermore, F.VIIa treatment can be used both in patients with hemophilia A and B regardless of the inhibitor titer.

Following the conversion of F.X to F.Xa and subsequent thrombin formation, F.Xa is inhibited by forming a complex with the tissue factor pathway inhibitor (TFPI). The F.Xa-TFPI complex then binds to the F.VIIa-TF complex resulting in inhibition of the extrinsic coagulation pathway. For this inhibition to occur, F.VIIa must be complexed to TF. In the presence of an abundance of F.VIIa, F.X activation may be facilitated on the phospholipid (PL) surface inducing local hemostasis, although F.VIIa-TF complexes are inhibited by TFPI. Consequently, F.VIIa therapy may be hemostatically effective in hemophilia in the presence of physiological levels of TFPI provided appropriate doses are administrated (see Figure 1). The half-life of infused recombinant F.VIIa in hemophiliacs is about 2.9 hours.

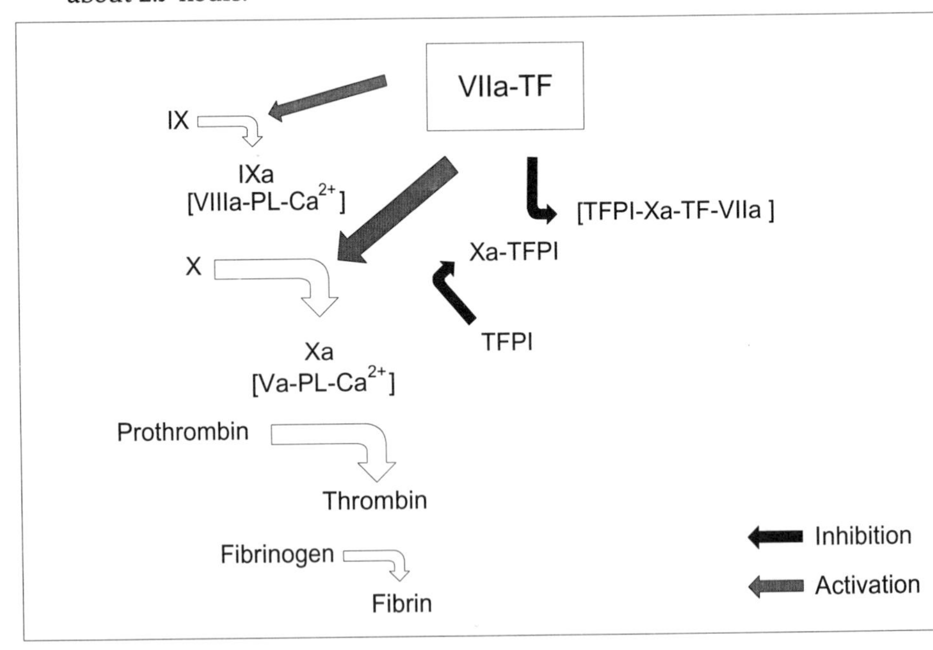

Figure 1. Recombinant activated F.VII as a bypassing agent

Clinical experience

The first published report of the efficacy of recombinant F.VIIa was described by Hedner and co-workers.[9] A hemostatic response was reported in synovectomy in a severely affected hemophilia patient with a high-titer F.VIII inhibitor. Clinical experiences with recombinant F.VIIa in other patients have been reported.[10-27] More than 240 patients have been treated with recombinant F.VIIa (NovoSeven™, Novostase™, Novo Nordisk, Denmark) for 1270 bleeding episodes.

Critical bleeds and essential surgery

Recombinant FVIIa has been given on a named-patient basis to 112 patients, 74 with hemophilia A, 8 with hemophilia B, 18 nonhemophilia patients with antibodies to F.VIII, 7 patients with F.VII deficiency, and 5 patients with other clotting disorders. Recombinant FVIIa was used on a named-patient basis on 638 occasions, of which 198 were critical bleeds or essential surgery where alternative therapy was ineffective or unsuitable. Seven patients were treated for more than 20 bleeding episodes.

Recombinant F.VIIa was usually implemented with doses of 70 to 100 µg/Kg bw every 2 to 3 hours and adjusted according to the clinical response. An excellent or effective response to recombinant F.VIIa treatment was reported by the physician in charge in 74 to 100% of cases depending on the type of bleed (see Table 1).

Table 1. Physician's global evaluation of clinical response

Type of bleed	No. of treatments	Documentation not available	Excellent/effective (%)		Partially effective	Ineffective
Major surgery	21	1	16/20	(76%)	3	1
Minor surgery	44	4	35/40	(88%)	3	2
Dental surgery	21	7	13/14	(93%)	-	1
Major muscle	31	8	17/23	(74%)	3	3
ENT & oral cavity	26	9	16/17	(94%)	1	-
CNS	17	5	12/12	(100%)	-	-
Internal/ Retroperitoneal	38	8	24/30	(80%)	3	3
Joint	268	121	127/147	(86%)	18	2
Prophylaxis	11	1	10/10	(100%)	-	-
Miscellaneous	161	18	107/143	(75%)	27	9
Overall	638	182	377/456	(83%)	58	21

In order to evaluate the hemostatic response at 8 and 24 hours after the initiation of recombinant F.VIIa therapy, a computerized algorithm was designed for specific categories of critical bleeds (see Table 2). An effective response was defined as definite pain relief or unequivocal reduction in bleeding site size. If there was no improvement in pain or no change/worsening in bleed site size, the response to treatment was defined as ineffective. Joint and mild muscle bleeds were not included in the analysis. In the first 55 consecutive bleeds evaluated according to the algorithm, 91% and 90% were judged to be effectively controlled at

8 and 24 hours, respectively. Using similar criteria to evaluate efficacy, 71% of life- or limb-threatening bleeds reported to be effectively controlled with porcine F.VIII.[28] In four clinical studies,[29-32] unactivated factor IX (F.IX) concentrates and activated prothrombin complex concentrates have been reported to be effective in about 50 to 60% of bleeds after 12 hours. Furthermore, about 70% of the bleeds treated in these four trials were joint bleeds.

On 86 occasions, recombinant F.VIIa was used in association with major and minor surgery and dental procedures. On 3 occasions (2 herniotomy and 1 hemipelvectomy), excessive bleeding was reported despite high doses of recombinant F.VIIa.[13,18] When a computerized algorithm was applied to the first 27 cases of surgery reported, postoperative oozing was noted in 15% and 22% of procedures at 8 and 24 hours, respectively, which is in accordance with published data in hemophilia patients without inhibitors treated with factor concentrates.[33] It remains to be clarified if postoperative oozing can be reduced by optimizing the dosage schedule or by a more rigorous use of concurrent antifibrinolytic therapy.

Table 2. Clinical response according to computerized algorithm—Critical bleeds

Bleeds	No. of bleeds	Effective response* 8 hrs.		24 hrs.	
Major muscle	11	7/7	(100%)	10/10	(100%)
ENT & oral cavity	12	10/10	(100%)	10/10	(100%)
CNS	10	6/9	(67%)	6/7	(67%)
Internal/ retroperitoneal	14	12/12	(100%)	12/13	(92%)
Wound	8	5/6	(83%)	6/7	(86%)
Overall	55	40/44	(91%)	44/49	(90%)

* 11 and 6 bleeding episodes were not assessable at 8 and 24 hours respectively and have been excluded from the efficacy analysis.

Monitoring of therapy

F.VII clotting activity (FVII:C)

In the treatment of hemophilia with factors VIII/IX, the goal is to increase the baseline plasma level of factors VIII/IX to a hemostatic level of 0.30 to >0.50 IU/mL depending on the type of bleed. When using recombinant F.VIIa, a totally different concept is used. Hemophiliacs have an essentially normal level of F.VII (1.0 U/mL plasma); however, by administering an abundance of activated F.VII the extrinsic coagulation pathway is pushed to induce hemostasis.

During treatment with recombinant F.VIIa, plasma samples were drawn for determination of peak and trough FVII:C at a Novo Nordisk core laboratory. The FVII:C was measured in a one-stage coagulation system containing F.VII-deficient plasma (immunodepleted, Novo Nordisk) and rabbit brain thromboplastin (type C, Manchester Comparative Reagents, UK). Coagulation was started by adding thromboplastin and Ca^{++}. Pooled citrated plasma from healthy subjects was used as calibrator and was assigned an arbitrary potency of 1 U/mL. Although it has not been logistically possible to investigate the correlation between peak and trough levels of FVII:C and hemostatic effect, descriptive data is presented in Table 3. Furthermore, there were no laboratory assays available at the time these patients were tested with recombinant F.VIIa to discriminate between F.VII zymogen and F.VIIa. Therefore, the amount of FVII:C that can be attributed to endogenous F.VII during treatment with recombinant F.VIIa could not be accurately assessed.

About 60% of trough values measured were between 3 and 5 U/mL

Table 3. Peak and trough values in plasma FVII:C

	Peak Values, U/mL					Trough Values, U/mL			
No. of determinations	Range	≤10	11-20	>20	No. of determinations	Range	<3	3-5	≥6
Hemophilia A N=89	7.1-38.0	10/89 (11%)	64/89 (72%)	15/89 (17%)	Hemophilia A N=158	2.2-9.6	20/158 (13%)	92/158 (58%)	46/158 (29%)
Hemophilia B N=67	10.0-35.0	1/67 (1%)	49/67 (73%)	17/67 (25%)	Hemophilia B N=4	2.6-9.5	1/4	1/4	2/4
Non-hemophilia N=13	14.0 27.0	0/13	9/13 (69%)	4/13 (31%)	Non-hemophilia N=29	2.4-9.3	2/29 (7%)	17/29 (59%)	10/29 (34%)

but were usually >5 mL during the initial 24 hours of treatment. Trough values of <3 U/mL were usually associated with milder bleeds, which were treated with lower doses of recombinant F.VIIa. However, in two patients oozing around a central line was associated with trough levels of 2.8 and 3.1 U/mL, respectively. In the first case, oozing stopped when the dose was increased to 100 μg/Kg and in the second case when concomitant therapy with tranexamic acid was implemented. In a third patient oozing around a tracheostomy site was associated with a trough level of 2.1 U/mL which immediately stopped when the dose was increased to 90 μg/Kg.

Peak FVII:C levels varied between 11 and 20 U/mL in about 70% of determinations. In 12 hemophilia A patients, the recovery of recombinant F.VIIa varied between 25% and 67% following the initial dose. Later during the course of treatment the recovery varied between 19 and 67%. The intra-individual variation was much less than the inter-individual variation (see Table 4).

Table 4. Recovery of recombinant F.VIIa

No. of	No. of patients	Dose determinations	µg/Kg	Recovery% 1st dose	Recovery% Later during treatment
Hemophilia A	12	38	30-104	25-67	19-67
Hemophilia B	1	2	70	60-91	
Non-hemophilia with antibodies against FVIII:C	1	5	61.9	61-70	52-79

Prothrombin time (PT)

Basically the PT measures the extrinsic coagulation pathway and is influenced by changes in the prothrombin complex coagulation factors. Normally, the PT is also within the normal range in hemophiliacs. Addition of F.VIIa shortens the PT in normal[5,6] and in hemophilic plasma.[9] Infusion of recombinant F.VIIa shortens the PT in a dose-dependent way in hemophiliacs in a nonbleeding state[34] and in critical bleeds (see Figure 2). The PT seems to reach a minimum plateau level of 6 to 7 seconds corresponding to plasma levels of FVII:C >6-10 U/mL. In 147 trough observations of FVII:C in hemophiliacs experiencing critical

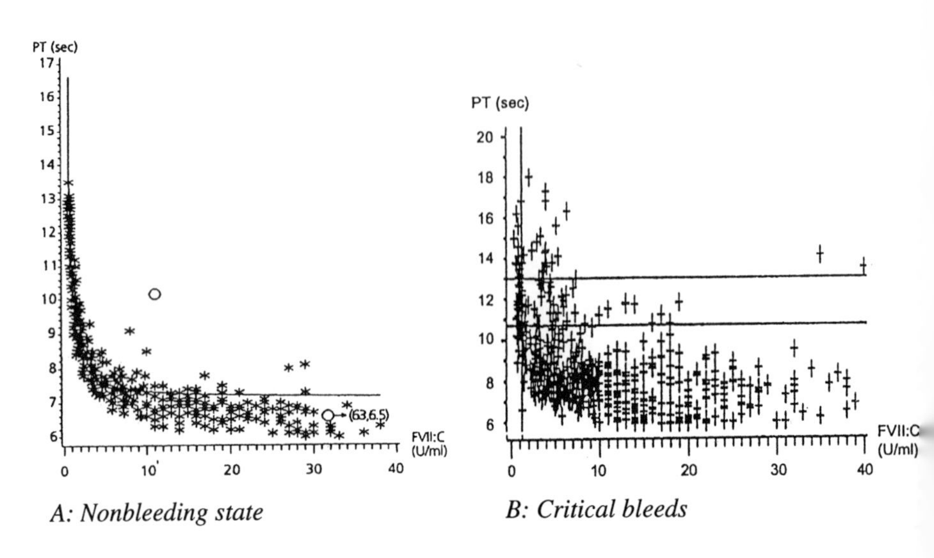

Figure 2. Relation between FVIIa coagulant activity and PT

bleeds, 52% and 48% had corresponding values of PT ≤8 seconds and >8 seconds, respectively. In 152 peak observations of FVII:C in critical bleeds, 83% and 17% had corresponding values of PT ≤8 seconds and >8 seconds, respectively.

Additional testing will indicate if the PT can be used to monitor clinical efficacy. The PT may also be useful to identify patients with low recovery of recombinant F.VIIa. Furthermore the PT assay is much easier to standardize than the FVII:C assay.

Activated partial thromboplastin time (APTT)

The APTT is assumed to basically be a measure of the intrinsic coagulation pathway. The test is markedly prolonged in plasma lacking factors VIII or IX. The APTT prior to recombinant F.VIIa treatment varied between 81.7 and 178 seconds and 79 and 144 seconds in hemophiliacs and nonhemophiliacs, respectively, who were tested at a Novo Nordisk core laboratory.

Surprisingly, the addition of recombinant F.VIIa to hemophilic plasma was reported to shorten the APTT *in vitro*[6,35] and in patients treated for critical bleeding in a dose-dependent way. F.VII is known to bind to phospholipids especially to phosphatidylserine, one of the components in the APTT reagent. Typically, a shortening of 15 to 25 seconds from baseline is measured following therapeutic doses; however, the inter-patient variation is quite large (range 3 to 104 seconds). Furthermore, the APTT never normalized, not even at plasma FVII:C levels >20 U/mL. No prolongation of the APTT during recombinant F.VIIa treatment was reported in 30 consecutive patients.

Baseline levels of APTT vary according to the patient and to the reagents; therefore, a comparison of the different reagents' ability to bind recombinant F.VIIa in vitro must be performed to evaluate to what extent APTT can be used to monitor treatment.

Safety

The postulated mechanism of action of F.VIIa provides theoretical support for its safety. Normally, F.VIIa complexes to tissue factor which is anchored to the vessel wall at the site of injury inducing local hemostasis.

An abundance of tissue factor could be supposedly exposed as a result of extensive tissue damage, some of which may find its way into the circulating blood. Circulating tissue factor could form a potential template for complex formation with F.VIIa inducing intravascular coagulation. Furthermore, many tissues and cells have been shown to possess tissue factor procoagulant activity. Tissue factor is expressed on the surface of monocytes following exposure to endotoxins of various origins. Tissue factor has also been demonstrated in arteriosclerotic plaques. Consequently, patients with crush injury, advanced arteriosclerotic disease, malignancy, and septicemia may be predisposed to adverse, coagulation-related events. Theoretically, this risk may be increased during treatment with recombinant F.VIIa.

As indicators of free proteolytic activity in the circulation, 30 consecutive patients treated for critical bleeds were monitored for changes in D-Dimer, ATIII and α_2-antiplasmin at a Novo Nordisk core laboratory. Platelets and fibrinogen were monitored by the hospital laboratories. With one exception, no clinically important observations were noted in these assays. Consumption depletion of coagulation factors was reported in one patient subjected to extensive resection of myonecrotic tissue.[18] Consumption of coagulation factors in the extensive wound area combined with the handling of necrotic tissue presumably provoked the release of a variety of enzymes capable of activating both the fibrinolytic and the coagulation systems. It is unclear if recombinant F.VIIa contributed to the condition. Seven additional patients with significant infections or septicemia were treated with recombinant F.VIIa without any signs of intravascular coagulation or fibrinolysis.

Transient cardiovascular events were reported in two nonhemophiliacs with atherosclerotic disease who were treated with recombinant F.VIIa.[2] Ataxia with clinical signs of a thrombotic event in the cerebellum was reported in a 5-year-old hemophiliac treated for an intraparenchymal hemorrhage.[27] Thrombosis could not be confirmed by CT or NMR.

Dosage and administration

The recommended dosage range of recombinant F.VIIa for the treatment of critical bleeding is 60-120 µg/Kg bw administered as bolus intravenous injections.

Critical hemorrhages

The recommended initial dose of recombinant F.VIIa is 90 µg/kg bw. The second dose should be given after 2 to 3 hours depending on the severity of the bleed. The dosage should then be repeated at 2- to 3-hour intervals for 1 to 2 days or until clinical improvement is observed. Thereafter, the dosage interval can be increased to 4 hours or the dosage decreased if continued therapy is indicated.

Surgery

An initial dose of 90 µg/kg bw should be given immediately before the procedure. The dose should be repeated after 2 hours and then at 2- to 3-hour intervals. In major surgery, the dosage should be continued at 2- to 3-hour intervals for 6 to 7 days. If continued therapy is indicated, the dosage interval can be increased to 4 hours for a further 6 to 7 days and then to 6 hours for the remaining period of treatment.

Preliminary experience indicates the concomitant use of antifibrinolytic therapy in surgery and dental procedures. Antifibrinolytic therapy is given to protect hemostatic plugs from the normally occurring local fibrinolysis. Antifibrinolytics (EACA, tranexamic acid) should be given intravenously in adequate doses immediately before the procedure. Treatment should be continued postoperatively for several days. Antifibrinolytics have been reported to reduce blood loss in association with surgery in hemophilia patients especially in regions rich in fibrinolytic activators such as the oral cavity.[36]

Preliminary clinical experience in thrombocytopenia

It is generally believed that the primary arrest of bleeding is essentially due to the development of a platelet plug at the end of transected vessels as occurs during the performance of the bleeding time test. The evidence has been derived largely from bleeding time incisions evaluated by light and electron microscopy.[37] The role of the coagulation mechanism during early stages of hemostasis is less clear; however, early fibrin formation has been reported to occur within the bleeding time wound.[38,39] These studies suggest F.VIIa-tissue factor-phospholipid activation of F.X occurs as an early event during the arrest of bleeding from bleeding-time incisions. Infusion of plasma-derived F.VIIa has been shown to shorten the bleeding time in rabbits with induced thrombocytopenia.[39]

Preliminary data is available from a recent study of the use of recombinant F.VIIa in thrombocytopenia.[40] The aim of this study was to investigate if infusion of recombinant F.VIIa can enhance fibrin formation in bleeding time wounds in patients with thrombocytopenia as reflected by significant shortening of the bleeding time. The effect of single intravenous bolus infusions on the bleeding time was studied in patients with impaired platelet production or immune destruction of platelets. The two groups of patients were subgrouped according to platelet count. The dose levels investigated were 50 and 100 μg/Kg bw and the Surgicutt bleeding time was measured before and 30 minutes after recombinant F.VIIa infusion. A total of 84 infusions were given. A reduction in bleeding time was found in 45 (53%) of the cases. Not surprisingly, the reduction was more pronounced when platelet count $>20 \times 10^9/L$.

Recombinant F.VIIa was administered to two thrombocytopenic patients with overt bleeding. The first case was a 55-year-old woman with advanced chronic lymphocytic leukemia, splenomegaly, and platelet count of $10 \times 10^9/L$. The patient suffered from extensive uterine bleeding with no effect of platelet transfusions and tranexamic acid. Following a single injection of 50 μg/Kg bw recombinant F.VIIa, the bleeding significantly decreased and almost stopped. The effect lasted for more than 24 hours. The Surgicutt bleeding time was reduced from 41.5 to 20 minutes.

The second case was a 70-year-old woman in blast crisis of chronic myelocytic leukemia and a platelet count of $32 \times 10^9/L$. After receiving a central venous catheter she continued to ooze from the neck incision. Local compression and platelet transfusions did not stop the bleeding.

Following a single injection of 50 μg/Kg bw the bleeding stopped immediately and did not recur. The Surgicutt bleeding time was reduced from 29.5 to 3 minutes.

Conclusion

Recombinant FVIIa is a highly effective alternative treatment of hemophiliacs and nonhemophiliacs with antibodies against F.VIII or F.IX. Treatment is not influenced by the inhibitor titer and appears to be equally effective in F.VIII- and F.IX-deficient patients. A hemostatic effect in critical bleeding is often reported within 8 hours after the initiation of treatment. The recommended dose schedule is bolus injections of 90 μg/Kg bw administered at 2- to 3-hour intervals until clinical improvement. There is a large inter-patient variation in the recovery of recombinant F.VIIa; however, therapeutic plasma levels of FVII:C have not been established. No specific assay has been established to monitor efficacy; however, the PT is consistently shortened during therapy with recombinant F.VIIa. It is recommended to use concomitant antifibrinolytic therapy in association with surgery.

Very few adverse events have been reported in inhibitor patients; however, experience in patients who may be predisposed to adverse, coagulation-related events is limited.

Recombinant F.VIIa may shorten the bleeding time in about 50% of patients with thrombocytopenia, in patients with impaired platelet production, or immune destruction of platelets. Single doses have significantly reduced overt bleeding in two patients. These anecdotal case reports should be confirmed in a controlled study in order to evaluate if recombinant F.VIIa therapy may be indicated to stop or reduce thrombocytopenic bleeding that cannot be controlled by conventional therapy.

References

1. U. Hedner, W. Kisiel, Use of human factor VIIa in treatment of two haemophilia A patients with high-titer inhibitors, *J Clin Invest.* 71:1836 (1983).
2. U. Hedner, S. Bjoern, S.S. Bernvil, L. Tengborn, L. Stigendahl, Clinical experience with human plasma-derived factor VIIa in patients with hemophilia A and high-titer inhibitors, *Haemostasis.* 19: 335 (1989).
3. F.S. Hagen, C.L. Gray, P. O'Hara, F.J. Grant, G.C. Saari, R.G. Woodbury, C.E. Hart, M. Insley, W. Kisiel, K. Kurachi, E.W. Davie, Characterization of a cDNA coding for human factor VII, *Proc Natl Acad Sci USA.* 83:2413 (1986).
4. L. Thim, S. Bjoern, M. Christensen, E.M. Nicolaisen, T. Lund-Hansen, A.H. Pedersen, U. Hedner, Amino acid sequence and posttranslational modifications of human factor VIIa from plasma and transfected baby hamster kidney cells, *Biochemistry.* 27:7785 (1988).
5. U. Hedner, T. Lund-Hansen, D. Winther, Comparison of the effect of factor VII prepared from human plasma and recombinant VIIa in vitro and in rabbits, *Thromb Haemost.* 58:270a (1987).
6. U. Hedner, J. Ljungberg, T. Lund-Hansen, Comparison of the effect of plasma-derived and recombinant human FVIIa in vitro and in a rabbit model, *Blood Coag and Fibrinol.* 1:145 (1990).

7. K.M. Brinkhous, U. Hedner, J.B. Garris, V. Diness, M.S. Read, Effect of recombinant factor VIIa on the hemostatic defect in dogs with hemophilia, hemophilia B, and von Willebrand disease, *Proc Natl Acad Sci USA.* 86:1382 (1989).
8. S.I. Rapaport, Inhibition of factor VIIa/tissue factor-induced blood coagulation: With particular emphasis upon a factor Xa-dependent inhibitory mechanism, *Blood.* 73:359 (1989).
9. U. Hedner, S. Schulman, K.A. Alberts, M. Blombäck, H. Johnsson, S. Glazer, K. Pingel, Successful use of rFVIIa in a patient with severe hemophilia A subjected to synovectomy, *Lancet.* ii:193 (1988).
10. G.B. Macik, H. Hohneker, H. Roberts, A.M. Griffin, The use of recombinant activated factor VII for treatment of a retropharyngeal haemorrhage in a hemophilic patient with a high-titer inhibitor, *Am J Hematology.* 32:232 (1989).
11. S.A. Grupp, S. Glazer, D.A. Williams, Recalcitrant epistaxis in a hemophiliac with inhibitors: Experience with recombinant factor VIIa. *Blood.* 74 (Suppl 1):3901 (abstract no. 1489) (1989).
12. M.L. Schmidt, H.E. Smith, S. Gamerman, S. Glazer, J.P. Scott, Prolonged recombinant activated factor VII (rFVIIa). Treatment for severe bleeding and surgical hemostasis in a factor-IX deficient patient with an inhibitor. Paper presented at a meeting of the Society for Pediatric Research, MA (1990).
13. A. Gringeri, E. Santagostino, P.M. Mannucci, Failure of recombinant activated factor VII during surgery in a hemophiliac with high-titer factor VIII antibody, *Haemostasis.* 21:1 (1991).
14. L. Tengborn, L. Stigendahl, P.O. Elfstrand, B. Kjellman, S. Glazer, Hemostatically effective treatment with recombinant factor VIIa (rFVIIa) of a hemophiliac boy with newly developed inhibitors to factor VIII. Abstract presented at the XIX International Congress of the World Federation of Hemophilia, Washington, D.C., August 14 to 19 (1990).
15. M.L. Schmidt, H.E. Smith, S. Gamerman, D. DiMichele, S. Glazer, J.P. Scott, Prolonged recombinant activated factor VII (rFVIIa) treatment for severe bleeding in a factor-IX deficient patient with an inhibitor, *Br J Haematol.* 78:460 (1991.)
16. U. Hedner, M. Feldstedt, S. Glazer, Recombinant FVIIa in hemophilia treatment, in: *Hemophilia and von Willebrand's Disease in the 1990s,* J.M. Lusher, C.M. Kessler, eds., Elsevier Science Publishers B.V. (1991).
17. J. Ingerslev, M. Feldstedt, S. Sindet-Petersen, Control of hemostasis by recombinant factor VIIa in a hemophilia A patient with inhibitors during major oral surgery. Abstract presented at the XIX International Congress of the World Federation of Hemophilia, Washington, D.C., August 14 to 19 (1990).
18. S.F. Stein, A. Duncan, D. Cutler, S. Glazer, Disseminated intravascular coagulation (DIC) in a hemophiliac treated with recombinant factor VIIa. Abstract presented at the American Society of Hematology 32nd Annual Meeting, Boston, MA, November 28 to December 4 (1990).
19. U. Hedner, Factor VIIa in the treatment of haemophilia, *Blood Coagul Fibrinolysis.* 1:307 (1990).
20. B.A. Bell, K. Birch, S. Glazer, Experience with recombinant factor VIIa in an infant hemophiliac with inhibitors to FVIII:C undergoing emergency central line placement. A case report, *Am J Pediatr Hematol Oncol.* 15:77 (1993).
21. S. Schulman, A therapeutic alternative for haemophiliacs with inhibitors, *Acta Pædiatr.* 81, paper 926 (1992).
22. J. Ingerslev, L. Knudsen, S. Sindet-Petersen, Haemostatic efficacy of recombinant factor VIIa in haemophilia A patients with inhibitors. Abstract presented at the XX International Congress of the World Federation of Hemophilia, Athens, Greece, October 12 to 17 (1992).
23. J. Ingerslev, L. Knudsen, S. Sindet-Petersen, Clinical experiences with recombinant factor VIIa in haemophilia A patients with inhibitors. Abstract presented at the 3rd Bari International Conference on Factor VIII/von Willebrand Factor, Foggia, Italy, June 7 to 13 (1992).
24. M. Makris, Recombinant factor VIIa: The UK experience. Abstract presented at the 24th Congress of the International Society of Hematology, London, UK, August 23 to 27 (1992).
25. U. Hedner, S. Glazer, Management of hemophilia patients with inhibitors, *Hematol Oncol Clin North Am* 6:1035 (1992).
26. M.L. Schmidt, S. Gamerman, H.E. Smith, J.P. Scott, D. DiMichelle, Recombinant

activated factor VII (rFVIIa) therapy for muscle and soft tissue hemorrhage in factor VIII and IX-deficient patients with inhibitors. Abstract presented at the 33rd Annual Meeting and Exposition of the American Society of Hematology, Denver, Colorado, December 6 to 10 (1991).

27. M.L. Schmidt, S. Gamerman, H.E. Smith, J.P. Scott, D. DiMichelle, Recombinant activated factor VII (rFVIIa) therapy for intracranial hemorrhage (ICH) in hemophilia A patients with inhibitors. Abstract presented at the 33rd Annual Meeting and Exposition of the American Society of Hematology, Denver, Colorado, December 6 to 10 (1991).
28. D.B. Brettler, A.D. Forsberg, P.H. Levine, et al, The use of porcine factor VIII concentrate (Hyate:C) in the treatment of patients with inhibitor antibodies to factor FVIII, *Arch Intern Med.* 149:1381 (1989).
29. J.M. Lusher, et al., Efficacy of prothrombin-complex concentrates in hemophiliacs with antibodies to factor VIII, *N Eng J Med.* 303:421 (1980).
30. L.J.M. Sjamsoedin, et al., The effect of activated prothrombin-complex concentrate (Feiba) on joint and muscle bleeding in patients with hemophilia A and antibodies to factor VIII. A double-blind clinical trial, *N Engl J Med.* 305:717 (1981).
31. J.M. Lusher, P.M. Blatt, J.A. Penner, et al., Autoplex versus proplex: A controlled, double-blind study of effectiveness in acute hemarthroses in hemophiliacs with inhibitors to factor VIII, *Blood.* 62:1135 (1983).
32. M. Hilgartner, L. Aledort, A. Andes, J.C. Gill, Efficacy and safety of vapor-heated anti-inhibitor coagulant complex in hemophilia patients, *Transfusion* 30:626 (1990)
33. C.K. Kasper, L. Boylen, N. Ewing, J.V. Luck, S.L. Dietrich, Hematologic management of hemophilia A for surgery, *JAMA.* 253:1279 (1985).
34. L. Johansen, D. Bucher, L.L. Hansen, H. Kastberg, C.M. Lindley, P. Madsen, W.T. Sawyer, Relation between factor VIIa coagulant activity and PT and between factor VIIa coagulant activity and APTT in human plasma, Abstract presented at the XIVth Congress of the International Society on Thrombosis and Hemostasis, New York, USA, July 4-9 (1993).
35. D.S.C. Telgt, B.G. Macik, D.M. McCord, D.M. Monroe, H.R. Roberts, Mechanism by which recombinant factor VIIa shortens the aPTT: Activation of factor X in the absence of tissue factor, *Thromb Res.* 56:603 (1989).
36. I.M. Nilsson, U. Hedner, A. Ahlberg, et al., Surgery of hemophiliacs–20 years' experience, *World J Surg.* 1:55 (1977).
37. R.P.C. Rodgers, J. Levin, A critical reappraisal of the bleeding time, *Semin Thromb Hemost.* 16 No. 1 (1990).
38. H.J. Weiss, B. Lages, Evidence for tissue factor-dependent activation of the classic extrinsic coagulation mechanism in blood obtained from bleeding time wounds, *Blood.* 629:635 (1988).
39. U. Hedner, D. Bergqvist, J. Ljungberg, B. Nilsson, Haemostatic effect of factor VIIa in thrombocytopenic rabbits, *Blood.* 66:1043 (1985).
40. J. Kristensen, A. Killander, E. Hippe, C. Helleberg, J. Ellegaard, M. Holm, U. Hedner, Recombinant factor VIIa (rFVIIa) reduces the bleeding time in patients with thrombocytopenia, Abstract submitted to The American Society of Hematology 35th Annual Meeting, St. Louis, USA, December 3-7 (1993).

Apheresis

Inga Marie Nilsson[1] and Christian Freiburghaus[2]

[1]*Department for Coagulation Disorders*
University of Lund, Malmö General Hospital
S-214 01 Malmö, Sweden
[2]*Excorim AB, Lund, Sweden*

Abstract

Treatment of hemophilia patients with inhibitors is most effective if circulating levels of factor VIII (F.VIII) or factor IX (F.IX) can be obtained. This is possible not only in patients with low inhibitor concentration, but also in those with high inhibitor levels after temporary removal of the inhibitors by means of apheresis methods.

Extensive plasma exchange combined with huge doses of F.VIII has occasionally been tried, but the method is of limited capacity, and exposure to plasma products can initiate an anamnestic response of the inhibitor before treatment with factor concentrates can be started. A better alternative is extracorporeal immunoadsorption for removal of antibodies against F.VIII or F.IX. Two methods developed at the center in Malmö are presented.

1) Group specific elimination using protein A. This is done by computer controlled extracorporeal adsorption to protein A Sepharose columns of the antibodies in plasma transferred on-line to the plasma treatment apparatus (Citem 10, Excorim, Lund, Sweden). The plasma depleted of inhibitor is returned to the patient together with the separated blood cells. At the Malmö center, high titer inhibitors have so far been removed in 10 hemophilia patients, and 4 patients with acquired hemophilia, on a total of 25 immunoadsorption episodes. Usually between 2 and 4 plasma volumes were processed in the course of 6 to 12 hours. The inhibitor decreased to zero or very low levels, enabling conventional replacement therapy to be given for control of severe bleeding or as cover for surgery. The treatment is well tolerated and no signs of activation of the coagulation, fibrinolytic, and complement systems were seen.

2) Specific elimination using coagulation factor. A method has been developed for the removal of F.IX antibodies directly from whole blood in a continuous extracorporeal system. A F.IX preparation with a specific activity of 92 units per mg protein was covalently coupled to Macro Beaded Sepharose, which allows passage of whole blood. One ml of the absorbent was capable of binding about 2100 BU. The F.IX antibodies could be specifically adsorbed directly from citrated whole blood with a minimal extracorporeal volume. In this way, high-titer antibodies could

Inhibitors to Coagulation Factors
Edited by Louis M. Aledort *et al.*, Plenum Press, New York, 1995

be removed from a boy with hemophilia B, by passing four times his blood volume through the F.IX column. No side effects were noted. In the future it will be possible to use immobilized F.VIII in a similar system for the removal of F.VIII antibodies.

Introduction

It is generally agreed that the optimal approach to the treatment of hemophilia patients with inhibitors is to achieve significant circulating levels of F.VIII (F.VIII) or IX (F.IX). In patients with initially low inhibitor titers (<10 BU, Bethesda inhibitor units) hemostasis can be obtained by giving only F.VIII/IX in sufficiently large doses. When the inhibitor titer exceeds 10 BU, the patient is not available for conventional replacement therapy. Thus, attempts have been made to temporarily remove the inhibitors prior to the infusion of factor preparations. Extensive plasma exchange combined with huge doses of F.VIII has occasionally been used, and successful clinical outcome has been reported for some patients.[1-3] However, the actual results with respect to inhibitor removal indicated plasma exchange to be inefficacious, and in most cases to require fairly extensive treatment periods to reduce inhibitor levels sufficiently. When the replacement fluids used have been restricted to albumin and crystalloid fluids, marked and progressive reductions have been reported in platelet counts and in the concentrations of various plasma constituents, particularly complement and coagulation factors.[4-6] In dealing with bleedings in patients with inhibitors to F.VIII or F.IX, it is essential that the hemostatic process is not further disturbed by depletion of various coagulation factors. The use of fresh-frozen normal plasma as a replacement fluid also has disadvantages, as the F.VIII or F.IX of the infused plasma can initiate an anamnestic response of the inhibitor before treatment with factor concentrates can be started.

A much better alternative than plasma exchange is extracorporeal immunoadsorption for the removal of inhibitors against F.VIII or F.IX. Two such methods have been developed at the center in Malmö, namely 1) group specific elimination using protein A, and 2) specific elimination using coagulation factor. Here we review these methods and our clinical experience in patients with hemophilia A and B complicated by high-titer inhibitors as well as in patients with acquired hemophilia.

Patients and methods

Patients

Five patients with severe hemophilia A and five patients with severe hemophilia B with high titer inhibitors (10-4200 BU) and four non-hemophilic patients with antibodies against F.VIII were treated.

Methods for extracorporeal immunoadsorption for removal of inhibitors against F.VIII or F.IX

1) Group specific elimination using protein A.[7-11] This method is based on the affinity of human immunoglobulin for protein A. Protein A binds predominantly to the Fc portion of human IgG antibodies of subclasses 1, 2, and 4 but also to various extent to IgG_3, IgA, and IgM. In all our patients with hemophilia A or B, as well as in non-hemophilic patients with inhibitors, it has been possible to adsorb the antibodies *in vitro* to columns of protein A Sepharose and then elute them completely. This principle has also been used for *in vivo* removal of inhibitors. In the clinical situation, huge amounts of IgG have to be removed, usually anything from 50 to 100 g, and this has prompted the development of a special device for *ex vivo* removal of allo- and autoantibodies. The clinical setup consists of two units, a plasma separator and the comput-

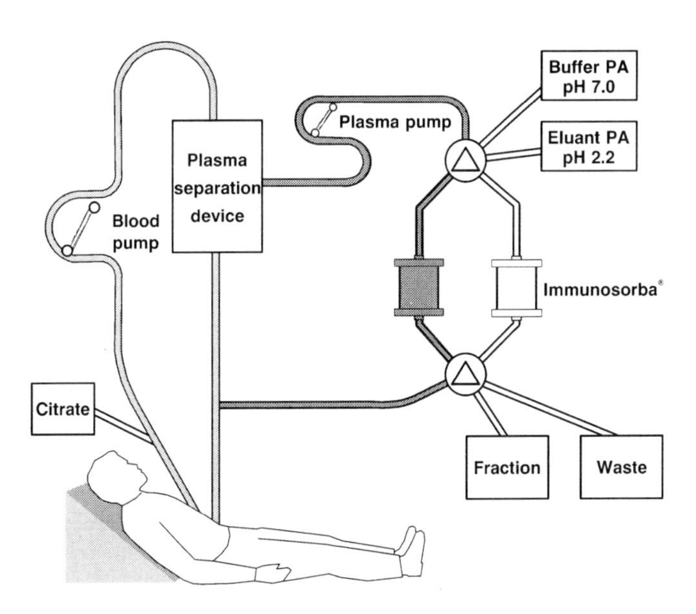

Figure 1. The principle of the continuous two-column immunoadsorption system

erized adsorption system (see Figure 1). Plasma is obtained using an ordinary cell centrifuge. Blood is drawn at a rate of up to 100 ml per minute and immediately citrated (0.13 M sodium citrate) at a ratio of 1:10. Vascular access for blood withdrawal is arterial (radial), whereas for return infusion an anticubital venous cannula is usually used. The patient himself is not heparinized. After separation in the cell centrifuge, the plasma is transferred on-line to the plasma treatment apparatus (Citem 10, Excorim, Lund, Sweden) monitoring the adsorption process. The adsorption unit includes two Immunosorba columns each containing 62.5 ml protein A Sepharose gel with a total binding capacity of 1.25 g of IgG. The monitor passes the plasma over the first column until this column is saturated with IgG. The plasma is then switched to the second column, while the first column is rinsed and the IgG eluted with a citrate-phosphate-gradient (pH from 7.0 to 2.2). The column is then neutralized again before new plasma enters the column. The plasma flow is switched from one column to the other every 10 minutes; and to compensate for the gradual lowering in levels of plasma IgG during treatment, the flow is successively increased to a maximum of 50 ml per minute. The effluent fluids from the columns are controlled by continuous pH and UV detectors. The data from these detectors are used in computer identification of the fluids as plasma, waste, or fraction. A series of sorting valves guide the fluids in the requested direction. The treated plasma depleted of inhibitor is returned to the patient together with the separated blood cells. The time required for a treatment depends on the inhibitor titer and the IgG level, of course; but anything from 6 to 12 hours is the average treatment time. The same columns can be used for repeated treatments. Between the treatment episodes, the columns are kept refrigerated and preserved by the addition of 0.1% Thiomersal.

2) Specific elimination using coagulation factor. A method for removal of F.IX antibodies directly from whole blood in a continuous extracorporeal system has been developed[12] (see Figure 2). A purified F.IX concen-

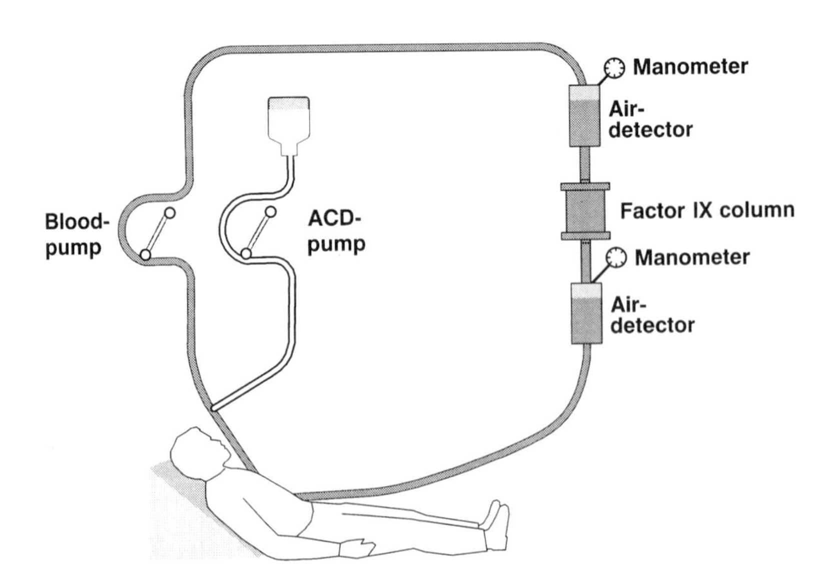

Figure 2. The principle of the continuous whole blood specific adsorption system

trate is required for a clinical adsorption system. A F.IX preparation essentially prepared with a method described by Andersson and coworkers[13] was used. It had a specific activity of 92 units per mg protein. The F.IX preparation was covalently coupled to Macro Beaded Sepharose, which allows the passage of whole blood, and packed in columns. One ml of the gel was capable of binding about 2100 BU of human anti-F.IX. The whole blood system is controlled by an Immunotherapy monitor (ITM10, Excorim, Lund). This consists of two pumps, one for blood, the other for the addition of citrate. The monitor is also equipped with pressure gauges and air detectors for the protection of the column as well as of the patient. Venous blood is drawn at a rate of 20 ml per minute and citrated directly. The blood is passed over the F.IX column containing 125 ml of adsorbant. During its passage over the column the blood is depleted of antibodies against F.IX; after adsorption it is retransfused to the patient. The extracorporeal volume of the system is only 200 ml. As the capacity of the adsorbant is so high, it is unnecessary to elute the column during the treatment procedure.

Assays

F.VIII coagulant activity (VIII:C) was determined both with a one-stage assay[14] and with a chromogenic assay (Coatest, Kabi), the results being expressed in international units (IU).

F.IX coagulant activity (IX:C) was measured with a one-stage assay[15] and the results expressed in international units.

F.VIII and F.IX inhibitor. A neutralization assay was used.[16] The inhibitory activity of the plasma is expressed as the number of units of F.VIII or F.IX inactivated by 1 ml of the plasma with 1 Malmö inhibitor unit (MU) corresponding to 3 BU/ml of plasma.[17]

Platelet count and coagulation tests were performed by standard methods.

Results and comments

Computer-controlled extracorporeal protein A adsorption has so far been used in 10 hemophilia patients (5 A and 5 B) and 4 patients with acquired hemophilia for removal of high-titer inhibitors, on a total of 25 immunoadsorption episodes (see Table 1). Usually between 2 and 4 plasma volumes were processed in the course of 6 to 12 hours. In cases of high titers, it has been found best if the removal is done in a two-day session, including a night´s rest to allow equilibration of intravascular and extravascular IgG. As expected, the inhibitor concentration decreased to zero or very low levels. In an extreme case, more than 40 liters of plasma (12.5 plasma volumes) had to be processed to lower the inhibitor concentration from 4200 to <10 BU per ml. This treatment does not prevent the anamnestic response on antigenic stimulation, but it enables conventional replacement therapy to be given for the control of severe bleeding or as cover for surgery. The IgG concentration was reduced to the same degree as the inhibitor. During the initial treatments a battery of tests were performed in connection with the adsorption technique, including hemoglobin, haptoglobin, platelet count, assays of fibrinogen, factor II + VII + X, factor V, VIII:C, IX:C, antithrombin III, FDP, fibrin monomers, ionized calcium, and C3 and C4 cell counts.The treatment is now performed on a routine basis with more than 600 patients treated all over Europe. There were no signs of hemolysis or of activation of the coagulation, fibrinolytic and complement systems. The patients tolerated the treatment well. Figure 3 shows the treatment course in a 5-year-old boy with hemophilia B and a high inhibitor concentration, 160 BU. After processing four liters of plasma over about four hours, the inhibitor had decreased to 6 BU. After a rest for 14 hours, it had increased to 30 BU, due to extravascular leakage. After processing a further 4 liters of plasma, the inhibitor decreased to zero. F.IX was given immediately, and his F.IX concentration increased to a high level. F.VIII or F.IX levels could usually be maintained for 5 to 8 days before the anamnestic response necessitated the withdrawal of factor infusions, except in those cases where the immunoadsorption was combined with induction of tolerance according to the Malmö treatment model.[10,18,19] In hemophiliacs actually bleeding at the start of immunoadsorption, the administration of factor concentrates resulted in prompt control of their bleedings. Similarly, in all hemophiliacs treated prior to surgery, postoperative hemostatic control was satisfactory.

Table 1. Extracorporeal immunoadsorption using protein A for removal of antibodies against F.VIII or F.IX

	5 hemophilia A	5 hemophilia B
Age (year)	5 - 66	5 - 44
Weight (kg)	30 - 65	17 - 48
Processed plasma volume (L)	6 - 52	2 - 12
Initial inhibitor (BU)	18 - 4200	18 - 160

Inhibitor decreased to zero or low levels enabling conventional replacement therapy.

However, the immunological activity in the inhibitor patient can complicate the overall outcome of an adsorption treatment. In January 1993 we received an overseas patient with hemophilia A and high-titer inhibitor. He had life-threatening abdominal and gastrointestinal bleeding and progressive respiratory distress requiring intubation. The patient had not responded to large doses of APCC, and several blood transfusions had been given. When he arrived in Sweden, with a military ambulance plane, his condition was extremely serious. He was immediately treated with immunoadsorption. The first day´s processing of about 3 plasma volumes lowered the inhibitor level from about 50 BU to around 5 BU. After a night´s rest, the titer was again over 40 BU, indicating the production of inhibitor since he had been boostered by the blood products given earlier. However, the second adsorption procedure was successful in removing the inhibitors, and the patient´s bleeding responded favorably to F.VIII replacement. Within a week, he returned home completely recovered.

Immunoadsorption (Immunosorba - Citem 10, Excorim, Sweden) is now available at many hospitals in Europe. This method has also been successfully used in the USA by Uehlinger and coworkers.[20]

Patients with acquired hemophilia have been treated with immunoadsorption both in the USA[20] and in Europe. In Sweden we have treated 4 patients on six different occasions (see Table 2). In all cases their autoantibody inhibitors were adsorbable to protein A *in vitro*. In acquired hemophilia, owing to the presence of circulating immuno-complexes and variation in the actual production of inhibitor, the

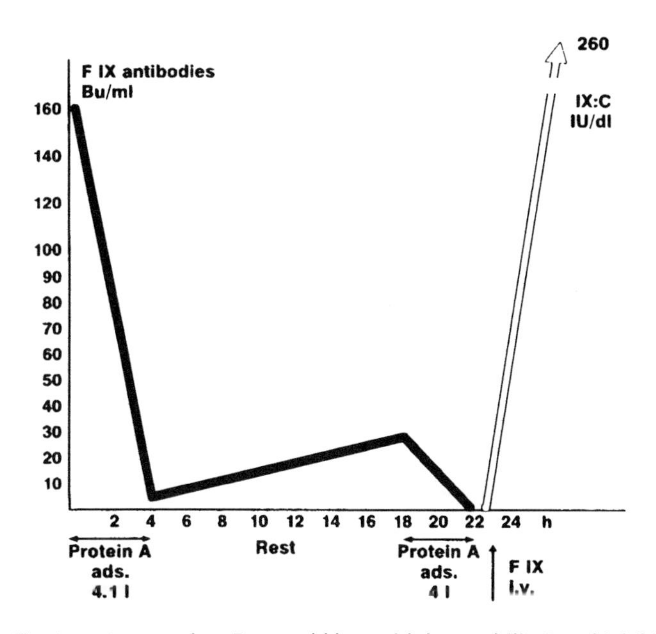

Figure 3. Treatment course in a 5-year-old boy with hemophilia B and inhibitors

outcome of treatment cannot be predicted. However, acute bleeding episodes in these patients could all be managed after extensive immunoadsorption followed by infusion of factor concentrates and/or DDAVP.

Another procedure is specific immunoadsorption for the removal of antibodies. Rather than removing all the immunoglobulins, a more ideal form of treatment of hemophiliacs with inhibitors would be the removal of antibodies by specific immunoadsorption, as even in patients with high inhibitor titers, only a minor portion of the total IgG is responsible for the inhibitor activity. A method for the removal of F.IX antibodies directly from whole blood in a continuous extracorporeal system has been used in a 10-year-old boy with hemophilia B and an inhibitor titer of about 120 BU (see Figure 4). On two consecutive days, four hours per day, approximately double the patient´s blood volume was passed over

Table 2. Extracorporeal immunoadsorption using protein A for removal of antibodies in acquired hemophilia

Patient	Age (y)	Weight (kg)	Plasma volume processed (L)	Initial inhibitor titer (BU)
1	34	90	14.8	36
2	67	108	17.0	38
3	55	85	19.7	63
4	71	60	50.0	160

Acute bleeding episodes could be managed after adsorption by infusion of factor concentrates and/or DDAVP.

the column. The inhibitor titer decreased from 120 to 21 BU per ml the first day. Following equilibrium of the intravascular and extravascular fluids, his inhibitor titer had increased to 27 BU after a night´s rest. After another four hours´ treatment the inhibitor decreased to about 5 BU. He was given F.IX, and an orthopedic correction could be performed without any complications. Normal hemostasis was maintained for eight days. There were no signs of activation of the coagulation, fibrinolytic or complement systems, or of hemolysis.

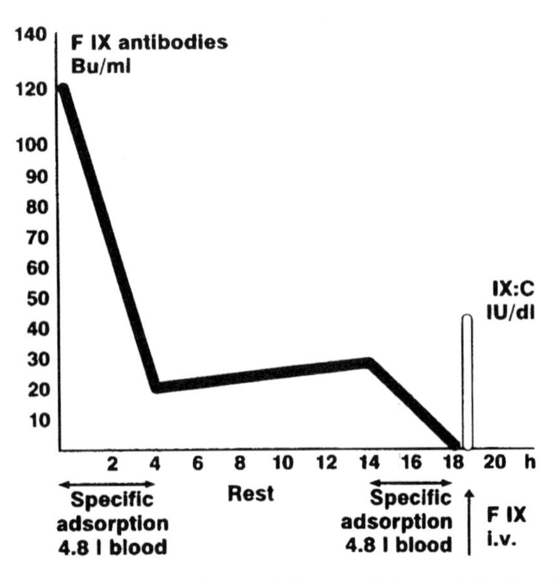

Figure 4. Treatment course in10-year-old boy with hemophilia B and inhibitors treated with specific adsorption.

The great advantages of the specific immunoadsorption system are the use of whole blood permitting small extracorporeal volumes, and that large amounts of inhibitors can be removed without loss of the normal IgG.

In the future, it will be possible to use immobilized F.VIII (recombinant F.VIII) or F.VIII fragments bearing those epitopes with which the antibody in the patient reacts in a similar system for the removal of F.VIII antibodies in patients with hemophilia A or acquired hemophilia.

Conclusion

Treatment of hemophilia patients with inhibitors is most effective if circulating levels of F.VIII or F.IX can be obtained. This is possible not only in patients with low inhibitor concentrations, but also in those with high inhibitor levels after temporary removal of the inhibitors by means of apheresis methods.

Extensive plasma exchange combined with huge doses of F.VIII has been tried, but the method has been shown to be of limited capacity. We have found immunoadsorption essential in the treatment of inhibitor patients. Both hemophilia A and B patients with inhibitors and patients with acquired hemophilia can be treated using computer-controlled extracorporeal adsorption of the antibodies in plasma to protein A Sepharose columns. Very high titers of inhibitors can be lowered to levels allowing conventional replacement treatment to manage acute bleeding episodes. Large volumes of plasma can be processed, which is often required.

A method for specific elimination of F.IX antibodies in a continuous extracorporeal system directly from whole blood has been developed.

Immunoadsorption should be followed by treatments aimed at minimizing, or even better, preventing the resynthesis of the inhibitor. The Malmö Treatment Model comprising cyclophosphamide, F.VIII or F.IX, and high dose intravenous IgG has been found successful in many cases in eliminating inhibitor production in hemophilia patients.

Acknowledgment

The study was supported by grants from the Swedish Medical Research Council (00087).

References

1. R. Edson, J.R. McArthur, R.F. Branda, J.J. McCollough, and S.N. Chou, Successful management of a subdural hematoma in a hemophiliac with an anti-factor VIII antibody, *Blood* 41:113 (1973).
2. R. Cobcroft, G. Tamagnini, and K.M. Dormandy, Serial plasmapheresis in a haemophiliac with antibodies to factor VIII, *J Clin Pathol.* 30:763 (1977).
3. M. Francesconi, C. Korninger, E. Thaler, H. Niessner, P. Höcker, and K. Lechner, Plasmapheresis: Its value in the management of patients with antibodies to factor VIII, *Haemostasis* 11:79 (1982).
4. A.J. Keller, A. Chirnside, and S.J. Urbaniak, Coagulation abnormalities produced by plasma exchange on the cell separator with special reference to fibrinogen and platelet levels, *Br J Haematol.* 42:593 (1979).
5. M.A. Flaum, R.A. Cuneo, F.R. Appelbaum, A.B. Deisseroth, W.K. Engel, and H.R. Gralnick, The hemostatic imbalance of plasma exchange transfusion, *Blood* 54:694 (1979).
6. J.B Orlin and E.M. Berkman, Partial plasma exchange using albumin replacement: Removal and recovery of normal plasma constituents, *Blood* 56:1055 (1980).
7. I.M. Nilsson, S.-B. Sundqvist, and C. Freiburghaus, Extracorporeal protein A-Sepharose and specific affinity chromatography for removal of antibodies, in: Factor VIII Inhibitors, p 225, Alan R. Liss, Inc. (1984).
8. L.-Å. Larsson, C. Freiburghaus, I.M. Nilsson, S.-B. Sundqvist, H. Thysell, P. Bygren, and T. Lindholm, Plasma regeneration system for extensive immunoadsorption, in: Progress in Artificial Organs, Y. Nosé, C. Kjellstrand, and P. Ivanonich, eds., p 902, ISAO Press, Cleveland (1986).
9. P. Gjörstrup, E. Berntorp, L. Larsson, and I.M. Nilsson, Kinetic aspects of the removal of IgG and inhibitors in hemophiliacs using protein A immunoadsorption, *Vox Sang.* 61:244 (1991).
10. I.M. Nilsson, The management of hemophilia patients with inhibitors, *Transfus Med Rev.* 6:285 (1992).
11. I.M. Nilsson, E. Berntorp, and C. Freiburghaus, Treatment of patients with factor VIII and IX inhibitors, *Thromb Haemostas.* 70:56 (1993).
12. I.M. Nilsson, C. Freiburghaus, S.-B. Sundqvist, and H. Sandberg, Removal of specific antibodies from whole blood in a continuous extracorporeal system, *Plasma Ther Transfus Technol.* 5:127 (1984).
13. L.-O. Andersson, H. Borg, and M. Miller-Andersson, Purification and characterization of human factor IX, *Thromb Res.* 7:451 (1975).
14. I.M. Nilsson, T.B.L. Kirkwood, and T.W. Barrowcliffe, In vivo recovery of factor VIII: A comparison of one-stage and two-stage assay methods, *Thromb Haemostas.* 42:123 (1979).
15. A. Wallmark, R. Ljung, I.M. Nilsson, L. Holmberg, U. Hedner, M. Lindvall, and H.-O. Sjögren, Polymorphism of normal factor IX detected by mouse monoclonal antibodies, *Proc Natl Acad Sci USA* 82:3839 (1985).
16. I.M. Nilsson, and U. Hedner, Immunosuppressive treatment in haemophiliacs with inhibitors to factor VIII and factor IX, *Scand J Haematol.* 16:369 (1976).
17. C.K. Kasper, L.M. Aledort, R.B. Counts, J.R. Edson, J. Fratantoni, D. Green, J.W. Hampton, M.W. Hilgartner, J. Lazerson, P.H. Levine, C.W. McMillan, J.G. Pool, and S.S. Shapiro, A more uniform measurement of factor VIII inhibitors, *Thromb Diathes Haemorrh.* 31:869 (1975).
18. I.M. Nilsson, E. Berntorp, and O. Zettervall, Induction of split tolerance and clinical cure in high-responding hemophiliacs with factor IX antibodies, *Proc Natl Acad Sci USA* 83:9169 (1986).
19. I.M. Nilsson, E. Berntorp, and O. Zettervall, Induction of immune tolerance in patients with hemophilia and antibodies to factor VIII by combined treatment with intravenous IgG, cyclophosphamide, and factor VIII, *N Engl J Med.* 318:947 (1988).
20. J. Uehlinger, G.R. Button, J. McCarthy, A. Forster, R. Watt, and L.M. Aledort, Immunoadsorption for coagulation factor inhibitors, *Transfusion* 31:265 (1991).

IVIg in the treatment of patients with factor VIII inhibitors

Y. Sultan,[1] U. Nydegger,[2] M.D. Kazatchkine,[3] F. Rossi,[3] M. Algiman,[1] G. Dietrich[3]

[1] *Centre des Hémophiles*
Hôpital Cochin
27 rue du Fg Saint Jacques
75014 PARIS - FRANCE
[2] *Division of Transfusion, Inselspital, Bern, Switzerland*
[3] *Unité INSERM 28, Hôpital Broussais*

Introduction

Ten years ago M. Kazatchkine, U. Nydegger, and Y. Sultan decided to treat patients with acquired hemophilia using intravenous immunoglobulins (IVIg). This was during a meeting in Denmark about autoimmune diseases and IVIg. In 1983, several authors had reported on the efficacy of IVIg in autoimmune diseases such as idiopathic thrombocytopenic purpura in children[1,2] and adults,[3] immune neutropenia,[2,4] and red cell aplasia.[5] Autoantibodies to factor VIII (F.VIII) occurring in either otherwise healthy patients or during the course of a more characterized immune disorder such as disseminated lupus erythematosus (DLE) and rheumatoid polyarthritis seemed to be good models. One major reason is that the neutralizing capacity of autoantibodies to F.VIII is easily and reproducibly quantified using the Bethesda assay, while this is not the case in most of the autoimmune diseases mentioned above.

Within three months, two patients with autoantibodies to F.VIII were treated using IVIg (Sandoglobulin^R) according to the protocol recommended for idiopathic thrombocytopenic purpura (ITP): 400 mg per kg body weight for five consecutive days.

Patients were a 29-year-old woman who had developed an inhibitor to F.VIII during pregnancy that persisted several months after delivery, and a 62-year-old male who had developed hematomas and acquired hemophilia without any concomitant disease. Inhibitor titers were 10.500 BU and 25.000 BU, respectively.

The patient observations, published in 1984,[6] showed that after infusion of IVIg a dramatic decrease in the antibody titer of 96% and 92% of the initial level occurred. The patients' follow-up demonstrated that additional infusion of IVIg did not improve the antibody titer,

Inhibitors to Coagulation Factors
Edited by Louis M. Aledort *et al.*, Plenum Press, New York, 1995

indicating that a residual amount of antibodies to F.VIII were not influenced by IVIg. It was also observed that after the initial treatment the antibody titer did not return to its initial level.

Observed immediately after IVIg infusion, the dramatic decrease suggested a direct interaction between IVIg and anti F.VIII autoantibodies, although the second effect suggested an effect on antibody synthesis.

The mechanism by which such a result had been obtained was totally unknown. The mechanism of action of IVIg in autoimmune diseases might be either Fc dependent or dependent on the variable region of Ig. Fc dependent mechanisms might be related to the blockade of Fc receptors on phagocytic cells; interaction with Fc receptors might result in modulation of T- and B-cell functions. Fc dependent mechanisms might be related to modulation of synthesis or release of pro-inflammatory monocytic cytokines or might inhibit the binding of activated complement proteins to targets of complement activation. Fc fragments might induce changes in structure and solubility of immune complexes.

The immunomodulatory functions of IVIg in acquired hemophilia occur through the variable region. Early effects are explained by neutralization of circulating autoantibodies, by passively transfused anti-idiotypic antibodies, and long-term effects are explained by selection of immune repertoires mediated by the variable region.

It was demonstrated that, initially, in both patients the direct interaction between IVIg and autoantibodies to F.VIII was related to the presence of anti-idiotypic antibodies to F.VIII inhibitors. The presence in polyspecific immunoglobulins prepared from a large pool of plasma of anti-idiotypic antibodies of such specificity was demonstrated by several kinds of experiments.

Anti-idiotypic antibodies in IVg

One crucial observation was the *in vitro* reproducibility of the *in vivo* interaction. Mixing and incubating patient plasma with increasing concentrations of IVIg from 0.1 to 10 mg/mL resulted in a decrease of F.VIII autoantibody concentration.

The same results were observed when $Fab_{,2}$ fragments of patient Ig were mixed and incubated with $Fab_{,2}$ fragments of commercial preparation of IVIg, indicating that the interaction occurred by the antibody binding sites of both immunoglobulins. These experiments also showed that the maximum interaction occurred for a given ratio between patient immunoglobulins or $Fab_{,2}$ fragment concentration and IVIg or $Fab_{,2}$ fragments of IVIg preparations.

The second set of experiments was developed by F. Rossi et al.[7] using immunoaffinity chromatography. $Fab_{,2}$ fragments of IVIg were coupled to sepharose and $Fab_{,2}$ fragments of Ig from patients with either autoantibodies to F.VIII or alloantibodies developed in hemophiliacs after transfusions were circulated through the columns. After extensive washing, Fab$_{'2}$ fragments attached to the column were eluted. Fab$_{'2}$ fragments eluted from the column contained high neutralizing activity towards F.VIII coagulant activity, indicating a higher concentration of anti-F.VIII activity than in the material deposited on the column.

The third kind of experiment demonstrated the binding of patient Fab$_{'2}$ fragments on microtiter plates coated with Fab$_{'2}$ fragments of IVIg. These experiments did not demonstrate the specificity of interaction but the high degree of binding of Fab$_{'2}$ fragments prepared from Ig of patients with autoantibodies to F.VIII.

IVIg in acquired hemophilia

Since publication of the two initial cases, five new cases of acquired hemophilia were treated with IVIg. Two of them failed to respond; however, there were three other partial successes in decreasing the antibody titer in a patient with a postpartum inhibitor and a 78-year-old female with F.VIII inhibitor associated with thrombocytopenia; full success was observed in a patient with cancer of the kidney.

Several reports in the literature have confirmed the beneficial effect of IVIg in patients with acquired hemophilia. In 1984, Borradori et al.[8] reported the case of a 13-year-old boy with a complex autoimmune disease who developed a 16 BU inhibitor controlled by vincristine and corticosteroids. When vincristine was discontinued and steroids reduced, the F.VIII inhibitor reappeared. Infusion of IVIg every two weeks led to the decrease then disappearance of the inhibitor after 10 weeks. The same year, Zimmerman et al.[9] reported the effect of IVIg in two women, 64- and 70-years-old, respectively. The first woman, with a F.VIII level <1% developed a 141 BU inhibitor. Plasmapheresis was followed by a reduction of the inhibitor titer to 28 BU. Treatment with Sandoglobulin^R was accompanied by an increase in F.VIII to 3% on the second day, to 20% after two weeks, and 25% after four weeks. The second case presented with F.VIII level <1% and a 51 BU inhibitor. After IVIg infusion, the inhibitor titer fell to 3.8 BU and F.VIII rose to 35%. Though transient and incomplete, the effect of IVIg in both patients illustrated both mechanisms: the direct interaction and the action on autoantibody synthesis. Green and Kwaan[10] reported in 1987 the total disappearance of a spontaneous inhibitor in a 55-year-old woman after one course of 25 g per day of IVIg for five days. The inhibitor titer was 6.5 BU before and became undetectable after treatment.

Carreras et al.[11] in 1988 reported the case of a F.VIII inhibitor responsive to gammaglobulin. This inhibitor developed in a 67-year-old woman with a post-hepatitis cirrhosis. The pretreatment level of F.VIII inhibitor was 356 BU. A first course of 400 mg/kg per day for five days lowered the inhibitor titer to 45 BU. Two weeks later, a two-day course of IVIg was followed by a further decrease until the inhibitor was suppressed.

IVIg in the treatment of hemophilia patients with an inhibitor to F.VIII

Two patients with congenital hemophilia, initially treated with IVIg, developed an alloantibody to F.VIII and failed to respond.[6] Until now, no differences have been found between auto- and alloantibodies to F.VIII. They have the same IgG subclasses and recognize the same epitopes on the F.VIII molecule.

In affinity chromatography experiments, Fab'_2 fragments from alloantibodies were eluted indicating that IVIg contained anti-idiotypic autobodies recognizing anti-factor VIII alloantibodies.

A protocol was elaborated: adult patients received 20 g of IVIg for two days then 10 g once a week; children received 15 g for two days then 7.5 g once a week. Seven hemophilia patients with an inhibitor to F.VIII ranging from 4 to 360 BU were included in the study. Two patients failed to respond and the five others showed a decrease in the inhibitor titer from 15 to 75% of the pretreatment level.

Origin of anti-idiotypes in IVIg

The origin of anti-idiotype to F.VIII antibodies in preparations of polyspecific immunoglobulins is unknown. It was observed that anti-idiotypic antibodies were found in the plasma of patients who had spontaneously recovered from acquired hemophilia.[12,13] Immunoglobulins isolated from the post-recovery plasma neutralized F.VIII antibodies from their own plasma collected during the disease and partially neutralized autoantibodies to F.VIII from other patients with acquired hemophilia. IgG Fab'_2 fragments from this patient, bound to sepharose, were as capable as Fab'_2 fragments from IVIg to bind specific anti-factor VIII antibodies from patients with acquired hemophilia antibodies, which can then be eluted from the column.[12] One of the main questions about the origin of anti-idiotypes against F.VIII inhibitors in IVIg was to determine if each donor contributed to their presence or if some privileged donors were acknowledged for the presence of anti-idiotypes of such specificity.

With U. Nydegger, therapeutic Ig from selected groups of donors were prepared and compared to IgG prepared from unselected blood donors. A higher frequency of neutralizing antibodies tested against a panel of spontaneous F.VIII inhibitors was found in pools of IgG from multiparous women while Ig from male donors had identical inhibitory capacity as commercial preparations of Ig.[14] IgG and Fab'_2 fragments of IgG were prepared from single donor plasma. IgG from young male donors contained less neutralizing activity than IgG from aged donors. Pooling IgG from several donors with no individual activity resulted in the expression of a neutralizing capacity against F.VIII inhibitors.[15]

Batches of IVIg from selected donors might lead to more active preparations for the treatment of autoimmune diseases.

References

1. P. Imbach, S. Barandun, V. d'Apuzzo et al., High dose intravenous gammaglobulin, for idiopathic thrombocytopenic purpura in childhood, Lancet i:1228 (1981).
2. J.B. Bussel, R.P. Kimberly, R.D. Inman et al., Intravenous gammaglobulin treatment of chronic idiopathic thrombocytopenic purpura, Blood 62:480 (1983).
3. R.E. Schmidt, U. Budde, G. Schafer et al, High dose intravenous immunoglobulin for idiopathic thrombocytopenic purpura, Lancet ii:475 (1981).
4. S. Pollack, C. Cunningham-Rundles, E.M. Smithwick, S. Barandun, R.A. Good, High dose intravenous gammaglobulin for autoimmune neutropenia, N Engl J Med 307:243 (1982).
5. JP. Clauvel, W. Vainchenker, A. Herrera et al., Treatment of pure red cell aplasia by high dose intravenous immunoglobulins, Br J Haematol 55:380 (1983).
6. Y. Sultan, M.D. Kazatchkine, P. Maisonneuve, U.E. Nydegger, Anti-idiotypic suppression of autoantibodies to factor VIII (anti-haemophilic factor) by high-dose intravenous gammaglobulin, Lancet ii:765 (1984).
7. F. Rossi, Y. Sultan, M.D. Kazatchkine, Anti-idiotypes against autoantibodies and alloantibodies to factor VIIIC (anti-hemophilic factor) are present in therapeutic polyspecific normal immunoglobulins, Clin Exp Immunol 74:311 (1988).
8. G. Borradori, A. Hirt, A. Luthy, H.P. Wagner, P. Imbach, Haemophilia due to factor VIII inhibitors in a patient suffering from an autoimmune disease. Treatment with intravenous immunoglobulin, Blut 48:403 (1984).
9. R. Zimmerman, B. Kommerell, J. Herenberg, W. Eich, K. Rother, KL. Schimpf, Intravenous IgG for patients with spontaneous inhibitor to factor VIII, Lancet i:273 (1985).
10. D. Green, H.C. Kwaan, An acquired factor VIII inhibitor responsive to high dose gammaglobulin, Thromb Haemost 58:1005 (1987).
11. L.O. Carreras, G.N. Perez, D.L. Xavier, A.N. Blanco, L.B. Penalva, Autoimmune factor VIII inhibitor responsive to gammaglobulin without *in vitro* neutralization, Thromb Haemost 60:343 (1988).
12. Y. Sultan, F. Rossi, M.D. Kazatchkine, Recovery from anti-VIIIC (anti-hemophilic factor) autoimmune disease is dependent on generation of anti-idiotypes against anti-FVIIIc autoantibodies, Proc Natl Acad Sci USA 84:828 (1987).
13. C. Tiarks, L. Pechet, RE. Humphreys, Development of anti-idiotypic antibodies in a patient with a factor VIII autoantibody, Am J Hematol 32:217 (1989).
14. M. Algiman, G. Dietrich, U.E. Nydegger, D. Boyeldieu, Y. Sultan, M.D. Kazatchkine, Antibodies to factor VIII (anti-hemophilic factor) in healthy individuals, Proc Natl Acad Sci USA 89:3795 (1992).
15. G. Dietrich, M. Algiman, Y. Sultan, U.E. Nydegger, M.D. Kazatchkine, Origin of anti-idiotypic activity against anti-factor VIII autoantibodies in pools of normal human immunoglobulin G (IVIg), Blood 79, 11:2946 (1992).

V: Tolerance to coagulation factors

Immunologic tolerance is a specific unresponsive state induced by exposure to an antigen. Like antibody formation, it requires the interaction of the immune system with an antigen or an antigenic determinant. Although there is only incomplete information about the molecular and cellular mechanisms underlying the tolerant state, Scott's chapter emphasizes that tolerance can be induced in mature and immature lymphocytes of both the B-cell and T-cell lineages. He then describes in more detail two approaches that should be kept in mind as we consider the Factor VIII (F.VIII) inhibitor problem. Deletion of specific B-cell populations has been shown to be dependent on critical interactions involving oncogenes and cyclins, with the outcome being the absence of antibody formation. A second approach, epitope specific tolerance generated by exposure to an antigen-IgG fusion protein, points out that the way in which cells encounter antigens is critical and emphasizes the effect of immunoglobulin domains on protein processing.

In the case of antibodies to F.VIII, recent studies emphasize the need to better differentiate inhibitor formation by patients with severe hemophilia A who are first treated with what is for them a foreign protein (primary intolerance in Briët's terminology) from inhibitor development in patients who appear to be tolerant to F.VIII for a long time period. The nature of resistance to antibody formation needs to be determined through careful immunologic studies, and the critical properties of the "modified" immunogenic F.VIII must be identified. The recent experience in the Netherlands and Belgium emphasizes that certain manufacturing processes can change F.VIII in a way that "breaks" immune tolerance in individuals who have been successfully treated with other F.VIII preparations for many years. It is hoped that more detailed studies of this phenomenon will help us understand how to evaluate new therapeutic products. Until then, as Dr. Briët points out in his review, there is no *in vitro* or animal study that identifies the subtle changes in F.VIII that can generate neoantigens. Thus, careful clinical evaluation remains the only way to assess the safety of newly introduced products.

The chapter in this section by Mariani and colleagues summarizes an international registry of patients treated with large amounts of F.VIII in order to establish immune tolerance. The combined data from many centers allows the identification of some relatively consistent associations. The importance of F.VIII dosage and the importance of a low inhibitor titer at the start of treatment are emphasized in this review. The registry also is important in documenting that tolerance is usually persistent once achieved.

Tolerance and intolerance to factor VIII: A clinical perspective

Ernest Briët
Hemostasis and Thrombosis Center, Department of Hematology
University Hospital
Leiden, The Netherlands

In this chapter I will discuss four aspects of inhibitor formation in patients with hemophilia A: why most patients with hemophilia are and remain tolerant to factor VIII (F.VIII), why some patients who are primarily tolerant to F.VIII get inhibitors later, how many patients get inhibitors, and how we can (re-)induce tolerance in patients with inhibitors.

Primary intolerance

The most obvious reason why patients are tolerant to F.VIII is the circulation of F.VIII in their plasma. This is true for all patients with mild and moderately severe hemophilia and for the small numbers of patients with severe disease who produce nonfunctional F.VIII molecules. Patients with severe disease who do not have any F.VIII due to gene deletions or other mutations have a higher than average chance of inhibitor development, but even in this group most do not get inhibitors at all. A simple explanation for this striking fact is not available. Findings in kidney transplantation patients, however, suggest that materno-fetal transfusion may induce tolerance to maternal antigens: multitransfused transplant recipients often lack antibodies to maternal HLA haplotypes (see Table 1) and maternal transplants appear to survive longer than

Table 1. Leukocyte antibodies against non-inherited maternal antigens (NIMA) and against non-inherited paternal antigens (NIPA) in hyperimmunized kidney transplant candidates[1]

Antibodies	First study		Second study	
	NIMA	NIPA	NIMA	NIPA
Absent	21	2	17	6
Present	24	23	14	21

Inhibitors to Coagulation Factors
Edited by Louis M. Aledort *et al.*, Plenum Press, New York, 1995

paternal transplants.[1] These considerations suggest that there is a subgroup of patients who remain totally immuno-ignorant of F.VIII during gestation due to the combination of complete absence of gene product and absence of materno-fetal transfusion. This subgroup is at the highest risk of inhibitor development in the early stages of replacement therapy. It is this subgroup that will be identified during the short-term follow-up of product evaluations in so-called PUP-studies. Prevention of inhibitor development in these patients is a major challenge. Theoretically, this might be met by the design of functional recombinant F.VIII, which lacks the most immunogenic epitopes. Alternatively, it has been suggested that intrauterine transfusion of F.VIII in fetuses with severe hemophilia A might have a favorable effect on the probability of inhibitor development.

Inhibitors after many exposures

On the other hand, another group of patients is confronted with inhibitors after many F.VIII exposures. McMillan et al. found a median of 43 exposures to F.VIII prior to inhibitor development in the national cooperative study in the USA.[2] It seems likely that many of the patients with inhibitors were primarily tolerant to F.VIII and became intolerant later on. Worries have been expressed over the years that manufacturing processes might induce neo-antigens in the F.VIII molecule that could induce neutralizing antibodies, but formal proof that this has ever happened was not available until recently. In vitro tests or animal experiments to detect neo-antigens are not predictive of clinical inhibitor development. In the best case, they might be used to detect grossly abnormal F.VIII that should not even be tried in patients. Apart from the late-onset inhibitors in the USA study, the best evidence for the causal role of neo-antigens in F.VIII concentrates stems from a small epidemic of inhibitors in Belgium and the Netherlands during 1991.[3,4] This epidemic was linked to a single product, F.VIII CPS-P, produced in the Netherlands. In Belgium, patients were randomized to receive either this

Table 2. Exposure days prior to inhibitor development in the epidemic of inhibitors associated with factor VIII CPS-P compared with data from preceding episodes

Study / Patient Group*	≤250 exposure days	≥250 exposure days
B CPS-P	1	4
NL CPS-P	2	9
NL CPS	4	0
NL '84-'89	2	0
McMillan USA	240	0

* Indications as in legend to Figure 1.

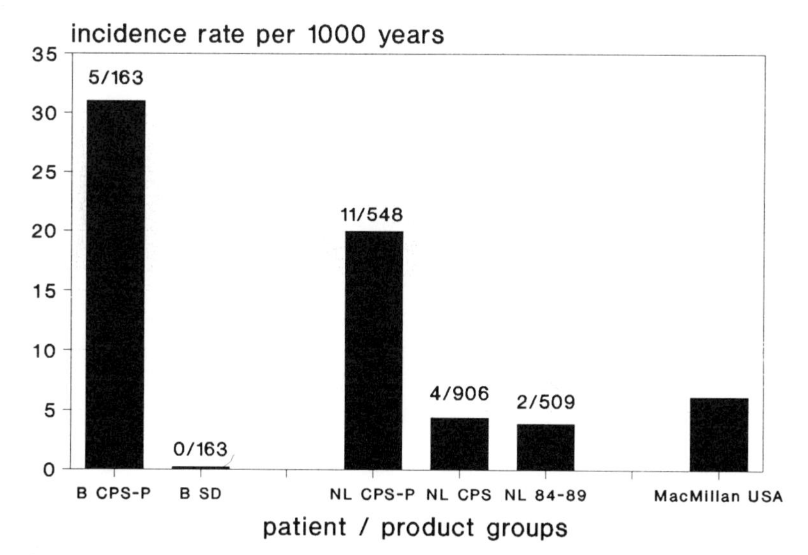

Figure 1. An epidemic of factor VIII inhibitors associated with F.VIII CPS-P. In Belgium 5 out of 163 patients randomized to receive factor VIII CPS-P (B CPS-P) developed an inhibitor, as opposed to none out of 163 patients receiving the solvent detergent treated product (B SD). Similarly, in The Netherlands the incidence of inhibitors associated with factor VIII CPS-P (NL CPS-P) was significantly higher than in preceding episodes during which other products were used (NL CPS and NL 84-89). The results of the USA cooperative study are given for comparison (McMillan USA).

or a solvent detergent treated product made in France. During the summer of 1991, 5 out of 163 patients using F.VIII CPS-P developed an inhibitor; none of the patients on the French product did so.[3] Similarly, in the Netherlands a significantly higher number of inhibitors occurred in patients on F.VIII CPS-P in comparison to other products and to the findings of McMillan et al.[2,4] The findings have been summarized in Figure 1. In support of the concept of secondary as opposed to primary intolerance to F.VIII, we found that the number of exposure days preceding the occurrence of these inhibitors was extremely large: almost all inhibitors arose after more than 250 exposures (see Table 2). These data emphasize the need to avoid damage to the structure of F.VIII during the purification process as well as during viral inactivation. While pure F.VIII has been the goal of all manufacturers, this goal is ideal only if the pure product is truely identical to F.VIII in its native state. Apart from the epidemic, it is currently unknown what proportion of inhibitors is due to neo-antigens in the product as opposed to primary intolerance of the recipient.

What is the "normal" incidence of inhibitors?

The previous discussion illustrates that patients and manufacturers of new products will be confronted with inhibitors whatever care they take to avoid neo-antigens in their product. Patients who are immuno-ignorant of F.VIII will detect the protein with their immune system and, more likely than not, will produce antibodies. Unfortunately, it is not currently possible to distinguish immuno-ignorant patients from those who have been exposed to F.VIII. Consequently, it will not be possible to do simple exposure studies aimed to demonstrate such neo-antigens by excluding immuno-ignorant patients. The best we can do is compare data obtained from patients on the new product with those obtained with an older generation product. Ideally, these comparisons should be prospective and randomized, but very large numbers of patients are required and currently such ambitious and costly projects have not been undertaken for any new product.

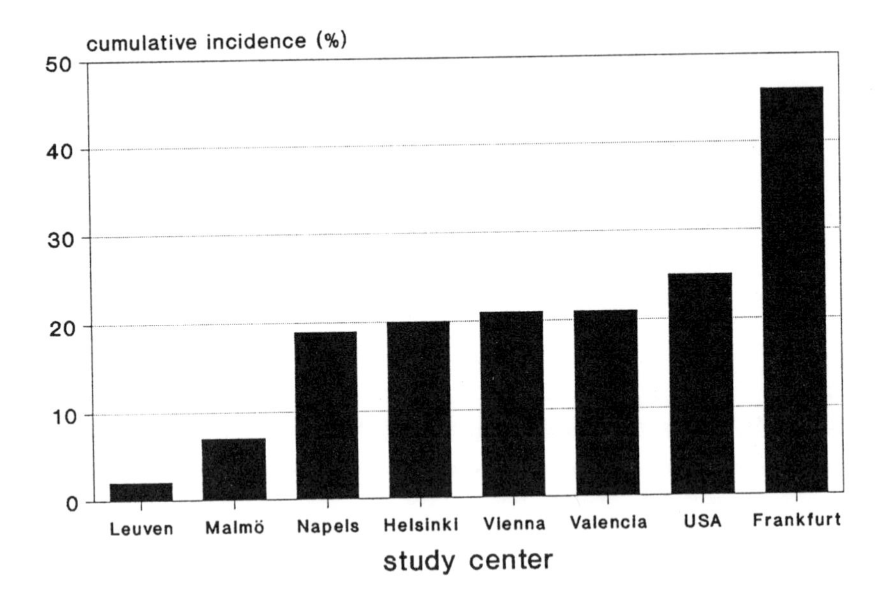

Figure 2. Cumulative incidences of high-titer inhibitors in severely affected patients from eight follow-up studies. All patients were followed from birth. The bars indicate the cumulative incidence at the plateau of the individual curves. Median follow-up ranged from 7 to 15 years.

In order to provide a useful estimate of the historical incidence of inhibitors, we have combined the data from eight follow-up studies and calculated a cumulative incidence for each of them with restriction to high-titer inhibitors in severely affected patients.[5] Unfortunately, even this "standardized" way of analysis results in widely disparate figures (see Figure 2). The cumulative incidences we found range from as low as 2% in Leuven to a staggering 46% in the Frankfurt experience. This leaves us with the disconcerting feeling that quantitative data from uncontrolled studies on new products will be close to meaningless, unless follow-up is very long and extreme results are found. Currently available data on ultra-pure plasma-derived concentrates and on recombinant products are based on studies that are both too small and have too short a follow-up to allow quantitative conclusions.[6-11]

Induction of immune tolerance

As early as 1972, Rizza and colleagues decided to continue F.VIII replacement treatment in spite of the presence of inhibitors.[12] In those days it was considered prudent to avoid any F.VIII to allow antibody titers to decrease and to reserve the F.VIII option for life threatening

Table 3. Immune tolerance induction schedules

Bonn[13-15]

- phase 1 until titer < 1 BU: factor VIII 100 U/Kg b.i.d. and FEIBA 50 U/kg b.i.d.
- phase 2 until T1/2 normal: factor VIII 150 U/Kg b.i.d.
- phase 3: tapering of factor VIII.
- phase 4: on demand treatment.

Malmö[16-19,32]

- protein A adsorption (day -1, 0).
- cyclophosphamide: 12-15 mg/Kg i.v. (day 1,2); 2-3 mg/Kg p.o. (day 3 - 13).
- i.v. IgG: 0.4 g/Kg (day 4 - 9).
- factor VIII: keep over 40% (day 0 - 21); 65 U/Kg 3x per week (day 21 - ...).

Bilthoven[20,21]

phase 1 until recovery > 30%: factor VIII 25 U/Kg on alternate days.
phase 2: tapering of factor VIII.
phase 3: on demand treatment.

hemorrhages. It was, however, the pioneering work from Bonn and Malmö that led to the current ideas about immune tolerance (see Table 3). It appeared from this early work that high doses of F.VIII administered on a daily basis could eradicate the inhibitors.[13-15] The major limitation of the Bonn approach was the astronomical cost involved. An alternative method was developed in Malmö, based on the simultaneous application of protein A adsorption to remove circulating antibody, intravenous IgG to neutralize antibody, F.VIII to stimulate the proliferation of specific immune response cells, and cyclophosphamide to kill these cells.[16-19] Although this method is technically demanding as well as expensive, the success rate is impressive. In addition, the equipment has become available in many countries, and the cost is limited since the success is obtained in a relatively short time. A much cheaper approach was worked out in the Netherlands where van Leeuwen et al. reasoned that the immune system might be tolerized by relatively low doses of F.VIII.[20,21] At present it is not possible to establish the roles for each of these approaches since the experience outside their places of origin is very limited.[22-31] The impression, however, seems to be justified that low dose regimens may suffice for low titer inhibitors, while more aggressive and expensive programs are necessary for inhibitors with high titers.

References

1. van Rood JJ, Claas FHJ. The influence of allogeneic cells on the human T and B cell repertoire. Science 1990; 248: 1388-1393.
2. McMillan CW, Shapiro SS, Whitehurst D, et al. The natural history of factor VIII:C inhibitors in patients with hemophilia A: A national comparative study. II. Observations on the initial development of factor VIII:C inhibitors. Blood 1988; 71: 344-348.
3. Peerlinck K, Arnout J, Gilles JG, Saint-Remy J-M, Vermylen J. A higher than expected incidence of factor VIII inhibitors in multitransfused haemophilia A patients treated with an intermediate purity pasteurized factor VIII concentrate. Thromb Haemost 1993; 69: 115-118.
4. Rosendaal FR, Nieuwenhuis HK, Van den Berg HM, et al. A sudden increase in factor VIII inhibitor development in multitransfused hemophilia A patients in the Netherlands. Blood 1993; 81: 2180-2186.
5. Briët E, Rosendaal FR, Kreuz W, et al. High titer inhibitors in severe haemophilia A. A meta-analysis based on eight long-term follow-up studies concerning inhibitors associated with crude or intermediate purity factor VIII products (letter to the editor). Thromb Haemost 1994; (submitted).
6. Lusher JM, Salzman PM, Monoclate study group. Viral safety and inhibitor development associated with factor VIIIC ultra-purified from plasma in hemophiliacs previously unexposed to factor VIIIC concentrates. Semin Hematol 1990; 27 (Suppl 2): 1-7.
7. Addiego JE, Jr., Gomperts E, Liu S-L, et al. Treatment of hemophilia A with a highly purified factor VIII concentrate prepared by anti-FVIIIC immunoaffinity chromatography. Thromb Haemost 1992; 67: 19-27.
8. Lusher JM, Arkin S, Abildgaard CF, Schwartz RS, and the Kogenate Previously Untreated Patient Group. Recombinant factor VIII for the treatment of previously untreated patients with hemophilia A-Safety, efficacy, and development of inhibitors. N Engl J Med 1993; 328: 453-459.
9. Bray GL, Courter SG, Lynes M, Lee M, Gomperts E, and the Recombinate Study Group. Safety, efficacy and inhibitor risk of recombinant factor VIII (Recombinate) in a cohort of previously untreated patients (PUP's) with severe hemophilia (abstract). Thromb Haemost 1993; 69: 1205.
10. Guerois C, Rothschild C, Laurian Y, et al. Incidence of inhibitors specific for factors VIII (FVIII) or factor IX (FIX) in severe hemophiliacs A and B only treated with very high purity FVIII or FIX SD concentrates (abstract). Thromb Haemost 1993; 69: 852.
11. Lusher JM, Arkin S, Abildgaard CF, et al. Inhibitor development in previously untreated patients (PUPS) with hemophilia receiving Kogenate: 4.5 year follow-up data, including response to immune tolerance (IT) with Kogenate (abstract). Blood 1993; 82, Suppl 1: 153.
12. Rizza CR, Matthews JM. Effect of frequent factor VIII replacement on the level of factor VIII antibodies in Haemophiliacs. Br J Haematol 1982; 52: 13-24.
13. Brackmann HH, Gormsen J. Massive factor VIII infusion in haemophiliac with factor VIII inhibitor, high responder (letter to the editor). Lancet 1977; 2: 933.
14. Brackmann HH, Egli H. Treatment of haemophilia patients with inhibitors. In: Seligsohn U, Rimon A, Horoszowski H, eds. Haemophilia. London: Castle House, 1981: 113-120.

15. Brackmann HH. Induced immunotolerance in factor VIII inhibitor patients. Prog Clin Biol Res 1984; 150: 181-195.
16. Nilsson IM, Sundqvist SB, Freiburghaus C. Extra-corporeal protein A sepharose and specific affinity chromatography for removal of antibodies. Prog Clin Biol Res 1984; 150: 225-241.
17. Nilsson IM, Berntorp E, Zetterval O. Induction of split tolerance and clinical cure in high responding hemophiliacs with factor IX antibodies. Proc Natl Acad Sci USA 1986; 83: 9169-9173.
18. Nilsson IM, Berntorp E, Zettervall O. Induction of immune tolerance in patients with hemophilia and antibodies to factor VIII by combined treatment with intravenous IgG, cyclophosphamide and factor VIII. N Engl J Med 1988; 318: 947-950.
19. Nilsson IM, Berntorp E, Freiburghaus C. Treatment of patients with factor VIII and IX inhibitors. Thromb Haemost 1993; 70: 56-59.
20. van Leeuwen EF, Mauser-Bunschoten EP, van Dijken PJ, Kok AJ, Sjamsoedin- Visser EJM, Sixma JJ. Disappearance of factor VIII:C antibodies in patients with haemophilia A upon frequent administration of factor VIII in intermediate or low dose. Br J Haematol 1986; 64: 291-297.
21. Mauser-Bunschoten EP, Nilsson IM, Kasper CK. Immune tolerance: A 1990 approach. In: Lusher JM, Kessler CM, eds. Hemophilia and von Willebrand's disease in the 1990's. Amsterdam: Excerpta Medica, 1991: 265-269.
22. White GC, Taylor RE, Blatt PM, Roberts HR. Treatment of high titre anti-factor VIII antibody by continuous factor VIII administration. Report of a case. Blood 1983; 62: 141-145.
23. Stenbjerg S, Ingerslev J, Zachariae E. Factor VIII inhibitor treatment with high doses of factor VIII. Thromb Res 1984; 34: 533-537.
24. Gomperts ED, Jordan S, Church JA, Sakai R, Lemire J. Induction of tolerance to factor VIII in a child with a high titre inhibitor: In vitro and in vivo observations. J Pediatr 1984; 104: 70-75.
25. Aznar JA, Jorquera JI, Peiro A, Garcia I. The importance of corticosteroids added to continued treatment with factor VIII concentrates in the suppression of inhibitors in hemophilia A. Thromb Haemost 1984; 51: 217-221.
26. Wensley RT, Burn AM, Redding OM. Induction of immune tolerance to factor VIII in hemophiliacs with inhibitors using low doses of factor VIII (abstract). Thromb Haemost 1985; 54: 227.
27. Wensley RT, Burns AM, Reading OM. Induction of tolerance to factor VIII in haemophilia A with inhibitor using low dose of human factor VIII (abstract). Ric Clin Lab 1986; 16: 104.
28. Scheibel E, Ingerslev J, Dalsgaard-Nielsen J, Stenbjerg S, Knudsen JB. The Danish Study Group. Continuous high-dose factor VIII for the induction of immune tolerance in haemophilia A patients with high responder state: A description of eleven patients treated. Thromb Haemost 1987; 58: 1049-1052.
29. Ewing NP, Sanders NL, Dietrich SL, Kasper CK. Induction of immune tolerance to factor VIII in hemophiliacs with inhibitors. JAMA 1988; 259: 65-85.
30. Mariani G, Solinas S, Pasqualetti D, et al. Induction of immunotolerance in hemophilia for high titre inhibitor eradication: A long term follow-up. Thromb Haemost 1989; 62: 835-839.
31. Gruppo RA. Immune tolerance induction with once weekly factor VIII infusions in patients with hemophilia A and inhibitors (abstract). Thromb Haemost 1991; 65 (Suppl): 1168.
32. Nilsson IM, Berntorp E. Induction of immune tolerance in hemophiliacs with inhibitors by combined treatment with i.v. IgG, cyclophosphamide and factor VIII or IX. Prog Clin Biol Res 1990; 324: 69-78.

Immune tolerance to Factor VIII: The international registry data

G. Mariani, M. Hilgartner, A.R. Thompson, J. Tusell, M. Manco-Johnson, A. Ghirardini, R. Bellocco and H.H. Brackmann for the participants in the International Registry of Immunetolerance Protocols

Other participants in the registry are: Miller R.T. (Los Angeles), Ewing N. (Los Angeles), Kasper C.K. (Los Angeles), Butler R. (Philadelphia), Morfini M. (Florence), Bell B.A. (Atlanta), Fressinaud E. (Angers), Gringeri A. (Milano), Kisker C.T. (Iowa City), Kessler C. (Washington), Gill J.C. (Milwaukee), Scharrer I. (Frankfurt/Main), Schimpf K. (Heidelberg), White G. (Chapel Hill), Growe G. H. (Vancouver), Nogao T. (Yokohama), Mauser-Bunschoten E. (Bilthhoven), Bosser C. (St. Alban Leyss), Thompson A.R. (Seattle), Inwood M.J. (London, Ontario), Brettler D.B. (Worchester), Magallon F. (Madrid), E. Scheibel (Copenhagen), Beardsley D.S. (New Haven), Markowsky L. (Buffalo), Poon M.C. (Calgary), Smith P.S. (Providence), Walker I. (Hamilton), Rodeghiero F. (Vicenza), Strawczynsky H. (Montreal), Rasi V. (Helsinki), Sultan Y. (Paris), Rubin S.H. (Moncton), Ble F. (New York), Guerois C. (Tours).

Introduction

Among the problems related to the treatment of classic hemophilia, the presence of inhibitors (alloantibodies) to foreign Factor VIII (F.VIII) is one of the most serious. These antibodies may arise at any age, but the cumulative risk of developing an inhibitor one year after the first treatment with a factor concentrate is between 22% and 25%.[1,2] The presence of such antibodies can render any further administration of F.VIII ineffective. On the basis of the speed and degree of immune response to F.VIII, patients have been divided into high- and low-responders.[3] This is an arbitrary classification, but it is related to important treatment characteristics. Low responders, that is patients whose maximum inhibitor titer is <10 Bethesda Units (B.U.), can be treated with F.VIII, albeit at higher doses; however, high-responders can seldom be treated with F.VIII and comprise the so-called "refractory hemophiliacs."

As a consequence, hemophiliacs with an inhibitor to F.VIII are difficult to treat, because hemostasis cannot always be achieved with

Inhibitors to Coagulation Factors
Edited by Louis M. Aledort *et al.*, Plenum Press, New York, 1995

certainty. A number of alternative treatments to F.VIII are available, but they are clearly less efficacious than F.VIII and may cause side effects, including anamnestic reactions and thrombosis.

In 1977 Brackmann and Gormsen proposed a treatment method capable of suppressing the inhibitor through the administration of large doses of F.VIII (200 IU/Kg b.w./day), and achieved tolerance in a high proportion of patients, mostly high-responders.[4,5] This type of treatment was subsequently used by other teams, with a wide array of F.VIII dosages (from <50 to >100), patterns of immune response, patient numbers, and success rates.[6-14]

Since Immune Tolerance (IT) is a treatment that is not sufficiently standardized, the F.VIII and F.IX Standardization Committee of the International Society of Thrombosis and Haemostasis proposed a Registry of the Protocols of Immunetolerance. This report stems from the registry and deals with patients treated from 1975 to 1993.

Materials and methods

Study participants

The Registry of Immunetolerance Protocols was set up in 1989. Hemophilia Care Centers (HCC) were invited to participate in the registry voluntarily and received an individual, standardized questionnaire on the following issues: a) demographic characteristics; b) levels of inhibitor at discovery, enrollment, and after treatment; c) dosage of F.VIII employed and other medications used, i.e., Activated Prothrombin Complex Concentrates (APCC), steroids; d) type of F.VIII concentrate (virus-inactivated, non-virus-inactivated); and e) clinical and laboratory data on Human Immunodeficiency Virus (HIV) infection. Each patient was identified by an individual code to avoid duplicate data on individuals cared for at more than one HCC. Forms to collect follow-up information were also sent.

Definitions of variable features

Inhibitor titers were expressed in "Bethesda Units" (B.U.).[15] "Maximum inhibitor titer" was defined as the highest titer recorded in that particular patient. The "cut off" value between "high" and "low responders" was 10 B.U.[3] Three classes of IT protocols were created: <50, 50 - <200, ≥200 I.U. of F.VIII/Kg b.w./day. Treatment outcomes were defined as follows: "failure," when there was no change in the type of immune response to F.VIII; "high-to-low," when the type of immune response shifted from that of a high- to that of a low-responder; and "success," when the inhibitor was no longer assayable and/or F.VIII recovery and half-life returned normal. "Relapses" were those patients in whom the inhibitor reappeared after one or more years and after the third "inhibitor 0" assay.

Collection and analysis of the data

A microcomputer-based program was developed for data collection and preliminary analysis, using the Epi Info programme.[16] The data were transferred to a Siemens 7-890 mainframe computer and BMDP[17] was used for statistical analyses. Differences in the distribution of variables were evaluated using the chi-square test (with Yates correction) and Fisher's exact test. Ninety five percent confidence intervals were calculated when applicable according to standard methods.[18] Stratified analysis was used to identify confounding and effect modifiers. Stepwise logistic regression was then used to assess the influence of various characteristics on the rate of success, taking into account the correlation between the explanatory variables. No interaction was found in the stratified analysis between the variables considered. The Kaplan Meier survival analysis was applied to evaluate the cumulative distribution function of the time patients remained free of inhibitor. The numbers of patients in the analyses do not always add up to the total number because of missing data for one or more variables.

Results

Two-hundred-four hemophilia A patients who had been or were currently on treatment with an IT protocol were enrolled in the registry. The registry included case reports from 40 HCCs (17 from the U.S.A., 6 from Canada, 16 from Europe and 1 from Japan). Figure 1 shows enrollment in the IT protocols by year.

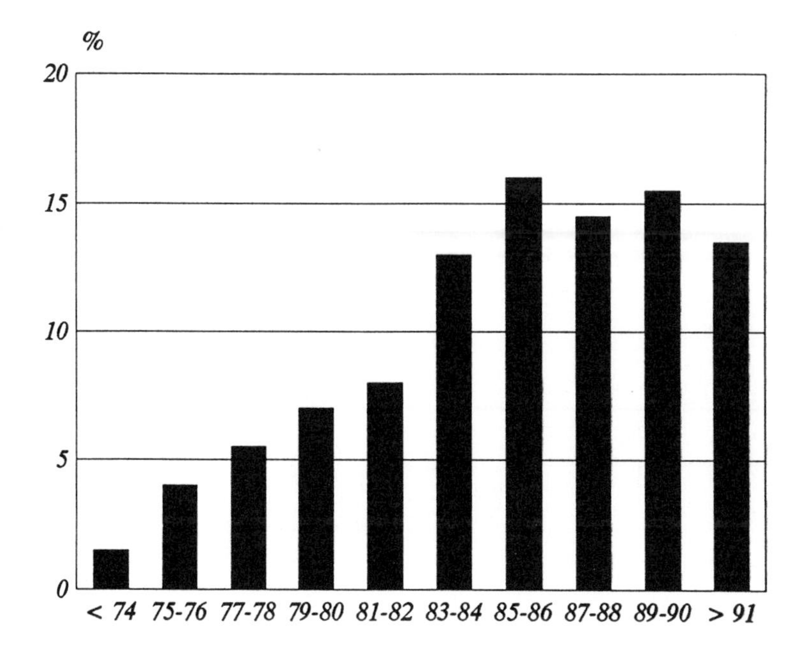

Figure 1. Patients enrolled (%) in immunetolerance protocols by year.

The mean age at inhibitor discovery was 11.6 years (median 4.5, range 1-63). In terms of immune response, 36 (17.7%) were low-responders and 168 (82.3%) high-responders. The cumulative frequency of inhibitor diagnosis by age is shown in Figure 2.

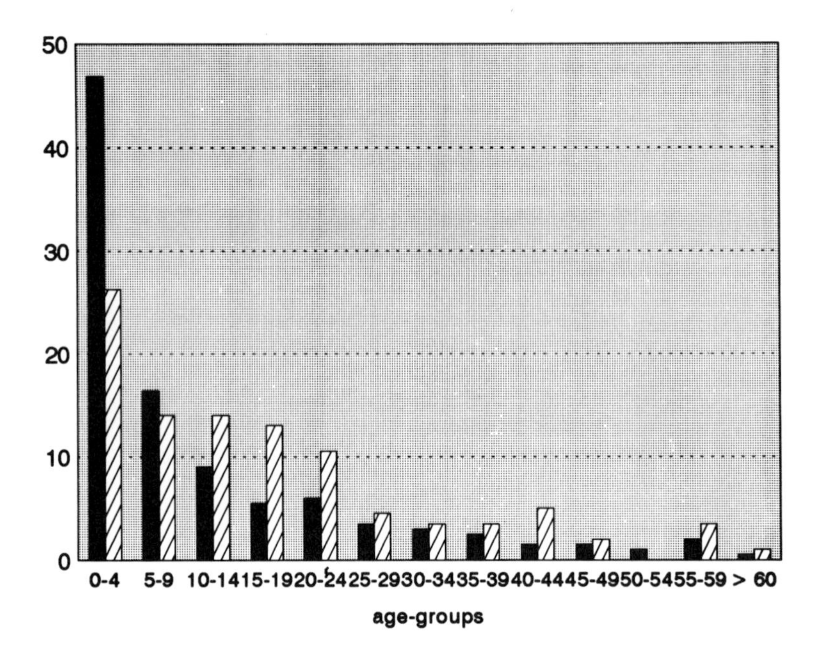

Figure 2. Age at inhibitor discovery (solid bars) and at enrollment (dashed bars) in IT protocols. % of total patients.

When treatment was started, 107 out of 204 patients had levels of inhibitor ≤ 10 B.U. Sixty-nine patients (33.8%) were treated with the highest F.VIII dosage (≥ 200 IU/Kg/b.w./day), 71 (34.8%) received dosages between 50 and < 200 IU, and 64 (31.4%) were given the lowest dosages (< 50 IU/Kg/b.w.).

Virus-inactivated F.VIII concentrates were used exclusively in 53 patients (26.2%); among the remaining 149 patients, 115 (56.9%) received non-virus-inactivated concentrates and 34 (16.9%) both types (Table 1). Fifty-four patients (26.7%) were given APCCs in association with F.VIII, 11 (5.4%) received steroids, and the remaining 137 (67.8%) were given F.VIII alone (see Table 1).

Table1. Immunetolerance protocols and other treatment modalities

Treatment	n.	%
F.VIII dosage (IU/Kg/b.w./day)		
<50	63	30.9
50 - <200	72	35.3
≤200	69	33.8
F.VIII concentrate used		
Virus-inactivated	53	26.2
Non-virus-inactivated	115	56.9
Both types	34	16.9
Protocols		
F.VIII alone	137	67.8
F.VIII+APCCs	44	26.7
F.VIII+ steroids	11	5.4

One-hundred-twenty-nine hemophiliacs (63.2%) were HIVAb-negative before the treatment; of the 118 HIV-screened after the treatment, 18 (15.2%) had seroconverted. The seroconversions occurred between 1977 and 1987. Treatment results are shown in Table 2.

Table 2. Results of Immunetolerance (patients who completed treatment. n= 158)

Outcome	n.	%
Success	107	67.7
High to low (<10 B.U.)	12	7.6
Failure	39	24.7

The remaining 46 patients were still on treatment as of September 1993. Multivariate logistic regression with observation of all the variables in the model was applied to a set of data regarding patients who had achieved success or failure (n=146). By using a stepwise procedure, the probability of success was shown to be independently associated with the dosage employed (F.VIII ≥ 100 vs. < 100 IU/Kg. b.w./day) ($p < .0001$) and with the level of inhibitor at enrollment ($p= .004$) (Table 3). The type of immune response, although showing a trend, did not reach formal significance (see Table 3).

Table 3. Stepwise regression analysis: Comparison of success vs. failure (n=146)

Variables	Relative risk	95% I.C.	"p"
F.VIII dosage (≥100 vs. <100 IU/Kg b.w./day)	13.7	4.5-41.5	<.0001
Inhibitor at enrollment (B.U.)			
(<10 vs. 10-100)	0.3	0.1-0.8	.004
(<10 vs. >100)	0.3	0.1-1.3	.004
Maimum inhibitor titer (B.U.)			
(<10 vs. 10-100)	1.7	0.5-6.0	.2
(<10 vs. >100)	0.6	0.2-2.4	.2

The rates of success with regard to the levels of inhibitor (maximum or at enrollment) are shown in Table 4.

Table 4. Outcome with reference to inhibitor levels at enrollment (n. of patients)

	Inhibitor titer							
	Maximum				At enrollment			
	<10	10-50	50-200	>200	<10	10-50	50-200	>200
Success	22	33	25	27	71	17	10	7
Failures	6	8	9	16	19	10	3	6

Dosage used to seek tolerance was then tested with reference to the levels of inhibitor at enrollment (see Table 5).

Table 5. Rate of success by type of protocol and levels of inhibitor at enrollment*

	Protocol (F.VIII I.U./Kg/b.w./day)			
	<100		>100	
	n.	%	n.	%
Inhibitor (B.U.)				
<10	36/52	69.2	35/37	94.6
10-50	6/15	40	11/13	84.6
>50	1/8	12.5	16/18	88.9
Total	43/75	57.33	62/68	91.2

* 3 missing data (2 among success, 1 among failures)

Mean duration of tolerance induction was not statistically different in success and failures (37.7 vs. 42.4 months, respectively; $p > .5$). The median duration of time without inhibitor has been calculated to be of 5 y (range 1 - 16.4) 1/107 patients relapsed.

Discussion

The data provided by the analysis of the registry allow us to draw some firm conclusions. The age frequency of inhibitor discovery confirms previous data (reviewed by Dr. Hoyer in this book): >50% of those who develop inhibitors do so before the age of 20 (see Figure 2). From an analysis of the years of enrollment, it is clear that this procedure has been used by many hemophilia treaters all over the world since its proposal in 1977.[4] Approximately half of the patients were enrolled before the age of 20, demonstrating that treaters prefer to seek tolerance in children or teenagers.

Of utmost importance is the fact that in the vast majority of the cases (82.3%), IT protocols were used in the high-responders, presumably because other therapeutic options were available in the low-responders. The levels of inhibitor at enrollment were, nonetheless, a concern for treaters: in more than half of the cases (52.4%) IT was started with low levels of inhibitor. On the multivariate analysis this feature became an important and independent predictor of success (see Table 3). In fact, 71/107 (66.35%) of the successfully treated patients had low levels of inhibitor at enrollment. This finding is in line with the findings of Nilsson et al.,[19] who, with a more complex protocol, obtained a high success rate (15/18 patients) only when treatment started with inhibitor levels below 10 BU through extracorporeal Ig absorption or when inhibitor levels declined below this level because there had been no recent anamnestic responses.

The other important, independent predictor of success has been demonstrated to be the dosage of F.VIII administered daily: 100 IU/Kg b.w. or more are far more likely to induce tolerance than are lower dosages. The rates of success analyzed by type of protocol and titers of inhibitor at enrollment (see Table 5) demonstrate that these two factors work together: by starting the treatment with low levels of inhibitor and giving high doses of F.VIII, success rates can be as high as 90%.

The long-term maintenance of tolerance is another important finding in this survey. Among the 107 patients in whom tolerance was achieved, only one (0.93%) relapsed after 47 months. In the patients who remained tolerant, the median inhibitor-free period was calculated at 5+ years (range 1 to 16). Another important consideration is that, by eradicating the inhibitor, one can return the patient to a less expensive, standard F.VIII treatment with predictable hemostasis. When IT procedures are proposed, this important aspect should be taken into account in the long-term analysis of the costs.

IT is, for the time being, only used empirically. We are awaiting further data from basic immunological studies and prospective clinical trials before we can make recommendations on specific dosages and treatment duration.

References

1. S. Ehrenforth, W. Kreuz, I. Sharrer, R. Linde, M. Funk, T. Gungor, B. Krackhardt, and B. Kornhuber, Incidence of development of factor VIII and factor IX inhibitors in hemophiliacs, Lancet 339:594 (1992).
2. J.M. Lusher, S. Arkin, C.F. Abilgaard, R.S. Schwartz and the Kogenate Previously Untreated Patient Study Group, Recombinant Factor VIII for the treatment of previously untreated patients with hemophilia A, N. Engl. J. Med. 328:453 (1993).
3. J. P. Allain, D. Frommel, Antibodies to factor VIII. V. Patterns of immuneresponse to factor VIII in hemophilia A, Blood 47:973 (1976).
4. H. H. Brackmann, J. Gormsen, Massive factor-VIII infusion in hemophiliac with factor VIII inhibitor, high responder, Lancet 2:933 (1977).
5. H.H. Brackmann, Induced immune tolerance in factor VIII inhibitor patients, Progr Clin Biol Res 150:181 (1984).
6. G.C. White, R.E. Taylor, P.M. Blatt, H.R. Roberts, Treatment of a high-titre anti-factor VIII antibody for continuous factor VIII administration: Report of a case, Blood 62: 141 (1983).
7. J.A. Aznar, J.J. Jorquera, A. Peiro, I. Garcia, The importance of corticoids added to continuous treatment with factor VIII concentrates in the suppression of inhibitors in haemophilia A, Thromb Haemost 51:217 (1984).
8. S. Stenbjerg, J. Ingerslev, E. Zacharie, Factor VIII inhibitor treatment with high doses of FVIII, Thromb Res 34: 533 (1984).
9. E. Scheibel, C. Feddersen, H. Hertz, Long-term high dose factor VIII treatment of 3 hemophiliacs with factor VIII inhibitor, Scand J Haematol 34: 378 (1985).
10. E. F. van Leewen, E.P. Mauser-Bunschoten, P.J. van Dijken, A. J. Kok, E.J.M. Sjamsedin-Visset, J.J. Sixma, Disappearance of factor VIII:C antibodies in patients with hemophilia A upon frequent administration of factor VIII in intermediate or low dose, Br J Haematol 64:291 (1986).
11. E. Scheibel, J. Ingerslev, J. Dalsgaard-Nielsen, S. Stenbjerg, J. B. Knudsen and the Danish Study Group, Continuous high-dose factor VIII for the induction of immune tolerance in haemophilia A patients with high responder state: A description of eleven patients treated, Thromb Haemost 58:1049 (1987).
12. N. Ewing, N.L. Sanders, S. L. Dietrich, C. K. Kasper, Induction of immune tolerance to Factor VIII in hemophiliacs, JAMA 259:65 (1988).
13. R.A. Gruppo, Immunetolerance induction with once weekly factor VIII infusions in patients with hemophilia A and inhibitors, Thromb Haemost 65:1168 (1991) (abstract).
14. G. Mariani, S. Solinas, D. Pasqualetti, A. Ghirardini, P. Verani, S. Butto, M. Lopez, T. Moretti, Induction of Immunetolerance in hemophilia for high titre inhibitor eradication: A long term follow-up, Thromb Haemost 62:835 (1989).
15. C.K. Kasper, L.M. Aledort, R.B. Counts, J.R. Edson, J. Fratantoni, D. Green, J.W. Hampton, M. W. Hilgartner, J. Lazeson, P.H. Levine, C.W. McMillan, J. Pool, S.S. Shapiro, N.R. Shulman, J. Van Eye, A more uniform measurement of factor VIII inhibitors, Thromb Diath Haemorrh 34:875 (1975).
16. A.G. Dean, J.A. Dean, A.H. Burton, R.C. Dicker, Epi Info, version 5: A word processing, database and statistics program for epidemiology on microcomputers. USD Inc., Stone Mountain, GA, 1990.
17. W.J. Dixon, 1985, "BMPD statistical software," University of California Press, Berkeley, CA.
18. J.L. Fleiss, 1973, "Statistical methods for rates and proportions," John Wiley & Sons, Inc., New York, NY.
19. I.M. Nilsson, E. Berntorp, O. Zettervall, Induction of immune tolerance in patients with hemophilia and antibodies to factor VIII by combined treatment with intravenous IgG, cyclophosphamide, and factor VIII, N Engl J Med 318:947 (1988).

Multiple mechanisms of immunologic tolerance: Novel approaches for unresponsiveness

David W. Scott, Ph.D.

University of Rochester Cancer Center and
Department of Microbiology and Immunology
School of Medicine and Dentistry
Rochester, New York 14642

Immunologic tolerance is defined as a specific unresponsive state induced by prior exposure to an antigen or antigenic determinant. It is now clear that tolerance can be induced in both immature and mature lymphocytes of the T- or B-cell lineages. The mechanisms involved appear to be multiple: anergy (that is, the failure to respond to specific signaling), deletion of specific clones (via apoptosis), or suppression of reactivity, all of which may exist in the same organism depending on the antigen. Procedures to facilitate tolerance induction in adults who have made undesirable immune responses (for example, to blood products or red-cell antigens) will require novel approaches. We report here on data supporting a deletional mechanism by apoptosis in both neoplastic and normal murine B cells, and further describe protocols for the induction and maintenance of unresponsiveness in T- and B-lymphocyte clones in adult hosts.

General properties of tolerance

Since the time of Ehrlich, immunologists have attempted to harness the failure of the body to respond against self-components in order to utilize this property of tolerance against undesirable immune responses. The production of inhibitors (antibodies) to coagulation factors in patients is a cogent example of such an undesirable immune response. As a background to our attempts to produce epitope-specific tolerance, it is useful to review the general properties and pathways of immunologic tolerance. (See Chris Goodnow and Marc Jenkins in this volume for further treatment of this subject.)

Tolerance is defined in negative terms: it is the failure to respond to an antigen. It is a specific, dose-dependent state of finite duration, and is developed in both T and B cells by prior exposure to a given epitope. As

Inhibitors to Coagulation Factors
Edited by Louis M. Aledort *et al.*, Plenum Press, New York, 1995

described by Peter Cresswell in this volume, this process requires that a given cell be able to "see" such an epitope. These lymphocytes "learn" tolerance during development when their receptors are engaged by an epitope, be it self-tissue antigen or a viral peptide introduced at a critical time or under inappropriate conditions. For T cells, we know that seeing antigen requires that it be expressed as part of a class I or class II major histocompatibility complex (MHC) antigen. No tolerance is likely to exist in CD4 T cells against self-antigens that are expressed within MHC class II negative cells since such T cells can only recognize an epitope in the context of MHC class II. Thus, "ignorance" of these antigens occurs, not tolerance.

Tolerance to many antigens can be induced in both types of CD4 T cells (TH1 and TH2), albeit the conditions for unresponsiveness in each may differ. It has been known for more than 20 years that T and B lymphocytes differ in the kinetics of tolerance induction, and that these cells differ in the waning of this unresponsive state and the doses required for achieving tolerance. Presumably this reflects several notable properties of T and B cells, especially their distinct life spans. Since tolerance is dose-dependent and B cells turn over rapidly, it is clear that as the amount of available antigen for tolerance decreases and as new B cells are made, there is a decreasing chance that enough antigen will remain to induce unresponsiveness in newly developed B-cell clones with rearranged receptors for that epitope. This proposal is independent of the mechanism of tolerance (deletion versus anergy) since replacement of lymphocyte clones is a normal property of the immune system.

Unresponsiveness is generally assumed to be a central property of the immune system. That is, tolerance is due to the absence of responsiveness, not the presence of an active suppression mechanism. Nonetheless, suppressor mechanisms have been described for some forms of unresponsiveness, namely oral tolerance, although these do not seem to be due to a unique cell type. Rather, suppression is likely to be mediated by cytokines which shift the type of T helper response from TH1 to TH2. Such an interplay is in fact characteristic of the checks and balances of the immune system.

Prior to the development of transgenic mouse systems to explore tolerance, it was difficult to determine whether the lack of responsiveness was due to the functional lack of responsiveness (anergy) or the absence of a particular antigen-specific clone (deletion). Our group had previously reasoned that the presence of antigen-binding cells (ABC) in tolerance hosts defined an anergic state, whereas a loss of ABC after tolerance induction supported deletion. Indeed, we reported just such a pattern for adult and neonatal tolerance, respectively, to the FITC hapten.[1] Nossal and Pike reported similar observations with an anti-IgM model of unresponsiveness: anergy was induced in bone marrow cells with low doses, but high concentrations of anti-IgM led to deletion.[2]

This was manifested by a lack of mitogenic responsiveness in cases of purported deletion.[1,2] Recently, Warner et al.[2] confirmed the induction of anergy in adult B cells with anti-IgM treatment; such cells failed to respond to antigen or mitogen (LPS) with antibody production, but still proliferated. Therefore, these B cells were not eliminated. However, under appropriate conditions, deletion of specific B cells can occur in the adult via IgM but not IgD receptors.[4]

A deletional model of B-cell tolerance: Role of BLKD

During the last decade, we have used B-cell lymphomas to examine the mechanisms of tolerance. Because these cells are transformed, they continue to grow *in vitro* and can be followed after their interaction with surrogate antigen, anti-μ. Three lymphomas, CH31, CH33 and WEHI-231, have proven to be elegant models for deletion: in the presence of low (submicrogram) doses of anti-μ, these cells are growth arrested and undergo apoptosis or programmed cell death. I will review the role of tyrosine phosphorylation and *c-myc* transcription in this process, before turning to a recent model for epitope-specific tolerance.

Crosslinking of surface IgM by anti-μ on these sensitive B-cell lymphomas in early G_1 leads to growth arrest that occurs late in G_1 near the G_1/S border; such growth-inhibited cells then undergo apoptosis.[5-8] Early studies of the molecular basis of signal transduction via B-cell membrane Ig receptors demonstrated that ligation of IgM initiates hydrolysis of membrane-associated phosphatidyl inositol 4,5-bisphosphate (PIP_2), yielding diacyl glycerol (DAG) and inositol 1,4,5-trisphosphate (IP_3). IP_3 then mediates the mobilization of calcium from intracellular stores, whereas DAG, acting in conjunction with elevated calcium, causes translocation and activation of protein kinase C.[9-12] One important outcome of these intracellular signals is the transiently increased expression of the *c-myc* and *c-fos* gene products. The first step in this process is the rapid tyrosine phosphorylation of a variety of intracellular proteins.[13,14]

Tyrosine (de)phosphorylation can regulate the activity of many intracellular catalytic enzymes and is utilized by different types of receptors in mediating signal transduction. Since neither the B-cell antigen receptor itself nor its associated components of the mb-1/B29 heterodimer possesses intrinsic kinase activity, the receptor complex must be functionally or physically coupled to certain kinases in order to transduce signaling. Recent studies have demonstrated involvement of the *src* family of nonreceptor protein tyrosine kinases (PTKs) in membrane Ig receptor-mediated signaling pathway in B lymphocytes. Importantly, ligation of membrane IgM causes activation of these receptor-associated PTKs, which in turn leads to tyrosine phosphorylation of intracellular substrates, most of which are catalytic enzymes whose activity is up-regulated by tyrosine phosphorylation. Presumably, they act directly or indirectly on cytosolically localized *trans*-acting factors, resulting in their translocation to the nucleus and regulation of critical nuclear-acting proteins that are required for cell cycle progression.[15,16] In both murine and human lymphomas, the role of tyrosine phosphorylation in initiating growth arrest has been established using either PTK-specific drugs or antisense oligos for a given *src* kinase.[17,18] The latter has been particularly effective with the CH31 B-cell lymphoma[18,19] reviewed below.

Initially, we screened a series of functionally different murine B-cell lymphoma lines, not only for patterns of tyrosine phosphorylation, but also for differences in the use of *src* family tyrosine kinases involved in this process. The patterns of tyrosine phosphorylation in nine different lines included a variety of common substrates, but also a number of unique targets.[18] However, there was no pattern that correlated with sensitivity or resistance to anti-μ.

We then began to concentrate on which kinases might be used in these different cell lines and found that the *blk* kinase was present and activatable in all anti-μ-sensitive lymphomas, whereas it was absent or in low abundance in most of the resistant lines. However, those anti-μ-resistant lines that possessed low *blk* activity also expressed the *fyn* PTK. All cell lines possessed both molecular forms of *lyn,* as well as *lck,* although no evidence for direct activation of these kinases by anti-μ was apparent. The pattern we saw was the consistent presence of *blk* in anti-μ sensitive cell lines (cf. ref. 18). Therefore, we proposed that the expression of activatable *blk* was required for growth inhibition by anti-μ, and that *fyn* might counterbalance this activity in resistant lines.

To test the role of *blk* in growth inhibition by anti-μ, we used antisense oligonucleotides to the *blk* PTK gene to reduce or eliminate expression of this kinase in the CH31 B-cell lymphoma. CH31 cells treated for 48 hours with antisense *blk* possessed less than 50% of the level of *blk* found in control cells and had significantly reduced *blk* kinase activation by anti-μ in *in vitro* assays.[18] Treatment of CH31 with antisense *blk* prevented both growth arrest and apoptosis induced by anti-μ, but had no effect on TGF-b inhibition.[18,19] Antisense oligos for the *fyn, lyn,* and *lck* PTK genes had no effect on anti-μ driven growth arrest. Hence, *blk* appears to be critically involved in initial signaling for growth arrest and apoptosis by anti-μ in this murine B-cell lymphoma. In contrast, we have been unable to render CH33 lymphoma cells resistant to anti-μ by transfection with *fyn* constructs, although the anti-μ-induced phosphorylation of *fyn* and other substrates was clearly evident.[18] Therefore, the hypothesis that *fyn* expression per se is sufficient to protect against anti-μ-initiated growth inhibition is indeed an oversimplification. Rather, *fyn* may be associated with other Ig isotypes (such as IgD) or may reflect redundancy of signaling in these lymphomas.

Interestingly, crosslinking IgM and IgD leads to the same initial and intermediate biochemical signals[14,20-25] but the ultimate results of such crosslinking differs for these isotypes. That is, although treatment with anti-μ leads to growth arrest and death of many B-cell lymphomas, anti-δ generally has no deleterious effect on the same cells,[21,22] a result confirmed using either m+d+ human or murine B-cell lymphomas.[26,27] Our working hypothesis has been that the phosphorylation of the IgM-associated and IgD-associated proteins[23,25] initiates a cascade affecting substrates that ultimately modify the transcription and/or translocation of *c-myc*/*myc* and cause cell cycle arrest and apoptosis.

Role of* myc *in deletional models of tolerance

It is now clear that the cell cycle is strictly controlled by the cyclins and a variety of oncogenes and anti-oncogenes.[15,16,28-38] Using B-cell lymphomas, several laboratories have focused on the roles of oncogenes or competence genes, such as *c-myc, bcl-2, egr-1,* as well as anti-oncogenes like the *retinoblastoma* gene product, pRB, as critical elements in the processes leading to apoptosis.[34-40] For example, Sonenshein and colleagues[31,34] demonstrated that *c-myc* transcription increased within 30 minutes in the growth inhibitable WEHI-231 B-cell lymphoma treated with anti-μ and that *myc* protein becomes phosphorylated within 1 hour under these conditions. Message levels for *c-myc* decrease thereafter to undetectable levels within 8 to 24 hours of treatment. Since *myc* protein is believed to function as a cell cycle controlling transcriptional element,[36,37] the changes in *myc* observed in these lymphoma lines are assumed to be causative of growth arrest. Interestingly, Tisch et al.[21] found that crosslinking of (transfected) IgD caused an increase in *c-myc* message, followed by a decrease to baseline or slightly elevated levels of *myc* message at 24 hours, a result that is consistent with the lack of growth inhibition elicited by anti-δ.[20,21,26,27]

Recent data have suggested an unexpected role of *myc* in apoptosis. Evan et al.[41] found that accumulation of *myc* protein at cell cycle borders signals apoptosis in rat fibroblasts, a result reproduced in activation-induced apoptosis in T-cell lines[42]. The role of *myc* in anti-μ-induced apoptosis in B-cell lymphomas was recently tested in our lab using antisense oligonucleotides by Fischer et al.[43] We treated WEHI-231 or CH31 cells with antisense oligos against the first coding sequences of *c-myc* and found that both growth arrest and apoptosis induced by either anti-μ or TGF-ß were prevented (see Figure 1). This effect was *c-myc* specific since nonsense *myc* and antisense *fos* had little effect on thymidine incorporation and none on apoptosis. Antisense *myc*-protected cells were also able to phosphorylate pRB normally.[43] Thus, the pathways of growth inhibition by these two agents may converge at a point involving *myc* proteins. Recent experiments with normal B lymphocytes undergoing apoptosis *in vitro* showed that antisense against three different *c-myc*

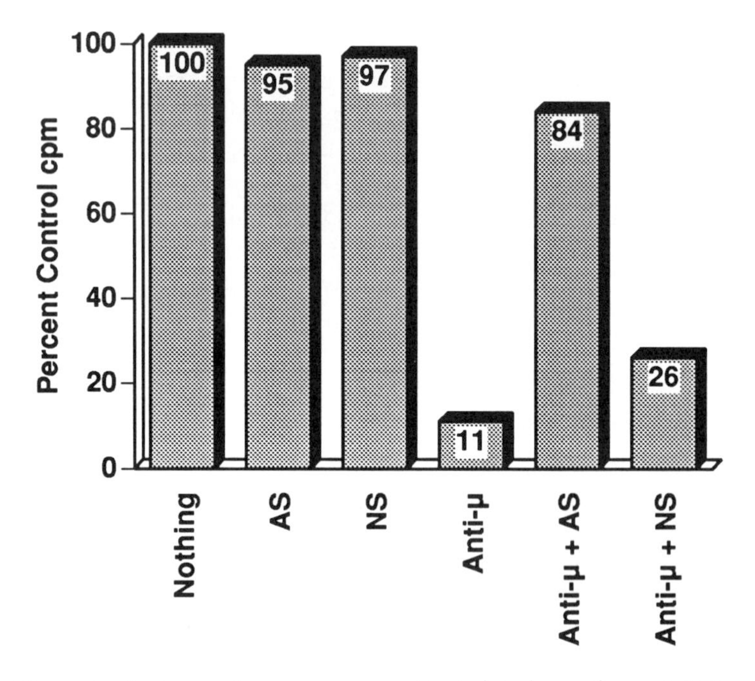

Figure 1. Effect on antisense *c-myc* on anti-μ induced growth arrest. CH31 murine B lymphoma cells were cultured for 24 hours ± 1 μg/ml anti-μ and the indicated oligos (at 2μM); they were then labeled with tritiated thymidine and harvested 4 hours later. Values represent percent control cpm compared to the no treatment control. AS = antisense *c-myc*, NS = nonsense *c-myc*. In an independent experiment, 45% of the anti-μ treated cells were apoptotic, while in the presence of antisense *c-myc* + anti-μ, only 4.7% of the cells were apoptotic.

sequences could prevent programmed death (Scott, D.W., Lamers, M., Köhler, G., and Carsetti, R., in preparation, 1993.). Thus, *myc* protein is a critical player in this pathway of tolerance. Studies in transgenic and knockout mice are planned to further test this hypothesis.

A novel IgG fusion protein approach for epitope-specific tolerance

Classic studies of immunologic tolerance require that an antigen or antigenic epitope be administratable in a tolerogenic *and* an immunogenic form. We decided to study tolerance to the epitope comprising residues 12-26 of the bacteriophage λ cI protein because it can be recognized by both T and B-cells, and it is the major immunodominant epitope of this protein in both H-2^d and H-2^k mice.[44] Initially, we expressed this unique epitope at the N-terminus of a mouse IgG construct as a tolerogen and have created a 12-26 challenge antigen by inserting it into a highly immunogenic, bacterial flagellin molecule. These novel fusion proteins establish a set of tolerogenic and immunogenic forms for expression of the 12-26 epitope for initial tolerance experiments outlined in Table 1. Our hypothesis was that isologous immunoglobulins (especially IgGs) would make efficient tolerogenic carriers because of their ability to crosslink B-cell Fc receptors and to persist in the circulation, as well as their lack of "intrinsic immunogenicity."

Table 1. Use of different challenge antigens to measure T-cell and B-cell tolerance to the 12-26 epitope

Tolerogen Pretreatment*	Challenge Antigen	T-cell Proliferation	Antibody Response
		Measures tolerance to 12-26 in:	
12-26 or 12-26-IgG	12-26-peptide in CFA	T_{H1} (IL-2) cells	T_{H2} + B-cells
12-26 or 12-26-IgG	12-26-flagellin in CFA	$T_{H1\ cells}$	B-cells
12-26 or 12-26-IgG	12-26-polymerized flagellin *in vitro*	–	B-cells

* 12-26 administered i.v./i.p. either as peptide or as 12-26-IgG fusion protein for tolerance induction; challenge is 4-7 days later. Responses measured are specific for 12-26 in indicated cell types. To analyze tolerance in T_{H2} cells, challenge is with 12-26-FITC and the anti-FITC IgG is determined.

The 12-26-flagellin construct has been useful for a number of reasons. Firstly, flagellin is highly immunogenic and can be polymerized to make a powerful *in vivo* and *in vitro* antigen.[45] Secondly, the use of the 12-26-flagellin will provide "carrier" epitopes for the B-cell responsiveness to 12-26. Thirdly, polymerized flagellin has been shown to be a type I "thymus-independent" antigen in *in vitro* responses and will, therefore, allow direct assessment of B-cell tolerance.[46] Moreover, *in vivo* (or *in vitro*) challenge with 12-26-flagellin can lead to anti-flagellin antibodies, serving as an internal positive specificity control. That is, mice (of any strain) can develop vigorous IgG and IgM antibody responses, measured by ELISA, to *both* 12-26 and flagellin epitopes when challenged with 12-26-flagellar protein. The flagellin epitopes also will serve as specificity controls for T-cell tolerance to 12-26 moieties. The details of their construction and preliminary data on the expression of 12-26 epitopes upon transformation of *Salmonella* will be reported elsewhere. Suffice it to say, both 12-26-flagellin and 12-26-IgG can be recognized by T cells and by antibody in proliferation assays and western blot/ELISAs, respectively. Finally, 12-26-polymerized flagellin is immunogenic for purified B-cells *in vitro*, thus allowing a direct readout of tolerance in the B-lymphocyte pool.

It is well established that high dose pretreatment with peptide injected i.v. or i.p. in saline or even emulsified in incomplete Freund's adjuvant can induce tolerance in both T_{H1} and T_{H2} cells upon subsequent immunization with peptide in complete Freund's adjuvant[44,47-50] (CFA). Indeed, Gefter and colleagues elegantly showed that unresponsiveness to 12-26 can be elicited by this protocol and that tolerance was not due to T-cell suppression.[47] To establish baseline data for tolerance, we pretreated CAF1 mice with 12-26 peptide and then challenged them seven days later with peptide in CFA or 12-26-flagellin in CFA as a test of T_{H1}, T_{H2} or B-cell tolerance (see Table 1). Our results confirmed that pretreatment with 12-26 peptide induced a form of unresponsiveness resulting in proliferative T-cell anergy to peptide, regardless of challenge (data not shown).

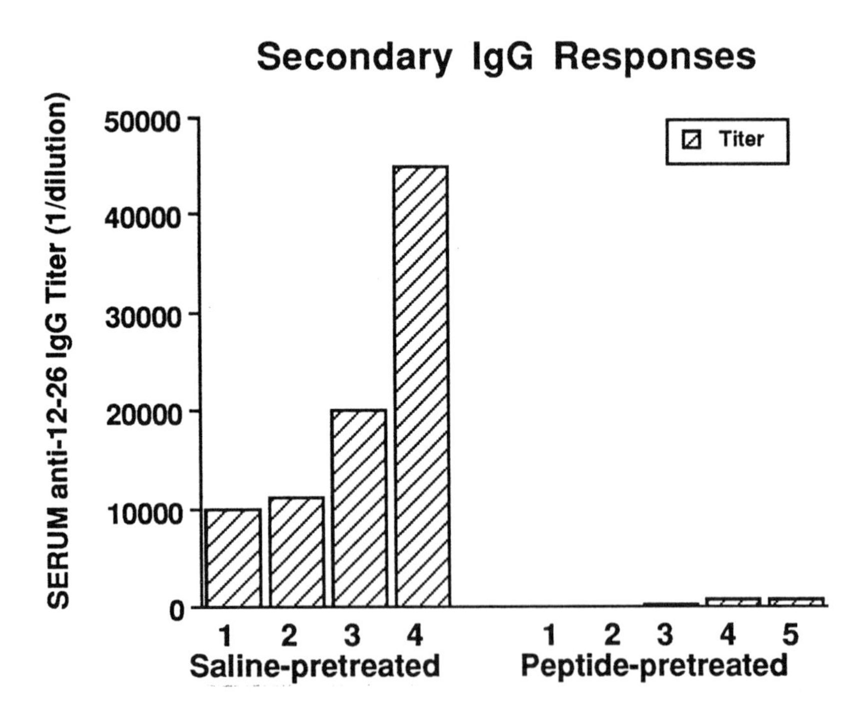

Figure 2. Induction of tolerance to 12-26 peptide. CAF1 mice were pre-treated with synthetic peptide or saline and were subsequently challenged with 50 mg peptide in CFA. Mice were boosted at 14 days with peptide in incomplete adjuvant and bled 10 days later. Sera were probed for anti-12-26 with a peptide-specific ELISA. Shown are individual secondary IgG titers for each animal (determined as the dilution at which O.D. levels returned to background pre-immunization titers).

Reduced antibody responsiveness was also observed in some experiments (see Figure 2), although this could represent either T_{H2} or B-cell tolerance. Challenge with 12-26-flagellin (which provides flagellin-derived helper epitopes and serves as a specificity control) demonstrated that B-cell hyporesponsiveness in terms of 12-26-specific IgG was apparently not induced (data not shown).

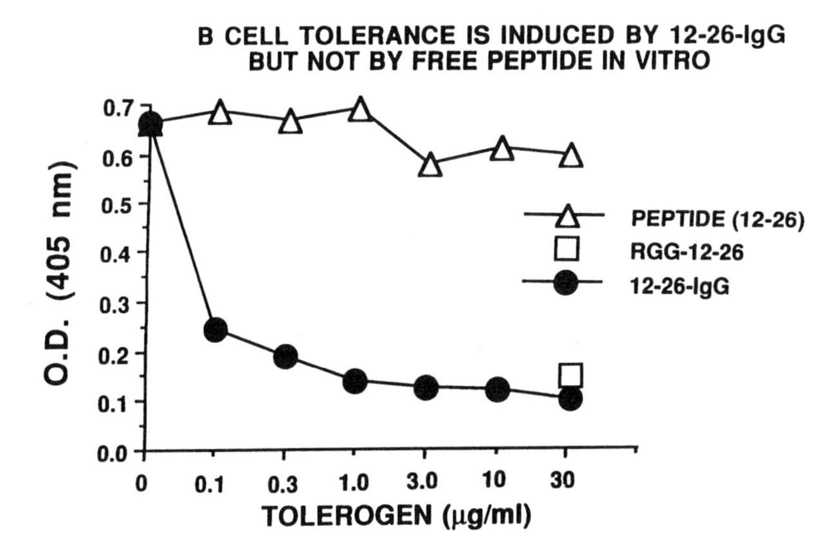

Figure 3. Induction of tolerance *in vitro* by 12-26-IgG_1 fusion protein, but not by free peptide. Spleen cells were cultured for 24 hours with the indicated concentrations of 12-26-IgG or free peptide, washed and challenged with 12-26-flagellin. Spleen cells treated with 12-26 chemically coupled to IgG were used as a tolerance control. Results are expressed as O.D. for IgM ELISA.

To further test whether the free peptide or the IgG fusion protein was capable of inducing B-cell tolerance, we incubated normal adult splenic lymphocytes with increasing doses of either 12-26 peptide or 12-26-IgG for 20 hours, washed and challenged these cells with either 12-26-flagellin or LPS (mitogen). As shown in Figure 3, only the IgG fusion protein was tolerogenic for B-cells *in vitro.*

In summary, we have devised a potent system to explore B- and T-lymphocyte tolerance to the same (overlapping) epitope. Efforts are under way to express this fusion protein in bone marrow cells so that tolerance once induced can be maintained by the continuous production of the tolerated epitope *in vivo.*

Therefore, our original hypothesis, that 12-26-IgG would be a superior tolerogen compared to free peptide, has been confirmed. We also believe that B-cell tolerance requires or is enhanced by bivalent expression of this epitope in an IgG molecule, which could also crosslink Fc receptors on B-cells, as we and others have shown.[51,52] Finally, the ability to maintain tolerance in the host without repeated injection of antigen will facilitate future therapy of undesirable immune responses, including those to factor VIII.

Acknowledgments

This work was supported by USPHS grants AI-29691 and CA55644. I thank Yao Xiao-rui, Gavin Fischer, Sally Kent, Elias Zambidis, Arti Gaur, and Anu Kurup for contributing their data, and Sandy Lynah for her secretarial assistance.

References

1. M. Venkataraman, and D.W. Scott, Cellular events in tolerance. VII. Decrease in tolerance spleens of clonable precursors stimulatable *in vitro* by specific antigens or LPS. *Cell. Immunol.* 47: 323 (1979).
2. G.J.V. Nossal, and B. Pike, Mechanisms of clonal abortion tolerogenesis. II. Clonal behavior of immature B cells following exposure to anti-μ chain antibody. *Immunology.* 37: 203 (1979).
3. G.L. Warner, and D.W. Scott, A polyclonal model for B-cell tolerance. I. Fc-dependent and Fc-independent induction of nonresponsiveness by pretreatment of normal splenic B cells with anti-Ig. *J. Immunol.* 146: 2185 (1991).
4. R. Carsetti, G. Köhler, and M.C. Lamers, A role for immunoglobulin D: Interference with tolerance induction. *Eur. J. Immunol.* 23: 168 (1993).
5. D.W. Scott, D. Livnat, C.A. Pennell, and P. Keng, Lymphoma models for B-cell activation and tolerance III. Cell cycle dependence for negative signaling of WEHI-231 B lymphoma cells by anti-μ. *J. Exp. Med.* 164:156 (1986).
6. D.W. Scott, B-lymphoma models for tolerance: The good, the bad, and the apoptotic. *ImmunoMethods.* 2: 105 (1993).
7. J. Hasbold and G.G.B. Klaus, Anti-immunoglobulin antibodies induce apoptosis in immature B-cell lymphomas. *Eur. J. Immunol.* 20: 1685 (1990).
8. L. Benhamou, P. Casenave, and P. Sarthou, Anti-immunoglobulins induce death by apoptosis in WEHI-231 B lymphoma cells. *Eur. J. Immunol.* 20: 1405 (1990).
9. J. Cambier, J. Monroe, M. Coggeshall, and K. Campbell, The biochemical basis of transmembrane signaling by B lymphocyte surface immunoglobulin. *Immunol. Today.* 6: 218 (1985).
10. M. Coggeshall, and J. Cambier, B-cell activation VIII. Membrane immunoglobulins transduce signals via activation of phosphatidylinositol hydrolysis. *J. Immunol.* 133: 3382 (1984).
11. G.G.B. Klaus, M. Bijsterbosch, A. O'Garra, M. Harnett, and K. Rigley, Receptor signaling and crosstalk in B lymphocytes. *Immunol. Rev.* 99: 19 (1987).
12. J. Imboden, The regulation of intracellular signals during lymphocyte activation. *Immunol. Today.* 9: 17 (1988).
13. J. Cambier, and K. Campbell, Membrane immunoglobulin and its accomplices: New lessons from an old receptor. *FASEB J.* 6: 3207 (1992).
14. M.-A. Campbell, and B.M. Sefton, Association between B-lymphocyte membrane immunoglobulin and multiple members of the Src family of protein tyrosine kinase mRNA expression. *Molec. and Cell. Biol.* 12: 2315 (1992).
15. L.R. Bandara, J.P. Adamczewski, T. Hunt, and N. La Thangue, Cyclin-A and the retinoblastoma gene product complex with a common transcription factor. *Nature.* 352: 249 (1991).
16. J. Milner, A. Cook, and J. Mason, p53 is associated with p34cdc2 in transformed cells. *EMBO J.* 9: 2885 (1990).
17. M. Beckwith, W. Urba, D. Ferris, D. Kuhns, C. Moratz, and D. Longo, Anti-Ig-mediated growth inhibition of a human B lymphoma cell line is independent of phosphatidylinositol turnover and protein kinase C activation and involves tyrosine phosphorylation. *J. Immunol.* 147: 2411 (1991).
18. X-r. Yao, and D.W. Scott, Expression of protein tyrosine kinases in the Ig complex of anti-μ-sensitive and anti-μ -resistant B-cell lymphomas: Role of the p55blk kinase in signaling growth arrest and apoptosis. *Immunol. Rev.* 132: 163 (1993).
19. X-r. Yao, and D.W. Scott, Antisense oligonucleotides to the blk tyrosine kinase prevent anti-μ-chain-mediated growth inhibition and apoptosis in a B-cell lymphoma. *Proc. Nat'l Acad. Sci. USA.* 90: 7946 (1993).

20. J-E. Alés-Martínez, G. Warner, and D.W. Scott, Immunoglobulin D and M mediate signals that are qualitively different in B cells with an immature phenotype. *Proc. Nat'l. Acad. Sci. USA.* 85: 6919 (1988).
21. R. Tisch, C. Roifman, and N. Hozumi, Functional differences between immunoglobulins M and D expressed on the surface of an immature B-cell line. *Proc. Nat'l. Acad. Sci. USA.* 85: 6914 (1988).
22. A. Burkhardt, M. Brunswick, J. Bolen, and J. Mond, Anti-immunoglobulin stimulation of B lymphocytes activates *src*-related protein-tyrosine kinase. *Proc. Nat'l. Acad. Sci. USA.* 88: 7410 (1991).
23. D. Hata, T. Kawakami, T. Ishigami, K.-M. Kim, T. Heike, K. Katamura, M. Mayumi, and H. Mikawa, Tyrosine phosphorylation of IgM and IgD-associated molecules of a human B lymphoma cell line B104. *Int. Immunol.* 4: 797 (1992).
24. M. Brunswick, A. Burkhardt, F. Finkelman, J. Bolen, and J. Mond, Comparison of tyrosine kinase activation by mitogenic and nonmitogenic anti-IgD antibodies. *J. Immunol.* 149: 2249 (1992).
25. J.-E. Alés-Martínez, E. Cuende, J. Casnellie, R.P. Phipps, G. Warner, and D.W. Scott, Crosslinking of surface IgM and IgD causes differential biological effects in spite of overlap of tyrosine (de)phosphorylation profiles. *Eur. J. Immunol.* 22: 845 (1992).
26. K.-M. Kim, T. Yoshimura, H. Watanabe, T. Ishigimi, M. Nambu, D. Hata, Y. Higaki, M. Sasaki, T. Tsutsui, M. Mayumi, and H. Mikawa, Growth regulation of a human mature B-cell line, B-104, by anti-IgM and anti-IgD antibodies. *J. Immunol.* 146: 819 (1991).
27. P. Mongini, C. Blessinger, D. Posnett, and S. Rudich, Membrane IgD and membrane IgM differs in capacity to transduce inhibitory signals with the same human B-cell clonal populations. *J. Immunol.* 143: 1565 (1989).
28. S. Mittnacht, and R.A. Weinberg, G1/S phosphorylation of the retinoblastoma protein is associated with an altered affinity for the nuclear compartment. *Cell.* 65: 381 (1991).
29. G. Draetta, Cell cycle control in eucaryotes molecular mechanisms of cdc2 activation. *Trends Biochem.* 15: 378 (1990).
30. T. Hunter, and J. Pines, Cyclins and cancer. *Cell.* 66: 1071 (1991).
31. J.E. McCormack, V.H. Pepe, B.R. Kent, M. Dean, A. Marshak-Rothstein, and G. Sonenshein, Specific regulation of *c-myc* oncogene expression in a murine B-cell lymphoma line. *Proc. Nat'l. Acad. Sci. USA.* 81: 5546 (1984).
32. V. Seyfert, V. Sukhatme, and J. Monroe, Differential expression of a zinc finger-encoding gene in response to positive vs. negative signaling through receptor immunoglobulin in murine B lymphocytes. *Mol. Cell. Biol.* 9: 2083 (1989).
33. G.L. Warner, D. Nelson, A. Gaur, J. Ludlow, and D.W. Scott, Anti-Ig induces TGF-ß and pRB hypophosphorylation in a model for B-cell tolerance. *Cell Growth and Differentiation.*3: 175 (1992).
34. S.M. Mashesaran, J.E. McCormick, and G. Sonenshein, Changes in phosphorylation of *myc* oncogene and *RB* anti-oncogene protein products during growth arrest of the murine lymphoma WEHI-231 cell line. *Oncogene.* 6: 1965 (1992).
35. C.A. Spencer, and M. Groudine, Control of *c-myc* regulation in normal and neoplastic cells. *Adv. Cancer Res.* 56: 1 (1991).
36. R. Hekkila, G. Schwab, E. Wickstrom, S.L. Loke, D. Pluznik, R. Watt, and L.N. Neckers, A *c-myc* antisense oligonucleotide inhibits entry into S phase, but not progress for G0 to G1. *Nature.* 328: 445 (1987).
37. E.M. Blackwood, and R.N. Eisenman, A helix-loop-helix-zipper protein that forms a sequence-specific DNA-binding complex with Myc. *Science.* 251: 1211 (1991).
38. V. Sukhatme, S. Kartha, F. Toback, R. Taub, R. Hoover, and C.A. Tsai-Morris, Novel early growth response gene rapidly induced by fibroblast, epithelial cell and lymphocyte mitogens. *Oncogene Res.* 1: 343 (1987).
39. J. Monroe, Molecular basis for unresponsiveness and tolerance induction in immature stage B lymphocytes. *Adv. Cell. Molec. Immunol.* 1b: 1 (1993).
40. V. Seyfert, S. McMahon, W. Glenn, A. Yellen, V. Sukhatme, X. Cao, and J. Monroe, Methylation of an immediate-early inducible gene as a mechanism for B-cell tolerance induction. *Science.* 250, 797 (1990).
41. G. Evan, A. Wyllie, C. Gilbert, T. Littlewood, H. Land, M. Brooks, C. Waters, L. Penn, and D. Hancock, Induction of apoptosis in fibroblasts by *c-myc* protein. *Cell.* 69: 119 (1992).
42. Y. Shi, J. Glynn, L. Guilbert, T. Cotter, R. Bissonnette, and D. Green, Role for *c-myc* in activation-induced apoptotic cell death in T-cell hybridomas. *Science.* 257: 212 (1992).

43. G. Fischer, S.C. Kent, L. Joseph, D.R. Green, and D.W. Scott, Lymphoma models for B-cell activation and tolerance X. Anti-μ-mediated growth arrest and apoptosis of murine B-cell lymphomas is prevented by the stabilization of *myc*. *J. Exp. Med.* in press (1993).
44. S. Roy, M. Scherer, T.J. Briner, J.A. Smith, and M.L. Gefter, Murine MHC polymorphism and T-cell specificites. *Science.* 244: 572 (1989).
45. G. Ada, G. Nossal, J. Pye, and A. Abbot, Preparation and properties of flagellar antigens from *Salmonella adelaide*. Austral. *J. Exp. Biol. Med.* 42: 267 (1964).
46. D.W. Scott, Role of self-carriers in the immune response and tolerance I. B-cell unresponsiveness and cytotoxic T-cell immunity induced by haptenated syngeneic lymphoid cells. *J. Exp. Med.* 144: 69 (1976).
47. M. Scherer, B. Chan, F. Ria, J. Smith, D. Perkins, and M. Gefter, Control of cellular and humoral immune responses by peptides containing T-cell epitopes. *Cold Spring Harbor Symp. on Quantit. Biol.* 54: 497 (1989).
48. D.M. Bitar, and C.C. Whitacre, Suppression of experimental autoimmune encephalomyelitis by oral administration of myelin basic protein. *Cell. Immunol.* 112: 364 (1988).
49. S.S. Zamvil, D.J. Mitchell, A.C. Moore, K. Kitamura, L. Steinman, and J. Rothbard, T-cell epitope of the auto-antigen myelin basic protein that induces encephalomyelitis. *Nature.* 324: 258 (1986).
50. A. Gaur, B. Wiers, A. Lin, J. Rothbard, and C.G. Fathman, Amelioration of auto-immune encephalomyelitis by myelin basic protein synthetic peptide-induced anergy. *Science.* 258: 1491 (1992).
51. M. Bijsterbosch, and G.G.B. Klaus, Crosslinking of surface immunoglobulin and Fc receptors on B lymphocytes inhibits stimulation of inositol phospholipid breakdown *via* the antigen receptors. *J. Exp. Med.* 162: 1825 (1985).
52. G.L. Warner, and D.W. Scott, A polyclonal model for B-cell tolerance I. Fc-dependent and Fc-independent induction of nonresponsiveness by pretreatment of normal splenic B cells with anti-Ig. *J. Immunol.* 146: 2185 (1991).

VI: New approaches to the inhibitor problem

In this section two separate issues are discussed, offering new insights into future areas of investigation.

In the management of inhibitors, IVIg has been part of the Malmö regimen and has been used by itself for patients with acquired inhibitors. Pioneers in this field have been the collaborative efforts of Drs. Sultan and Kazatchkine. The efficacy of IVIg has led this group to explore the elements in IVIg that might explain their mechanism of action. In elegant experiments, Dr. Kazatchkine demonstrates that gamma globulin derived from normal persons contains anti-idiotypic antibodies with specificity to F.VIII antibodies. These antibodies are more commonly seen in aged donors and from multiparous women. These findings give treaters renewed interest in pursuing the adjunctive advantages of IVIg in the management of inhibitors.

The second paper by Dr. Kay offers great hope that gene therapy for hemophilia may be a reality sometime in the future. Initially, retroviral vectors containing the F.IX gene were transduced into hepatocytes from partially hepatectomized-normal dogs. Those vectors were then transfused into the splenic veins of these dogs as early experiments. At a later stage, direct hepatic gene transfer was carried out in hemophilia B dogs using a retroviral vector. Small amounts of F.IX were produced over a nine-month period with improvements in laboratory data.

Use of adenoviral vectors eliminated the need for transduction of hepatocytes, thereby eliminating the need for hepatectomy. More F.IX was produced, but gene expression was transient. The transience of the expression may well be due to antibodies to adenoviruses.

These data are very exciting and offer great hope for the future, but once again inhibitor induction may be the limiting factor in the rapid advance of this technology.

Variable (V) region-mediated control of autoreactivity to factor VIII

Michel D. Kazatchkine,[1] Srinivas S. Kaveri,[1] Urs E. Nydegger,[2] and Yvette Sultan[3]

[1]*Service d'Immunologie and INSERM U28, Hôpital Broussais, Paris, France*
[2]*Center for Blood Transfusion, Inselspital, Bern, Switzerland*
[3]*Centre de Traitement des Hemophiles, Hôpital Cochin, Paris, France*

It is now recognized that autoreactive T and B cells are normally present in healthy individuals. Natural antibodies reactive with self-antigens are also present in normal serum and are frequent among hybridomas obtained from healthy individuals. Evidence accumulated in the last ten years indicates that autoreactivity is positively selected during ontogeny, that it provides the basis for an autonomous activity of the normal immune system independent of foreign antigens and that "physiological autoreactivity" is regulated within a network of interactions between variable (V) regions of antibodies and lymphocytes.[1]

Normal human serum IgG has been shown to react with a wide range of self-antigens including cell surface molecules, intracellular, nuclear, and plasma proteins.[2] Reactivity of normal IgG with self-antigens is higher when IgG is tested in a purified form rather than in whole serum. The autoantigens recognized by human natural autoantibodies include evolutionarily conserved molecules such as actin, tubulin, and myosin and molecules that may be targets for autoantibodies in autoimmune diseases, e.g., DNA, factor VIII (F.VIII), thyroglobulin, and intrinsic factor.[3] There are indications that in certain autoimmune diseases, autoantibodies are directed towards epitopes of the target molecule that are distinct from those recognized by natural antibodies and that the antibodies express specific idiotypic markers (e.g., in autoimmune thyroiditis). In other autoimmune diseases, however, neither the analysis of epitopic or idiotypic specificities of autoantibodies nor that of VH family usage and sequence of V regions has allowed discrimination between natural and "pathogenic" autoantibodies (see chapter by V. Pascual in this volume). As the boundaries between physiological and pathological autoimmunity appear more subtle, these issues become critical for delineating our concepts about autoimmunity and our

Inhibitors to Coagulation Factors
Edited by Louis M. Aledort *et al.*, Plenum Press, New York, 1995

approaches to the therapy of autoimmune diseases.[4]

Natural autoantibodies: Antibodies to F.VIII are found in normal human serum IgG

Antibodies to F.VIII in normal plasma were identified by screening heated plasma from healthy blood donors for its ability to neutralize F.VIII activity in a reference plasma pool by using the assay described by Kasper et al.[5] Eighty five of 500 plasma samples tested (17%) were found to contain F.VIII-neutralizing activity > 0.4 BU.[6] Dose-dependent F.VIII-neutralizing activity was present in purified IgG from plasma and in $F(ab')_2$ fragments of IgG. IgG antibodies to F.VIII were also detected in normal plasma by ELISA and Western blotting. Natural antibodies to F.VIII did not exhibit restricted isotypic heterogeneity. The absorbance values determined by ELISA did not correlate with anti-F.VIII activity measured in a functional assay, suggesting that natural anti-F.VIII antibodies recognize epitopes located both within and outside the functional sites on the F.VIII molecule. We also observed that mean levels of F.VIII activity in the plasma of healthy individuals with anti-F.VIII antibodies did not differ from the levels measured in individuals with no detectable F.VIII-neutralizing activity. In addition, purified IgG from the plasma of a single donor was found to neutralize F.VIII activity in the plasma of several other donors but not in autologous plasma, suggesting that some natural antibodies may be directed against allotypic determinants on the F.VIII molecule.

Probing the natural IgG autoreactive repertoire

No detailed study is yet available on the portions of the F.VIII molecule and epitopes that are recognized by natural anti-F.VIII antibodies and inhibitors from patients with anti-F.VIII autoimmune disease or hemophiliacs immunized against allogenic F.VIII. We have recently undertaken a comparative analysis of the natural and "pathological" autoimmune antibody repertoires by using protein extracts from homologous tissues as a source of antigens and a computer-aided quantitative immunoblotting method allowing the simultaneous comparison of the patterns of immunoreactivity of several antibody sources with hundreds of antigens in a tissue.[7] By using this technique, we observed that IgG from healthy individuals recognizes a restricted set of antigens in tissues such as thymus, liver, stomach, muscle, and endothelial cells. In addition, although differences in intensity of reactivity with a given protein band are seen between individuals, the pattern of immunoreactivity of natural IgG and IgM antibodies with each tissue extract is conserved among healthy donors.[8] We will use this approach to define the normal expressed autoreactive B-cell repertoire and compare it with that expressed in autoimmune diseases, whether the disease is systemic or, as in the case of spontaneous inhibitors to F.VIII, believed to be restricted to one or few epitopes of the target molecule or tissue. These studies should help to distinguish between pathological autoimmune responses of clonal (oligoclonal) nature and those which represent a defect in the autoreactive repertoire extending beyond the unique or few "abnormal" specificities detected by usual diagnostic assays. In the latter case, autoimmunity would result from altered selection of expressed autoreactive repertoires rather than it would represent a "normal" immune response to an abnormal antigen or abnormally presented normal antigen.

V region-dependent "connectivity" between autoantibodies: Anti-idiotypic antibodies against anti-F.VIII antibodies are present in normal human serum IgG

Analysis of natural antibodies, particularly those of the neonatal period, has revealed that antibodies which react with self-antigens are highly "connected" through V regions with other antibody molecules of the same individual. Pooled normal human IgG (intravenous immunoglobulin, IVIg) and IgG purified from plasma of a single healthy donor can be separated into "connected" and "nonconnected" fractions by subjecting IgG or $F(ab')_2$ fragments of IgG to affinity chromatography on itself and eluting at acid pH from the column, the fraction that bound with high affinity to V regions present within the same IgG preparations.[9] The connected fraction of IgG was shown to express higher reactivity with self-antigens (e.g., F.VIII) than nonconnected IgG; no difference was observed between the connected and nonconnected fractions in the case of the reactivity with foreign antigens, further documenting that a high degree of connectivity is characteristic of natural autoantibodies.[9] We and others have suggested that the network structure of the normal immune system is ensured by such multiconnected antibodies and homeostasis of autoreactivity under physiological conditions requires those critical levels of connectivity, probably regardless of the precise epitopic (idiotypic) characteristics of individual antibodies in the system.[10] In this view, autoimmune disease would result from primary or secondary (e.g., to an infection) alterations in connectivity that would allow for uncontrolled expansion and, possibly, mutations of otherwise normal autoreactive clones. As discussed below, we believe that a major mechanism by which IVIg is effective in autoimmune disorders, may be that it would restore normal levels of connectivity. In this respect, spontaneous inhibitors to F.VIII have provided an *in vivo* and *in vitro* model for studying V region-dependent connectivity between pathogenic autoantibodies and therapeutic IgG (IVIg).

The suggestion that IVIg may directly interact with circulating autoantibodies in the patients came from the observation that, in patients with inhibitors to F.VIII, autoantibody levels decreased rapidly following administration of IVIg (i.e., within 24 hours of infusion).[11] Such interaction between inhibitors to F.VIII and IVIg and its dependency on V regions was demonstrated *in vitro* by the finding of a dose-dependent inhibition of anti-F.VIII activity of patients' IgG or $F(ab')_2$ fragments of IgG by IVIg and $F(ab')_2$ fragments of IVIg. The following lines of evidence then demonstrated that IVIg contains anti-idiotypic antibodies against anti-F.VIII autoantibodies: (1) anti-F.VIII activity in purified IgG from patients' plasma or in $F(ab')_2$ fragments of IgG was specifically retained on an affinity chromatography column of Sepharose-bound $F(ab')_2$ fragments of IVIg;[12] (2) IVIg was found not to contain detectable amounts of antibody activity against the most common allotypes located in the $F(ab')_2$ region of human IgG;[13] (3) IVIg was found to inhibit the binding to $F(ab')_2$ fragments of an anti-F.VIII antibody of a mouse monoclonal antibody directed against the antibody-binding site of the anti-F.VIII antibody, indicating that IVIg contains antibodies recognizing the same idiotopes on anti-F.VIII antibodies as those recognized by the heterologous anti-idiotypic reagent.[14] Analogous experiments in other antibody systems have demonstrated that IVIg contains V region-connected and anti-idiotypic antibodies against a wide range of pathogenic or "disease-related" autoantibodies.

By examining the capacity of IgG from single donors or pooled IgG from several donors to neutralize anti-F.VIII activity of spontaneous inhibitors to F.VIII, we found that IgG from aged donors and from multiparous women contained inhibitory activity against the autoantibodies at a higher frequency than IgG from young adult males. These studies (see chapter by Y. Sultan in this volume) also demonstrated that anti-idiotypic activity of pooled IgG (IVIg) results from a synergistic participation of anti-idiotypes from each donor contributing to the pool.[15]

Implications for therapy of anti-F.VIII autoimmune disease

A physiological significance for natural autoantibodies may be inferred from their ability to neutralize biologically active molecules such as anti-F.VIII antibodies, to select B-cell and antibody repertoires early in life, regulate the production of other serum antibodies, and to prevent occurrence of autoimmune diseases in genetically susceptible animals.[16, 17] Further evidence for the existence of a functional network controlling IgG autoreactivity in healthy individuals came from the finding of nonrandom, conserved patterns of spontaneous fluctuations of serum levels of natural autoantibodies in healthy individuals.[18] These patterns were clearly distinct from those observed in patients with autoimmune diseases. Recent evidence from our laboratory indicates that infusion of IVIg into an exemplary patient with autoimmune thyroiditis restored the dynamic pattern of spontaneous antibody fluctuations characteristic of healthy individuals.[19] Based on these views, we have postulated that because IVIg originates from healthy individuals, it should contain the whole set of regulatory molecules that normally participate in maintaining immune homeostasis. Thus, if autoimmune disease is associated with primary or secondary deficiency or dysfunction of the regulatory network, administration of IVIg would reconstitute the defective idiotypic regulation of effector functions (for example, as it is seen with the immediate neutralizing effects of IVIg) and the ability to select a normal immune repertoire (as seen with the long-term effects of IVIg therapy). These concepts are supported by the *in vivo* observations of short-term and prolonged suppression of spontaneous inhibitors to F.VIII by administration of IVIg[20] and by the experimental data summarized above demonstrating V region connectivity between anti-F.VIII autoantibodies and IVIg.

References

1. F. Varela and A.C. Coutinho, Second generation immune networks, *Immunol Today.* 12: 159 (1991).
2. S. Avrameas, Natural autoantibodies: From "horror autotoxicus" to "gnothi seauton," *Immunol Today.* 12: 154 (1991).
3. V. Hurez, S.V. Kaveri, and M.D. Kazatchkine, Expression and control of the natural autoreactive IgG repertoire in normal human serum, *Eur J Immunol.* 23:783 (1993).
4. A.C. Coutinho, and M.D. Kazatchkine, "Autoimmunity: Physiology and Disease," Wiley-Liss, New York (1993).
5. C.K. Kasper, L.M. Aledort, R.B. Counts, J.R. Edson, J. Fratantone, D. Green, J.W. Hempton, M.N. Hilgartner, J. Lazarson, P.H. Levin, C.N. Macmillan, J.C. Pool, S.S. Shapiro, N.R. Schulman, and J. Eyes, A more uniform measurement of F VIII inhibitors, *Throm Diath Haemorrh.* 34:869 (1975).
6. M. Algiman, G. Dietrich, U.E. Nydegger, D. Boieldieu, Y. Sultan, and M.D. Kazatchkine, Natural antibodies to Factor VIII (anti-haemophilic factor) by high dose intravenous gammaglobulin, *Lancet.* II: 765 (1984).
7. A. Nobrega, M. Haury, A. Grandien, E. Malanchere, A. Sundblad, and A. Coutinho, Global analysis of antibody repertoires. II. Evidence for specificity, self-selection and the immunological "homunculus" of antibodies in normal serum, *Eur J Immunol* (1993), in press.
8. L. Mouthon, N. Ronda, S.V. Kaveri, and M.D. Kazatchkine, Natural human IgM and IgG antibodies recognize a restricted set of antigens in homologous tissues, *FASEB J.* (Abstract). (1994), in press.
9. G. Dietrich, S.V. Kaveri, and M.D. Kazatchkine, A V region-connected autoreactive subtraction of normal human serum immunoglobulin G, *Eur J Immunol.* 22:1701 (1992).
10. M.D. Kazatchkine, and A.C. Coutinho, Are lymphocytes concerned with our definition of idiotypes? *Immunol Today.* 14:513 (1993).
11. Y. Sultan, M.D. Kazatchkine, P. Mainsonneuve, and U.E. Nydegger, Anti-idiotypic suppression of autoantibodies to Factor VIII (anti-haemophilic factor) by high dose intravenous gammaglobulin, *Lancet.* II:765 (1984).
12. Y. Sultan, F. Rossi, and M.D. Kazatchkine, Recovery from anti-VIII:C (anti-hemophilic factor) autoimmune disease is dependent on generation of anti-idiotypes against anti-VIII: C antibodies, *Proc Natl Acad Sci USA.* 84: 828 (1987).
13. F. Rossi, Y. Sultan, and M.D. Kazatchkine, Anti-idiotypes against autoantibodies and alloantibodies to VIII:C (anti-hemophilic factor) are present in therapeutic polyspecific normal immunoglobulins, *Clin Exp Immunol.* 74:311 (1988).
14. G. Dietrich, P. Pereira, M. Algiman, Y. Sultan, and M.D. Kazatchkine, A monoclonal anti-idiotypic antibody against the antigen-combining site of anti-factor VIII antibodies defines an idiotope that is recognized by normal human polyspecific immunoglobulins for therapeutic use (IVIg), *J Autoimmunity.* 3: 547 (1990).
15. G. Dietrich, M. Algiman, Y. Sultan, M.D. Kazatchkine, Origin of anti-idiotypic activity

against anti-Factor VIII autoantibodies in pools of normal human polyspecific immunoglobulin G (IVIg), *Blood.* 79: 2946 (1992).

16. S.V. Kaveri, G. Dietrich, V. Hurez, and M.D. Kazatchkine, Intravenous immunoglobulins (IVIg) in the treatment of autoimmune diseases, *Clin Exp Immunol.* 86: 192 (1991).
17. L. Mouthon, S.V. Kaveri, and M. D. Kazatchkine, Immune modulating effects of IVIg in autoimmune diseases, *Transfus Sci.* (1993) in press.
18. F. Varela, A. Anderson, G. Dietrich, A. Sundblad, D. Holmberg, M.D. Kazatchkine, and A.C. Coutinho, The population dynamics of antibodies in normal and autoimmune individuals, *Proc Natl Acad Sci USA.* 88: 5917 (1991).
19. G. Dietrich, F. Varela, V. Hurez, M. Bouanani, and M.D. Kazatchkine, Selection of expressed B-cell repertoires by infusion of normal immunoglobulin G in a patient with autoimmune thyroiditis, *Eur J Immunol.* (1993) in press.
20. Y. Sultan, M.D. Kazatchkine, U.E. Nydegger, F. Rossi, G. Dietrich, and M. Algiman, Intravenous immunoglobulin in the treatment of spontaneously acquired Factor VIII inhibitors, *Am J Med.4:* 5A35S (1991).

Hepatic gene therapy for hemophilia B

Mark A. Kay

Investigator, Markey Molecular Medicine Center
Assistant Professor of Medicine
Division of Medical Genetics RG-25
University of Washington
Seattle, Washington 98195

A number of different cell types including keratinocytes, hepatocytes, fibroblasts, myoblasts, and endothelial cells have been targeted for factor IX (F.IX) gene replacement therapy.[1-7] Following gene transfer, these cells have produced functional F.IX because of their ability to gamma-carboxylate the appropriate "gla" domains and secrete the functional protein. Most efforts to date have employed an *ex vivo* gene transfer method in rodents and involve transfer of the F.IX gene into the cells via a recombinant retroviral vector and subsequent transplantation of the cells back into the animal. Two major problems encountered in these studies have been the rather low concentration of circulating F.IX and/or the loss of F.IX expression over time. The inability to achieve long-term expression is primarily the result of loss of the transduced cells or inactivation of the expression vectors.

The liver produces virtually all of the circulating F.IX and a significant proportion of factor VIII (F.VIII); thus, it will undoubtedly be an important target for gene therapy of the hemophilias. The liver represents an excellent tissue for gene therapy for several reasons: (1) Many genetic disorders are the result of hepatic enzyme deficiencies which require liver specific enzymes or cofactors that are needed for proper metabolic function. (2) The liver is a major synthetic organ and is responsible for production of a large number of circulating proteins including F.VIII, F.IX and other coagulation factors. (3) Gene transfer into this organ may be the best way for achieving high-level production of hematogenously delivered proteins. (4) The liver is the only organ in the body besides bone marrow that can regenerate itself after partial ablation. (5) The liver provides a large cellular target for direct *in vivo* gene transfer using intravenous infusion of cell-specific vectors. With the availability of a F.IX-deficient canine model that can be used for preclinical trials, the liver represents an excellent target organ for gene therapy for hemophilia.

Inhibitors to Coagulation Factors
Edited by Louis M. Aledort *et al.*, Plenum Press, New York, 1995

We have approached the hepatocyte as a target for F.IX gene therapy, and I will discuss our efforts toward this goal. Both rodent and canine animal models have been used in preclinical trials. The mouse was selected because of its small size and easy handling; the dog was selected because of the existence of a well-characterized hemophilia B canine model.[8] Three different approaches have been tried. The first is transplantation of autologous canine hepatocytes after retroviral-mediated gene transfer referred to as the *ex vivo* approach. The second approach is direct *in vivo* transfer of recombinant retroviral vectors into the livers of mice and dogs. The last approach is direct *in vivo* hepatic gene transfer using recombinant adenoviral vectors.

Ex vivo *hepatic gene therapy*

To develop this method for the potential treatment of hemophilia B dogs, initial trials were performed in normal dogs.[9] The animals were subjected to a 20% partial hepatectomy and hepatocytes were isolated and cultured. About 2 to 3 x 10^9 hepatocytes could be isolated from each 5 kg animal. Transplantation methods were developed that allowed transplantation of at least 5% of the original hepatic mass. Hepatocytes cultured from two animals were subjected to gene transduction with a recombinant retroviral vector that encodes the human alpha-1-antitrypsin (hAAT) cDNA under the transcriptional control of the human intermediate-early CMV promoter/enhancer. The hAAT serves as a serum marker that is quantitated by a specific immunologic ELISA assay that specifically measures the human protein. Optimal conditions for hepatocyte culture and gene transfer were developed such that about 30% of the hepatocytes could be transduced with this or other retroviral vectors. Hepatocytes transduced with this retroviral vector secreted large quantities of hAAT into the media. The hepatocytes were harvested 2 to 3 days after isolation following gene transduction and were transplanted into the portal vasculature via a catheter that was positioned in the splenic vein (near its entry into the portal vein). The proximal portion of the catheter (port) was placed in the subcutaneous space so that the transplantation of 4 to 7 x 10^8 hepatocytes in each of two animals was performed without a second surgical procedure.

Following gene transfer, the animals were monitored periodically for serum protein and found to have hAAT concentrations of up to 5 ug/ml. However, the serum hAAT level began to fall, and after about 1 month no human protein was detected in the bloodstream. We were able to demonstrate that loss of hAAT protein was not due to antibody formation to the human protein. By obtaining liver biopsies from the animals at 2 time periods up to 4.5 months after gene transfer, we were able to demonstrate a constant number of proviral DNA copies per hepatocyte. This implies that the transduced cells survived long-term but that gene expression was extinguished in transduced cells. Inactivation of the

CMV promoter enhancer in the retroviral vector was hypothesized to be responsible for inactivation of gene expression because transgenic mice contain transcripts from the CMV promoter in most tisssues except the liver.[10] Additionally, similar extinction of CMV driven gene expression had been demonstrated in transplanted fibroblasts in a rodent model.[11] Clearly, the expression of transgenes from vectors in cultured primary hepatocytes do not reflect the level of gene expression *in vivo.*

The *ex vivo* method mentioned above is extremely labor intensive and it would not be feasible to test a large number of vectors in a significant number of animals to determine which expression sequences would be most useful for gene transfer trials. Thus, it was important to develop a relatively simple method for testing different gene constructs *in vivo.*

In vivo *hepatic gene transfer of recombinant retroviral vectors*

We developed a more direct method of hepatic gene transfer in the mouse that was accomplished by partial hepatectomy followed by portal vein infusion of viral particles.[12] Partial hepatic ablation is needed to stimulate cell division, which is a requirement for proviral DNA integration into the host chromosome. In order to determine the frequency of gene transfer into hepatocytes a retroviral vector that encodes the E.coli beta-galactosidase gene was used as an intracellular reporter protein. After gene transfer, the hepatocytes were isolated and stained with 5-bromo-4-chloro-3-imdoyl-6-D-galactopyranoside (x-Gal) . Hepatocytes expressing the recombinant enzyme stain blue, and we found that 1% to 2% of hepatocytes were transduced by this method. We estimate that about 1 x 10^6 viral particles were infused into an animal with about 2 x 10^7 hepatocytes after partial hepatectomy. A transduction frequency of 1% to 2% translates into an efficiency of gene transfer in a range of 5% to 10%.

A number of different vectors that used different promoters to drive hAAT production were constructed. When we infused the same vector (CMV promoter-hAAT cDNA) in mice that was used for the *ex vivo* therapy in dogs, transient hAAT expression was observed similar to that observed in the dogs. When vectors containing different promoters were used for hepatic gene transfer, the recipient animals had constitutive serum hAAT concentrations (about 1 ug/ml) for the life of the animal. This data further supports the hypothesis that the CMV promoter enhancer is inactivated in the liver after somatic gene transfer. Most importantly, by swapping promoters one could achieve long-term transgene expression.

The relative simplicity of this method lent itself to testing a number of different retroviral vectors in the mouse before returning to *ex vivo* gene therapy in the dog. Moreover, we hypothesized that it may be possible to perform *in vivo* delivery of retroviral vectors into a large animal model. Thus, we performed a number of studies of direct *in vivo* hepatic gene transfer into the hepatocytes of dogs. Similar to the mouse experiments, our first step was to use the retroviral vector that expressed the beta-galactosidase gene. A 2/3 partial hepatectomy was performed in normal dogs and at the same time a catheter was placed into the splenic vein of the animal so that retrovirus could be infused without additional surgery. Recombinant virus containing the beta-galactosidase gene virus was infused at 24h, 48h and 72h post-hepatectomy. The frequency of hepatocyte transduction varied between 0.3% and 1% and was similar to what had been observed in the mouse.

Hepatic gene therapy for canine hemophilia B

Direct hepatic gene transfer was adapted for trials in the Chapel Hill hemophilia B canine model.[8] These animals have no detectable F.IX antigen or activity that results from a missense mutation in the catalytic domain.[13] Three animals that varied in weight from 5.3 to 7.5 kg (ages 11 to 14 weeks) were infused with a retroviral vector that contained the canine F.IX cDNA under the transcriptional control of the mouse Moloney LTR promoter enhancer.[14] About 1.5 x 10^9 viral particles were infused into the portal vasculature, as was done in the normal dog studies described above. The plasma canine F.IX concentrations were determined by an immunologic ELISA assay and by a biological assay. Assuming a wild-type animal has 11.5 ug/ml, we were able to detect F.IX concentrations in the range of 2 to 10 ng/ml. Hemostatic parameters included the measurement of the whole blood clotting times (WBCT) and partial thromboplastin times (PTT).

The whole blood clotting time in hemophilia B dogs is about 45 to 50 minutes whereas the treated animals had reductions in the whole blood clotting times of more than 50% to 18 to 22 minutes. The WBCT in normal animals is 6 to 8 minutes. The PTT was also significantly shortened in these animals. Thus, with as little as 0.1% of the normal concentration of F.IX there was a dramatic improvement in biochemical parameters of hemostasis. Most importantly we were able to demonstrate that production of recombinant F.IX persists. In the first animal treated, the plasma F.IX concentration and hemostatic parameters have remained stable over 9 months.

Adenoviral-mediated hepatic gene therapy

Recombinant adenoviral vectors will transduce nonreplicating cells, hence there is no requirement for partial hepatectomy for gene transfer. These vectors have the advantage of being able to be concentrated to high titers up to 10^{12} particles/ml. However, the adenoviral genome remains extra-chromosomal, a fact that has major implications for long-term stable gene expression. Studies were initiated to evaluate the adenoviral vector as a means of gene transfer into the liver of animals.

Using a recombinant vector that encodes the beta-galactosidase gene, we were able to transduce virtually all hepatocytes in the mouse after portal vein infusion of about 100 particles per cell without evidence of toxicity.[15] Moreover, a dose response challenge revealed that the proportion of transduced cells was directly related to the amount of adenovirus infused. If the dose was increased to about 500 particles per cell, the mortality rate rose to greater than 90%. In spite of the high transduction rates, the expression from transduced hepatocytes diminished over time in the mouse. At 16 weeks posttreatment only 0.5% to 10% of the hepatocytes contain beta-galactosidase, and this is due in part to loss of the adenoviral genome over time.

Adenoviral-mediated hepatic gene therapy for hemophilia B

The high efficiency of gene transfer makes this vector attractive for the treatment of hemophilia. We have constructed adenoviral vectors that express canine F.IX and used these to treat hemophilia B dogs. We infused 2.4×10^{12} viral particles into the portal vasculature of three hemophilia B dogs weighing 14 kg to 17 kg.[16] The animals produced 2 to 3 times the wild-type level of F.IX, which normalized their hemostatic abnormalities as determined by monitoring the WBCT, PTT, and secondary bleeding times. As predicted from the mouse experiments, expression was transient and the hemostatic parameters slowly returned to their pre-treatment levels over a period of 2 months concomitant with a fall in plasma F.IX. Multiple infusions may be possible; however, there is preliminary evidence that animals develop a block to secondary transduction with repeat administration.

Conclusions

Retroviral-mediated F.IX gene transfer into deficient dogs has resulted in constitutive expression of low levels of F.IX which has led to persistent improvement of clinically relevant parameters such as the WBCT and PTT. Conversely, *in vivo* adenoviral mediated delivery of the F.IX cDNA into hepatocytes of hemophilia B dogs has resulted in greater than wild-type plasma concentrations of clotting factor with complete, albeit transient normalization of hemostasis.

In order for the retroviral-mediated gene transfer approach to be useful in clinical trials, it will be necessary to improve F.IX production by 10- to 100-fold. This may be achieved by increasing the amount of protein factor made in each of the transduced cells by altering cis acting DNA elements and/or by developing methods to increase the proportion of transduced cells.

It will be necessary to either develop noninvasive methods for repeat adenoviral gene transfer or to develop a method which maintains the recombinant adenoviral genome in transduced hepatocytes for this method of gene transfer into hepatocytes to be useful in the clinic for the treatment of hemophilia.

Gene therapy and inhibitor formation

Gene transfer for the treatment of hemophilia does not directly address the issue of inhibitor formation for at-risk patients. It is possible that constitutive expression of moderate levels of F.IX will allow patients with inhibitors to develop tolerance and diminish the risk to new patients. Many issues regarding ectopic expression of F.IX and inhibitor formation will need to be addressed. For example, if ectopic expression from non-hepatic cells is possible, will the factor become more or less immunogenic? If F.IX is expressed from selected populations of cells derived from bone marrow progenitors will this induce tolerance or increase the likelihood of inhibitor formation? Although these issues are not known, it is clear that with our ability to deliver genes to various somatic tissues *in vivo*, these questions and others can now be directly addressed.

References

1. Brinkhous, K.M. (1992) Thromb. Res. 67: 329-338.
2. Thompson, A.R., Palmer, T.D., Lynch, C.M., and Miller, A.D. (1991) Curr. Stud. Hematol. Blood Transfus. 58:59-62.
3. Dai, Y., Roman, M., Naviaux, R.K., and Verma, I. (1992) Proc. Natl. Acad. Sci., USA. 89: 10892-10895.
4. Miyanohara, A., Johnson, P.A., Elam, R.L., Dai., Y., Witztum, J.L. Verma, I. and Freidmann, T. (1992) New Biol. 4:238-246.
5. Yao, S.N. and Kurachi, K. (1992) Proc. Natl. Acad. Sci., USA. 89:3357-3361.
6. Gerrard, A. J., Hudson, D.L., Brownlee, G.G. and Watt, F.M. (1993) Nature Genet. 3:180-183.
7. Armentano, D., Thompson, A.R., Darlington, G. and Woo, S.L.C. (1990) Proc. Natl. Acad. Sci., USA 87:6141-6145.
8. Brinkhous, K.M., Davis, P.D., Garris, J.B., Diness, V., and Read, M.S. (1973) Blood. 41:577-585.
9. Kay, M.A., Baley, P., Rothenberg, S., Leland, F. Fleming, L., Parker-Ponder, K., Liu, T.J., Finegold, M., Darlington, G., Pokorny, W., and Woo, S.L.C. (1992) Proc. Natl. Acad. Sci., USA. 89:89-93.
10. Schmidt, E.V., Christoph, G., Zeller, R. and Leder, P. (1990) Mol. Cell. Biol. 10:4406-4411.
11. Scharfmann, R., Axelrod, J.H. and Verma, I.M. (1991) Proc. Natl. Acad. Sci., USA. 88:4624-4630.
12. Kay, M.A., Li, Q.T., Liu, T.J., Leleand, F., Toman, C., Finegold, M., Woo, S.L.C. (1992) Human Gene Therapy. 3:641-647.
13. Evans, J.P., Brinkhous, K.M., Brayer, G.D., Reisner, H.M., High, K.A. Proc. Natl. Acad. Sci., USA. 86:10095-10099.
14. Kay, M.A., Rothenberg, S., Landen, C., Bellinger, D.A., Leleand, F., Toman, C., Finegold, M., Thompson, A.R., Read, M.S., Brinkhous, K.M. and Woo, S.L.C.. (1993) Science. 262:117-119.
15. Li, Q.T., Kay, M.A., Finegold, M., Stratford-Perricaudet, L.D., Woo, S.L.C. (1993) Hum. Gene Ther. 4:403-409.
16. Kay, M.A., Landen, C.N., Rothenberg, S.R., Taylor, L.A., Leleand, F., Wiehle, S., Fang, B., Bellinger, D., Finegold, M., Thompson, A.R., Read, M.S., Brinkhous, K.M. and Woo, S.L.C. (1994) Proc. Natl. Acad. Sci., USA. 91:2353-7.

Panel discussion on the development of inhibitors: Who and why?

Editors' note: The transcript is incomplete because of tape recording problems. Also, we were not able to identify all the participants in the discussion. Those quoted had an opportunity to review and edit their comments to insure accuracy.

Chairman
Gilbert C. White II
The University of North Carolina at Chapel Hill

Panelists
Ernest Briët, M.D.
University Hospital, Leiden
Jean Guy G. Gilles, Ph.D.
Universite Catholique de Louvain
Leon W. Hoyer, M.D.
Holland Lab, American Red Cross Blood Services
Michel Kazatchkine, M.D.
Hôpital Broussais, Paris
Robert R. Montgomery, M.D.
The Blood Center of Southeastern Wisconsin
Francis E. Preston, M.D.
Royal Hallamshire Hospital
Howard M. Reisner, Ph.D.
The University of North Carolina at Chapel Hill

Participants
Louis M. Aledort, M.D.
Mt. Sinai Medical Center
David Aronson, M.D.
George Washington University
Christopher Goodnow, Ph.D.
Stanford University Medical Center
Marc Jenkins, Ph.D.
University of Minnesota
Mark Kay, M.D., Ph.D.
University of Washington School of Medicine
Dougald M. Monroe, Ph.D.
University of North Carolina at Chapel Hill
Dorothea Scandella, Ph.D.
Holland Lab, American Red Cross
David W. Scott, Ph.D.
Holland Lab, American Red Cross Blood Services

Dr. White: I think for the past two days we've heard some very stimulating discussions on what we know about the formation of inhibitors, both in hemophiliacs and nonhemophiliacs. We've heard some stimulating presentations and discussions on what we know about the immune system and the molecular mechanisms involved in immunity. I think the purpose of this panel discussion will be to try and bring these two topics even closer together, as close, if you will, as an antigen processing cell and a T-cell receptor. So, I think as we go through this panel I'd like to see discussion from both immunologists and people interested in hemophilia. Let's try and keep in mind that what we are trying to do here is to develop new ways to try and look at and take care of patients with inhibitors. This first panel discussion is going to focus on who and why and the second panel will focus on treatment.

I'd like to open it up with rather a broad question. I think one of the things that this panel is perhaps designed to address is, by the title of the panel, who develops inhibitors. Can we predict patients who are going to develop inhibitors? Are there ways that we can look at F.VIII molecules to tell whether they are conformationally different from plasma-derived F.VIII or not? Before doing that, let me take one short step backward, and let's talk about why it's important to be able to tell who will develop inhibitors. And I'd like to ask Dr. Hoyer to open the discussion with his thoughts on why we need to know who will develop inhibitors.

Dr. Hoyer: Thank you. From my perspective, the most important reason for understanding which patients are at risk for inhibitor development is to make possible the testing of interventions that might, in fact, prevent or modulate antibody formation. Of course, there may be differences in our ideas about who is at risk for inhibitor formation. I'd like to hear comments from people about how we can develop ways to recognize those individuals whose molecular defects are consistent with production of a sufficient amount of F.VIII protein that they are naturally tolerant. That may or may not be the same as having F.VIII protein in the circulation that can be identified by immunoassay. We have, I think, had an over-simplistic view that if you can measure F.VIII antigen, there is synthesis, and if you can't, there is no F.VIII protein produced. Some molecular defects may be associated with a failure of secretion or very rapid clearance from the circulation. So, we need to know which patients are at risk. Can we correlate that with the molecular defects? It is most important to know if there is a population that is not at risk because of their specific molecular defect. As we begin evaluation of intervention, it is only prudent to include only those patients that are at high risk.

Dr. White: Any other comments from any other panel members on that? Or from the audience?

Dr. Reisner: I'd just like to look at it in a slightly different way. We've heard in a series of magisterial talks this morning that there really are possible approaches, things we should think about. For example, perhaps starting in young adults, children who don't have an antibody. Perhaps intervention to try and downregulate some of the particular CD-defined antigens that are important in interaction. These are not going to be trivial exercises and before we want to go to even planning clinical trials of ways of preventing inhibitors, we really have to know who it is we are preventing inhibitors in. If, early intervention, if early treatment or exposure to F.VIII will indeed prevent or ameliorate the inhibitor response, this is going to be associated with cost and undoubtedly some risk. I think we have to know the population to properly target, for ethical and for very practical reasons.

Dr. Briët: We should not forget that even patients with mild hemophilia who have apparently normally functioning molecules in their circulation may get inhibitors, so the best you can hope for is to get a group with a higher risk. It is very unlikely that you can identify a group with no risk.

Dr. Hoyer: Doesn't that help us distinguish between two kinds of inhibitor populations?

Dr. Briët: Maybe.

Dr. Hoyer: ... One group is at the risk you would expect if they lack the protein and are infused with it. They then respond to it. Perhaps earlier, perhaps later in some cases. Individuals who are, in fact, reacting to a modified F.VIII protein are a different group, and are not protected by their previous exposure to endogenous or exogenous F.VIII.

Dr. Briët: I agree with that.

Dr. White: Are there immunologic ways to distinguish between those two? Do we have the immunologic methods to distinguish between those two?

Dr. Kazatchkine: I don't see any easy method or approach to that but I'd like to followup on what I guess was underlying Dr. Hoyer's comments and was stated earlier this morning by Dr. Goodnow. I think rather than focusing on the antigen here, since all people–those who will develop and those who will not develop an inhibitor–are somehow exposed to the same sources of antigens. I guess maybe we should focus our attention more than we have done before on why are nonresponders nonresponders. Isn't it really that, indeed, those individuals become tolerant as they are infused with F.VIII? Then identify immunological parameters of induced tolerance and then see whether these are missing in some patients that thereby could be identified as being at risk. There we can think of various ways of looking at precursor cells reacting with some F.VIII, with F.VIII epitopes, or the presence of various types of antibodies or anti-antibodies circulating in the patients. What I mean is better characterizing the immunological studies of the nonresponder individuals.

Dr. Preston: I feel that whereas this is interesting, I'm not convinced it will be clinically useful. It will certainly generate an enormous amount of data. But I'm not at all convinced that at the end of that day we will be able, in absolute terms, to identify those individuals who are at risk. Although we may be able to identify a small subset. I've been very impressed, in the context of inhibitor formation, by the reports from Belgium and the Netherlands about the relationship between the introduction of a new concentrate and the appearance of inhibitors. I wonder whether we ought to pay more attention to the concentrates rather than trying to identify all patients at risk. I think we should redirect our focus.

Dr. Gilles: Yes, but I think that all patients should be considered at a high risk. And, all patients, including the patient who has no inhibitor at the beginning of the treatment should be considered at risk. Look at what has effectively happened in Belgium and in Holland. The modification in the F.VIII structure could be a very small modification only in the presentation of one or two epitopes. It involved a very high immune response in some of these patients. So, effectively, I think that the F.VIII molecule in the different products actually on the market should be very carefully monitored. It's important to have controls in the production of F.VIII in an attempt to avoid the type of problem we have in Holland and Belgium.

Dr. Hoyer: We should certainly try to avoid problems like those that developed in the Netherlands and Belgium. Do you have any suggestions as to how that outcome could have been predicted?

Dr. Gilles: Yes, of course. If we are looking at the immune response of all of these patients, we see that there is very specific immune response against one or two domains of the F.VIII molecule. So if we are able to produce and to have assays which are not only following the functional activity of the F.VIII molecule but also from the complete structure and its environment, which is also very important, we are able to have, at the end of the process, quality control. So we may be able to avoid such a problem. But, actually, it is only when we are looking at the control for the production of this F.VIII we realize some very crude control, a functional assay that is our chromogenic assay. And so that's very important to have a series of other assays. If we are able to detect such possible modification–that could be done by ELISA, but we can also have a method perhaps using the mouse SCID model. By injecting the serum products which have been made, we can control if there is modification in the immune response or not.

Dr. Jenkins: I'd like to say that, in fact, it may be difficult to identify any predisposing factor that will correlate very well. I think you could build sort of a stochastic model based on what the status of your immune response is at the time you get your first F.VIII infusion and that could be controlled by many different variables that would be difficult to identify. So it could be that the 30 percent that get primed are unfortunate enough to have an ongoing inflammatory reaction at the time that they get their F.VIII. So if you look for some genetic basis for that, you'll never find it.

Dr. White: Because it's too diverse...there are just too many variables?

Dr. Jenkins: It's just bad luck. (laughter)

Dr. Monroe: I'm interested in the question as to whether or not there is now in products that are formed recombinantly and pasteurized, treated differently than they have been in the past, is there essentially a different antigen that people have been getting. That's a theoretical question. It seems to me that there are a couple of concrete things that can come out of it. One is, has there been an increase in the incidence of inhibitor formation in the last three years which was when this product would have become widespread? You have patients who have transitioned from one to the other. Are there more inhibitors? Can we tell from that from the data base? The other thing would be, not looking at individual patients, but just statistically, in the next five years are we more likely to see inhibitors in patients than we were ten years ago? Whereas there was a stable incidence of 20 percent to 30 percent, is it going to be 30 percent to 45 percent over the next ten years?

Dr. White: Let me just interject so that everybody is on the same wavelength here. I suspect that some of the immunologists are not aware, perhaps completely, of the issue here. Again, we have talked about hemophiliacs and once they've had perhaps 250 exposure days to F.VIII they are at very low likelihood to develop inhibitors. But there are patients who receive pasteurized concentrate, both in the Netherlands and in Belgium, a new concentrate, who had been previously exposed to a lot of F.VIII who began to develop inhibitors [on treatment with the new concentrate]. Not all of them, not a certain percent of them, but the idea was that none of them probably should have developed inhibitors. There was a concern that these new F.VIII concentrates were somehow altered. Ernest [Briët] or Jean Guy [Gilles], would you like to respond to Dr. Monroe's question?

Dr. Briët: I think that the data base, as I have seen it, does not enable us to answer your question. Is it true that switching from one product to another increases the likelihood of getting inhibitors later on? With the exception of the experience we had in the Netherlands and in Belgium, there are no reports supporting this idea. For that reason, in the Netherlands we have set up a nationwide registry of side effects of hemophilia treatment after we had experienced this small epidemic. In that way we hope to correlate the use of product and the types of product and hope to correlate eventual changes in product to the occurrence of side effects of inhibitors and also other things. So I don't believe that the data we have now are good enough to answer the question, but I think that if other countries also are going to make registries like that it might be possible in the future to monitor these events.

Dr. White: Jean Guy [Gilles] you have looked at some of the epitopes, at least in the patients who developed those antibodies, you looked at the epitope specificity of some of the antibodies in those patients. Were there differences from other patients with inhibitors?

Dr. Gilles: Yes, the immune response of these patients was completely different from the immune response of patients who developed inhibitor [to other products] or had a long-standing inhibitor. We have studied two different groups, the first one with long-standing inhibitor. We have seen that the immune response was directed against functional epitopes and also against nonfunctional epitopes. And they were from the four different IgG subclasses with a high level of IgG4 and IgG1 as described in the research. But when we looked at the other immune group, we see that the immune response is restricted to one or two regions of the F.VIII molecule and strictly restricted to this region. Also, the IgG subclasses were completely different because we do not see any increase of the IgG4 subclass. And, the third thing, the affinity of this antibody is probably different and we have also measured if it was a type 1 or type 2 inhibitor and all of them were type 2 inhibitors. So that was a very classical immune response against a normal antigen, if we can say that. What we do not forget is that when we switch from one product to another it involves a better immune response. These patients were injected with pasteurized F.VIII for more than one year without any immune response problem. This problem appeared very suddenly. And so it is probably a problem with the production of this product and not modification of the immune response. Because if you inject a patient or an animal, you do not observe an immune response one year or two years after the first injection, of course. In all of these patients the increase of the inhibitor appears approximately in the same period, in two or three months. So it's indicated very well that it's probably the production of this product which is involved, not necessarily the method of this production.

Dr. Montgomery: I understand if you denature the F.VIII protein and look at it, there would be no difference in reactivity whether it was against native F.VIII or the F.VIII in the product. Do you know if there are confirmational differences? Could these antibodies be, in fact, in any way used to describe an immune alteration of the product?

Dr. Gilles: Unfortunately, we were not able to have a sufficient amount of product to clearly analyze the product. What we can do is only to analyze the immune response of these patients by using normal recombinant F.VIII or plasma-derived F.VIII. But we were not able to use it in assay, by immunoassay, or by using the classical assay. We have this F.VIII. We were only able to analyze this immune response of a normal F.VIII. But we see that the immune response a large amount of this antibody from these patients were directed against two regions of F.VIII.

The pattern is completely different from the other inhibitor patients.

Dr. Scandella: I'd like to present just a slightly different perspective relative to the epitope specificity. I agree that in the one case of the heat-treated F.VIII there were problems that look like they might be different. But we've looked at patients that have been treated with a variety of different F.VIII products. The 28 patients that I talked about the other day have been treated with everything from cryoprecipitate up to the more intermediate purity products. We've also looked at a group of 20 hemophiliacs treated only with recombinant F.VIII. And, surprisingly, we tend to get very similar epitope specificity. The functional inhibitors also seem to be very similar. I think it's even more interesting and more relevant probably that spontaneous inhibitors are also the same and they've never seen any foreign F.VIII. Their F.VIII therefore cannot be affected by how it's purified. So I'd just like to raise a point which represents the experience that I've had, even though there is one isolated case where there may be differences. We also plan to look at epitope specificity in some of these hemophiliacs that developed inhibitors after the heat-treated F.VIII product was given.

Dr. Aledort: I'd like to not let the immunologists off the hook here. We've been listening for a couple of days to some very exciting things. We're learning about surface receptors and T cells and B cells, and their interactions, and B7 and CD28, and we've got a lot of models of inhibitor development. It's not just hemophiliacs who are on new products. We've got twins and brother pairs where one has an inhibitor and one doesn't. We've got animals that when infused with either F.VIII or F.IX develop inhibitors to human F.VIII. And I'd like to hear some of the immunologists tell us with those human and animal models what can we look at in terms of B and T cells rather than the amount of energy that we've now put on the antibody itself. The final antibody seems pretty similar from what Dorothea [Scandella] tells us all the time. But not only will it help us understand who might get it, it might help us understand how we could block some of it and I'd like to hear them help us understand what we could do or do collaboratively in this area.

Dr. White: Maybe we can get a response from David [Scott] or Chris [Goodnow]?

Dr. Aronson: I just want to make one comment coming back to the production control. The only model I think has any substance at all is one developed by Mike Fournel at Cutter, transgenic mouse with a human AHF gene and I think that would be a wonderful model to look at the Dutch material. I think that is the only way, before you go into clinical trials, but I think constant surveillance is absolutely the final answer.

Dr. White: I think that brings up a good point. There is also a newborn mouse model at Genetics Institute. I think either one of those would be a good or complementary technique for differences in F.VIII.

Dr. Goodnow: I guess that one thing that might be interesting to do would be whether to see if you could compare individuals who have not developed inhibitors with those who have by some of these new methods to really sample what cells are present in the circulation. Some of the things that Paolo Cassali has done where you can activate polyclonally all the B cells [at limiting dilution] and then enumerate how many cells are anti-F.VIII. And if you do the same kind of thing in T cells you might get some sort of baseline information as to whether in the tolerant individuals you actually are eliminating the autoreactive cells. But those experiments are so difficult when you're dealing with a polyclonal situation. Even in experimental inbred mice it's been very hard to, once you get down to very low frequency cells, to do much cellular immunology.

Dr. Scott: I wonder if I can reserve my right to not appear on the next panel if I speak now. (laughter) I think you have to not ignore some of the ethical issues before we start coming up with some basic scientists jumping into the field. When you are dealing with the issues, where as I understand it, you cannot a priori predict who is or who will or will not form inhibitors. If you have a high enough frequency, and I think 20 percent to 30 percent is a high enough frequency, and you start early enough intervention, that combined with some of the things we've heard today from Chris, from Mark, and some of the approaches we are using, should lead to a significant change in the response pattern and you should be able to get significant successes. While I do feel that the human neonate is immunologically competent, you are certainly better off starting as early as possible in those intervention therapies. But there is an ethical question in terms of whether you are going to treat everybody. I probably would lean on the side that you've got a 20 percent chance that it's going to happen, you're going to be giving F.VIII to patients in one form or another, do it in a way that is going to facilitate tolerance production and facilitate the maintenance of tolerance.

Dr. Hoyer: I had a question for either David [Scott] or Chris [Goodnow]. Is there any *in vitro* way that you can demonstrate tolerance to a given antigen using the peripheral blood of a patient or an experimental animal? If you could, you could determine if a normal newborn is tolerant of F.VIII. The patient who is not at risk of inhibitor formation, because he is making a small amount of F.VIII, ought to be like the normal and the hemophiliac at risk ought to be different.

Dr. Scott: Chris [Goodnow] should elaborate on this, but basically there's some evidence for looking at precursor frequencies and reading out whether there are cells present that are making high-affinity responses. As Michel [Kazatchkine] said there are lots of autoantibodies, autoreactive cells present T cells and B cells that you can detect. It's the ones that potentially have deleterious effects which will be of higher affinity, so not only do you need to be able to polyclonally stimulate them and bypass immunoglobulin and T-cell receptor signaling but you are going to have to be able to read them out and discern which are high affinity and which are low affinity and which are relevant potentially inhibitory responsive cells. And if you can do that in a polyclonal system, then you might be able to discern who is already tolerant, but I think that may not be sufficient.

Dr. Goodnow: I think David's point about the low-affinity cells really plague that kind of analysis. But it has been done successfully. The one example I am thinking of is a group has done it with human peripheral blood for the ABO blood group antigens showing that if you do limiting dilution repertory analysis activating and sifting through thousands of individual wells of 96 well plates, you can show that a person who lacks blood group A has, say, one cell in 10,000 that can make some reasonable affinity blood group A specific antibody. Whereas someone that's type A positive, it's down by a factor of 10. So you could do something like that. But it's incredibly laborious and probably what you really want to do more than anything else is look at the T-cell repertoire, and I'm not really sure how much progress has been made really at being able to sample the T-cell repertoire in unimmunized individuals.

Dr. Kazatchkine: There is contradictory evidence to this. A recent paper in the European Journal of Immunology by Zubler from Geneva showed there is an equal frequency of anti-A and anti-B cells in A and B blood group individuals. And this comes back to the issue that Dr. Scandella was raising. I am not too surprised that there is no difference in epitopic specificity between alloantibodies and autoantibodies or those that are found in healthy individual's spontaneous inhibitors and acquired inhibitors. As we've heard from Dr. Pascual yesterday, sequencing variable regions of autoantibodies in diseased individuals and healthy individuals has not allowed us to differentiate between pathogenic and nonpathogenic clones. So there are things that we do not yet understand that will differentiate the healthy individual who will be able regulate clonal extension of self-reactive clones and the individual who will have uncontrolled expansion either of self-reactive clones or uncontrolled expansion of alloreactive clones. My point is that we should not necessarily focus all of our attention on eliminating precursor cells, but perhaps blocking the response at the periphery by better understanding how the healthy individual blocks his own natural autoantibody responses.

Dr. Goodnow: I think though that paper in your journal of immunology is a good example of how one could potentially proceed for the B-cell repertoire to at least try that approach for F.VIII. Because the key thing they used was discriminating between the cells that were specific only for blood group A, which did disappear in the people where that was expressed as self versus reactive with everything even the plastic dishes which are the other ones [natural polyspecific reactivity]. The other thing that I thought would perhaps be useful for the T-cell analysis would be to choose patients where the genetic defect is understood in detail, where you know that, say, the C terminal half of the protein isn't made at all and so they should definitely not be tolerant to those T-cell epitopes. Then, at least that way, one should be able to have essentially a positive control where you should be able to detect T cells that can see those epitopes in the context of their MHC.

Unidentified speaker: When you suggest it might be useful to establish a T-cell repertoire what does it actually mean? Are you talking about T-cell subsets, T-cell function? Can you tell us exactly what you're recommending?

Dr. Goodnow: It may be totally impractical. The classic example would be to do what several groups have done in a particular inbred mouse strain. They take, say, myelin basic protein. For example, Lee Hood's group did this in a real tour de force that no one else would ever attempt. They make synthetic peptides staggered one amino acid apart, marching from one end of the protein to the other, sampling potential T-cell epitopes. Then what they show is that in a mouse that lacks the capacity to make myelin basic protein because it's got a genetic defect, that there is a particular peptide somewhere in the C terminal half that is able to be recognized by a certain fraction of T cells when the T cells are exposed to that peptide and appropriate presenting cells *in vitro,* and they can just enumerate that either in bulk assays or potentially, although probably Marc [Jenkins] has better experience with this, or potentially by plating the cells out at limiting solution for individual wells that give a response. But the trouble is, of course, that's one MHC and one protein; and when one is dealing with an outbred population of humans, the peptides that are going to pop up are going to depend entirely on their HLA DR and DQ and it will be a different pattern for each one. So you are only talking about a tiny fraction of susceptible patients, but it might still at least elucidate some general principles if one focused on the most frequent DR types.

Dr. White: I think if Dr. Hood does that in his lab, maybe if everybody in the room worked on that problem we could find a solution. (laughter)

Unidentified speaker: I don't like to contribute to this area but wish to finish this type of discussion and otherwise proceed with the more practical issue.

Dr. Jenkins: Let me add one point to that. If you look at a lot of fields where autoimmunity is being looked at, the field will progress from an intensive study of what antibodies are there and where they bind the autoantigen only to find out that the actual pathogenesis is often the T cells. I think that needs to be looked at. Having said that, doing that is extremely difficult. We don't have assays even in mice or whatever to look at, in a reproducible way, the precursor frequency of reactive cells before we immunize. And so even if you have the potential to respond, we can't read that out unless we prime and so in a person we are not going to have a positive control. The other point about even thinking about this is, I think it is critical to identify people who either are or are not making any of the protein. I think if you've got people who are making even low levels, they could be clonally deleting all of their reactive lymphocytes and these people will never have this inhibitor problem. The molecular basis of the defect needs to be identified. The amount of protein that is there needs to be found. The people that you have to worry about for peripheral tolerance are the people who have none. And that protein is going to come in, and it's going to be just as foreign as KLH. I think those are important points.

Dr. Kay: As a geneticist, I'd like to address the immunologists in the following way: Obviously the process of inhibitors involves many genes, it's a polygenic pattern. If there are good rodent models to study inhibitors, one could use DNA linkage. Would that be profitable in trying to distinguish the genes that are actually responsible for making the inhibitors? And, if they lie in the immune system, then one could ask more directed questions about immunologic function. This approach has been taken for polygenic inheritance of hypertension and some of the autoimmune disorders.

Dr. White: Is that on the same topic?

Dr. Gilles: There is maybe one thing that we should not forget, speaking of the decreasing of the production of the inhibitor. Everyone is looking after the production of the inhibitor with a different assay. What we have looked at in the characterization of the immune responsiveness of our group that still presents the anti-idiotype antibody and we are able to have found in the normal number of groups. In all the donors we found a large amount of inhibitor, also against functional epitope and nonfunctional epitope. Most of them are able to inhibit chromogenic assay, about 70 percent of these patients are going to inhibit this chromogenic assay when you purify this anti-F.VIII antibody. In the circulation you do not detect in the plasma or in the serum, you do not detect any activity or very low activity because they are probably neutralized by anti-idiotypic antibody. We have assay that are able to detect the presence of this anti-idiotypic antibody. There is probably a very important neutralization and equilibrium between these two type of antibodies. Secondly, we have also studied patients who have been treated and where the treatment succeeded. We are able to demonstrate that after the treatment the quantity of anti-idiotype antibodies were completely increased in all the patients where the treatment succeeded. We are also able to find, again after separation of this complex (anti-idiotype-anti-idiotype), an anti-F.VIII activity. So, it seems indicated that when there is treatment and you have succeeded you do not decrease the production of IgG but you have increased the production of anti-idiotype antibody. And it is a completely different way, but I think we don't do things only to decrease production of IgG because in all patients there is IgG against F.VIII, but we have to look for a method that is able to increase the production of this anti-idiotype antibody and that is a completely different way than that we are seeing.

Unidentified speaker: I would like to discuss how much damage we do to the molecule if we manufacture and purify it. I think Dr. Hoyer has demonstrated in his presentation very well two extremes. On the one side, the incidence experienced by the use of cryoprecipitate and, on the other hand, the recent paper from Ehrenfuhrt with a rather high incidence of inhibitors, and please don't forget these were mainly pasteurized products. As we have heard there are pasteurized products made under certain conditions, as the Dutch-Belgian experience shows, that introduce quite a high rate of inhibitors. So, what I would like to express is rather than categorize concentrates into intermediate, high, and ultra-high purified preparations, it is better to look to the manufacturing method itself. Because there might be the point where we damage the molecule, where we unfolded and the refolding might not be as good as assumed. And it is old immunological experience which states now more than one hundred years old that if we heat antigens they might become more antigenic.

Tape recording of panel ends here.

Panel discussion on the treatment of inhibitors: Today and tomorrow

Editors' note: We were not able to identify all the participants in the discussion. Those quoted had an opportunity to review and edit their comments to insure accuracy.

Chairman Margaret W. Hilgartner, M.D.
New York Hospital

Panelists Louis M. Aledort, M.D.
Mt. Sinai Medical Center

H.H. Brackmann, M.D.
Universität Bonn

Carol K. Kasper, M.D.
University of Southern California

Jeanne M. Lusher, M.D.
Wayne State University School of Medicine

Frederick R. Rickles, M.D.
The Centers for Disease Control and Prevention

David W. Scott, Ph.D.
Holland Lab, American Red Cross Blood Services

Participants Edward Alderman, Ph.D.
Genetics Institute

Joan Gill, M.D.
Great Lakes Hemophilia Foundation

Jean Guy G. Gilles, Ph.D.
International Institute of Cellular and Molecular Pathology

Marc Jenkins, Ph.D.
University of Minnesota

Michel Kazatchkine, M.D.
Hôpital Broussais, Paris

Guglielmo Mariani, M.D.
University of Rome

Robert Montgomery, M.D.
The Blood Center of Southeastern Wisconsin

Inga Marie Nilsson, M.D.
Malmö General Hospital

Francis E. Preston, M.D.
Royal Hallamshire Hospital

Howard M. Reisner, Ph.D.
The University of North Carolina at Chapel Hill

Gilbert C. White II, M.D.
The University of North Carolina at Chapel Hill

Dr. Hilgartner: ... I would hope that many of you in the audience will [in this session] step to the microphone and remind us of the things that we have heard in the last couple of days that you think may be applicable in devising our tolerance programs. I would hope that the immunologists will enter into the dialogue and say, "Yes, that's possible to translate what's been known in the mouse to the human, or it's not possible." So, I would open the floor then to anyone who feels like beginning the dialogue. What do you think we could add to our tolerance programs now that might indeed help us with inducing tolerance?

Dr. Jenkins: I think some animal models would be a reasonable thing to try to develop. I assume that people are trying to knock out some of these proteins in mice to get a model, and then experimentation can begin on what happens when you inject a protein into the knock out. I think that's a reasonable starting point. I think another key point is not only to have a rodent model where you can do very elegant work, but you need to start to think about large animal models because the reality is that a lot of the very elegant rodent transplantation tolerance protocols that have been developed do not succeed in larger animal models. So, a successful rodent protocol, I think, has to be then tested in a larger animal model like the dog or the pig. I think if it will work in a dog or a pig, then it's likely that it will work in a human. But, being a basic scientist, I think it would be critical to develop some kind of animal models to start looking at the immunology.

Dr. Hilgartner: What do we do with these models once we get them?

Dr. Jenkins: You need to know how much of these proteins are required to get central tolerance, to get clonal deletion of immature cells. Then you might get some feel for how much is it going to take for us to not worry about a subsequent immune response because all the clones are gone. If you had an animal model you could begin asking how do we make the protocol better without taking huge risks. I think that just makes sense to me. But again, the caveat is that you often have to do everything you did in a rodent in a larger animal model at huge expense. We do some pig work and, from Dr. Kay's talk, I sympathize with how difficult it is to work in a large animal model, but I think that that's unfortunately what has to happen.

Dr. Lusher: Considering immune tolerance, to take this a step back further, I was hoping that following the previous panel discussions we might have a better indication of how we might identify fetuses at risk. We would certainly anticipate that if they have a family history of other affected individuals who are CRM negative, perhaps known family members with gene deletions, perhaps even nonsense mutations or those with close relatives who have high-titer inhibitors, that those fetuses then would be at particular risk for developing an inhibitor, and perhaps we should start immune tolerance in utero. But the questions remaining–at least as a non-immunologist and perhaps our immunology colleagues can help us with this–how soon would we have to start giving F.VIII to the fetus? Assuming that venous access problems were taken care of in some way, what would be the ideal time to start it in order to prevent that individual at great risk from ever developing a high-titer inhibitor? Now, I think we know that starting it at birth is probably too late because I think we've all had experiences with children who were treated at birth or on the first day of life–for circumcision bleeding, or for whatever–and have still developed high-titer inhibitors after the next few exposures. So, at least at birth seems to be too late. How soon in utero should this be started, and how long would it have to be continued once one does institute it? If any of our immunology colleagues could be of some help with questions concerning timing, then we could focus on the practicalities of the access and so forth.

Dr. Reisner: Well, I think sometimes relatively simple-minded experiments really can yield information. For example, Dorothea [Scandella] showed beautifully that the epitopes are probably limited and that probably we know an awful lot about the actual immune response in terms of important epitopes already. We've shown, although certainly not beyond the possibility of error, that it looks like at least one relatively rare epitope–the response to that rare epitope–is clearly associated in a particular family and may be (and, certainly, it could be a chance observation) associated with the HLA complex. It seems to me that being able to very precisely map epitopes, as has been done in Dr. Hoyer's and other labs, combined with family studies and looking for some of the things that we have pretty good prior reason to believe may be associated with the immune response are probably worthwhile, particularly considering the tremendous sophistication in terms of the molecular analysis of these sites. It can be done today, it can be done at a relatively low cost. While I wouldn't predict whether we will be able to get any sort of tremendously valuable information out of it, we can at least ask the question and find out. I think the time is probably now to put together that type of a study and to do it.

Dr. Hilgartner: What do you think about Jeanne's [Lusher] suggestion that this process be begun before birth?

Dr. Reisner: I am coming back to what you said. I don't think you can undertake that until you can define the patient who is at high risk. There is a real ethical concern. You might be willing to undertake the experiment–even perhaps with in utero treatment, which is certainly possible–but I certainly wouldn't want to bring that up to my committee unless I could say that I have a good reason to believe that this patient will be at risk of a very, very debilitating condition. I think we really have to make our best attempt at trying to figure out who is at risk.

Dr. Lusher: For example, if you have a family where there is a fetus–known to be a male fetus, known to be affected, with a family history, say with two brothers who have gene deletions and have high-titer inhibitors. Okay, so we know that infant is at risk. Putting it in those terms, immunologically, when would it be optimal to start tolerance in utero?

Dr. Reisner: I think in humans as early as possible. I think as early as you can, in fact, expose to F.VIII technically. I think the only limitation would probably be the technical limitation of dealing with the fetus in being able to infuse. That would be my own opinion.

Dr. Lusher: How early is "as early as possible"?

Dr. Hilgartner: You know we do infuse via the cord for a number of products these days and that can begin in the last trimester.

Dr. Reisner: I certainly would say, for starters—if we are staying away from molecular types of intervention and we are talking about the more physical—I think infusion via cord in the last trimester is certainly something that should be contemplated given that you can identify the sort of case that you mentioned, someone who is at a very high prior probability of developing an inhibitor.

Dr. Lusher: And then, would one dose suffice, two doses, the entire last trimester? How would the immunologists advise this be done?

Dr. Alderman: Accepting that for the most part humans aren't mice, we did establish a model in Cambridge [Massachusetts], and in the mouse we found that 30 hours postpartum was too late. I don't know exactly how that relates to a human. We found again in the mouse that this single dose of 25 micrograms—and I don't know how that relates to units off the top of my head—was sufficient to induce a tolerant state that was relatively long-lived in the animal model that we used. However, we should point out that these animals still had their own factor VIII (F.VIII). They were not murine F.VIII knockouts. So we envision that there is probably some kind of a suppressive event that's going on after an initial tolerant event. I think the most important thing is to look for the antigen. If you can find any F.VIII in the fetal blood sample, there is a good probability, from our perspective, that you can induce tolerance successfully. If you don't find it, then it's time to intervene.

Dr. Hilgartner: Then you would suggest that in all of the newborn patients with hemophilia one of the first things we should do is to characterize the deficiency more and find out just exactly how much they do have in the way of F.VIII antigen present.

Dr. Alderman: I would say that's one of the most compelling things to do.

Dr. Lusher: But it's mainly the CRM negative ones who develop the inhibitors.

Dr. Alderman: In that case, move very rapidly for intervention.

Dr. Lusher: If you find them to be CRM positive, they are certainly not going to be a high risk group.

Dr. Scott: I agree with the notion of starting as early as possible—

avoiding the ethical questions–because I think Howard [Reisner] has raised the point that these are important ethical questions to be resolved. You need to identify the earliest point at which you have that potential to induce tolerance. You have to realize that the immune repertoire develops in a stepwise fashion. We don't have this magical moment in the third trimester when you can respond to every antigen. There are some antigens that you don't respond to until you are several years of age; other antigens you can respond to very early. There are some elegant studies by Silverstein in sheep that show this and these have been reproduced in mice; the same holds for the human repertoire. I don't know what the earliest time is at which you can respond to F.VIII but certainly, if it were ethically possible, you'd want to go into the second trimester. I think that you have to realize that tolerance, whether you're looking at T- or B-cell tolerance, can be long lasting but is not forever. Long lasting in a mouse is several months, probably equivalent to about 10 years in human life. Is it sufficient to give F.VIII and hope you don't have to come again for another five years, or should you try the genetic manipulations? Once we get around some of these ethical questions, if you had a way in which you could start relatively early, identify those at risk, do gene therapy–and I would personally lean towards putting in an amount of hematopoietic stem cell or you probably can't get it into the earliest stem cells, but certainly go towards the amount of hematopoietic lineage, get it so that the patient continues to produce it–you'll be killing two birds with one stone. But, that's theoretical.

Dr. Jenkins: I think there actually is some information, though. When Medawar was doing his experiments with neonatal tolerance, when he went from mice to some larger animal species that had a longer gestational period, he found that at birth what was okay in a mouse was not okay in some of these larger animals. In fact, he had to go much, much earlier to get tolerance to work. So, I think that that's a daunting thing.

Dr. Scott: Fetal sheep will reject allografts.

Dr. Jenkins: You don't want to be in the position where you are trying to induce peripheral tolerance in already mature cells. You want to induce central clonal deletion of the reactive cells. They're gone. To do that you are going to have to introduce this very early, and you are going to need a continuous source of the antigen or new clones that are produced later in life will not be exposed to the toleragens. I think that some sort of gene therapy approach very early on would be the ideal solution.

Dr. Hilgartner: Before we leave this question, I'd like to ask the immunologists one last question. If we go into this business of trying to delete certain clones, will we in any way interfere with the response that we expect and that we rely upon when we vaccinate all babies with the eight vaccines that we use now? Will we, in any way, interfere with those responses if we do go ahead and try to knock out a certain clone? There is no cross-reactivity? We would not interfere with the immunization process as a whole? What do you think?

Dr. Jenkins: There is absolutely no reason to believe that that should be a problem.

Dr. Scott: Can I just add one additional thing? When we realize we come up with a real-life situation which is beyond the theoretical ones we are talking about now–you have an individual who has formed an inhibitor or you even cannot intervene early on in life but you want to knock out the ability to form that inhibitor–Marc [Jenkins] has very elegantly discussed how one can modulate the CD4 response in a so-called TH1 type clone. In that system, understanding costimulation for these cells is absolutely critical. However, we should note that what we want to knock out are TH2 cells that are the helpers for an antibody response or for certain classes of antibody. I think we need to focus a little more on understanding exactly how that's done and how that may apply in terms of human T-cell subsets.

Dr. Aledort: One of the immunologists mentioned how important it might be to remove the bulk of antibody before starting a tolerance program in those people who have an antibody. And, if one looks at the data that's been presented in the last few days, those with high-titer antibodies need more F.VIII over longer periods of time than those with low antibody titers. If one looks at a cost-effective methodology, despite the costly staph A columns. Dr. Nilsson presented data to show that when you do remove antibody with these columns, the cost and time to achieve tolerance was markedly shortened. I wonder if the immunologists would comment: with our clinical experience–coupled with the data that was presented here in the animal model–should they not encourage us to be very cautious about initiating therapy when the titers are very high and that we'd be far better off removing those titers and push harder for us to be able to get this to happen.

Dr. Jenkins: I think it just comes down to doing it. I think your field is at the point where you have–even, as you point out, in people that have an incredibly high, ongoing immune response–you already have evidence that this can be modulated with an immune tolerance protocol. So the question is: Are you going to use a prophylactic sort of regimen to make sure that you never get to the immunized state? Seems to me the only way to do it is to do some kind of a double-blinded clinical trial. Since you already have a frequency of at least 30 percent of people are going to get these inhibitors, it seems to me that you could start early, agree on a protocol and give it a try and ask do you have a statistically significant effect on the outcome. You are doing exactly what Dresser did in 1960 in terms of giving a de-aggregated protein antigen without an adjuvant intravenously. In terms of inducing peripheral tolerance, I think this is the best protocol for doing it. So, you are already doing it in a very nasty clinical situation.

Dr. Lusher: The clinically significant inhibitors are not 30 percent. That's the overall [percent] including all the mild and transient ones.

Dr. Jenkins: I think your window is big enough though that you could probably design a study that could answer the question.

Dr. Aledort: I think the assumption that we are giving this pure F.VIII has to be modified by the fact that the F.VIII product has albumin in it, and may have aggregates and thus they're not pure single molecules of albumin. How they play a role in augmenting or making it more likely to develop inhibitors is really unknown. But we do not have, at the moment, PUP studies with a F.VIII that has no other material in it other than F.VIII. And we know that albumin when heated has many aggregates, and we have been shown many of those data at the Bureau of Biologics for years. And I don't know what the actual aggregate content is and whether that plays a role. I wonder if you might comment. We really don't have that pure system. People were given the impression at this meeting that we have this nice pure F.VIII molecule with no other substances in it.

Dr. Jenkins: And those would all be very vilifying, horrifying concerns except for the fact that you've already done it in the worst possible situation. What I'm saying is: if I walk into this room and you hadn't already tried it and you said well, we're going to give prophylactic immune tolerance protocols to people, I'd say you were absolutely insane without a large animal model. But you've already essentially done the experiment. Now, I think you have to ask yourself, and I didn't understand the data well enough to really know, whether you are convinced that it is immune tolerance as opposed to giving so much of the actual antigen that you essentially clear all of the antibody out of the system.

Dr. Aledort: Most of the time they don't come back, in the Brackmann experience, in the Malmö experience. But there are cases where what you are saying is correct and they do disappear. But we have actually–a group of us in the audience–have proposed–although not at birth and not in utero but within a very short time after birth being in the first year or so–putting a program together of prophylaxis for different reasons, on demand with intermittent therapy, versus regular therapy for five or six years. [The program] is going to look at prospectively the likelihood of developing inhibitors. So, we have already asked that question. Whether or not we get funded is another question.

Dr. Lusher: Already we have data, of course, from our Swedish colleagues who have been using prophylaxis starting at a very early age (1 to 2 years of age) and, in fact, have a very low prevalence of inhibitors. So, that does not seem to be a problem.

Editors' note: A portion of the discussion is missing.

Dr. Scott: The Malmö protocol does involve a number of different things. It's sort of what I call a "bulldozer effect" but it works. In terms of doing studies to figure out how we can incrementally improve it, I think it's important to understand it. With regard to the IVIg, I don't know how that's acting in this particular situation/combination. I was going to simply point out, however, that there's a well-recorded phenomenon that occurs in experimental animals. Elegant work was done, I think over 20 years ago, by Jonathan Uhr and Jean-Claude Bystren. They essentially plasmapheresed rabbits that had been immunized against one or more antigens and specifically removed one of the antibodies on a column. They got beautiful depletion of that antibody and then they got the de novo synthesis of new antibody that came back. So, at some time, there is going to be a rebound. You are coming in with a high-dose tolerance regimen and Cyclophosphamide and really socking it to them. You're essentially creating a situation that would be an excellent situation. I think if you took the Cyclophosphamide out of there it wouldn't work; that's just my theoretical, quick look at what I have just learned about. I do think it's important to think about the way in which the IVIg protocol can be improved from the point of view of an anti-idiotype network. I think that's where, perhaps, I would suggest taking a good look at those patients who already are inhibitor positive and making some rational decisions about pooled sources of known idiotype and anti-idiotype components and doing a good study on that.

Dr. Lusher: If I might just comment on your statement that if you took the Cyclophosphamide out of the Malmö regime that it probably wouldn't work. I agree that the results with the Malmö regime are certainly superior to those of others in terms of the degree and rapidity of response, but a number of other protocols which do not employ Cyclophosphamide have worked very well too. In fact, the reason I asked Dr. Mariani about the age of the patients in his immune tolerance registry yesterday is that our experience in young children put on immune tolerance with F.VIII alone has been very good. Many of the inhibitor concentrations have come down to nondetectable levels within two to four weeks. Several of these children are now back on episodic (as necessary) treatment with F.VIII and have not had a recurrence of their inhibitor. So, I think there are a lot of things that we still don't fully understand about individuals' responses to immune tolerance regimens and age-related phenomena.

Dr. Scott: I was simply trying to analyze the very elegant data from Sweden through the retro-spectroscope and, from the point of view of if you are going to remove the antibody there is going to be a rebound. We don't have to discuss the reasons but there's going to be some kind of a rebound if you do nothing else. And, under those conditions, the Cyclophosphamide-induced tolerance model is an elegant one. You can certainly induce high-dose tolerance with no drugs, there's no doubt about it. We've all done it.

Dr. Lusher: Especially when we are dealing with infants and young children. Many of the inhibitors make their first appearance in the first few years of life, and many of us are a little hesitant to use Cyclophosphamide in children without malignant disease.

Dr. Kasper: In this regard, I'd like to ask the immunologists to comment more on the use of immunosuppressive drugs during induction of immune tolerance. We heard earlier that cyclosporine might be counterproductive. Many pediatric hematologists are wary of Cyclophosphamide. Dr. Kazatchkine has said that we need to improve the health of the immune system rather than smash it. What do the immunologists suggest about the use of immunosuppressive medications such as prednisone, Cyclophosphamide and so on?

Dr. Scott: There are certain situations where I think it's indicated. As a general approach, I would like to avoid it. But I think in the situation where you're going to get a rebound effect, I think you may want to think about employing that as a part of the protocol. Certainly it's something that you can experimentally verify whether it is or is not necessary in the appropriate animal model. We are talking about doing gene therapy at midgestation human infants with a certain risk factor. There are big ethical issues here.

Dr. Kasper: What about right now–tomorrow–when I go back to work and I have an inhibitor patient who's 3 years old, 5 years old, 10 years old. Is there a rationale for using prednisone, for using Cyclophosphamide, for using any other drug together with the F.VIII , perhaps together with intravenous gammaglobulin?

Dr. Scott: If I was going to do anything, I would rather use a cytokine. I mean I would generally try to avoid immunosuppressive drugs. I don't think they are going to be very helpful but it's part of a protocol that works. So, you have to go with what you've already done which, as Marc has so elegantly and candidly said, you've already done it. And, if you came to us 10 years ago and said we're going to do this we would have been saying "don't get me involved in it, I don't want to be involved in doing that to patients."

Dr. Hilgartner: Marc [Jenkins], do you want to add to that?

Dr. Jenkins: Briefly. I think it's worth at least playing with the idea of an overlay of immunosuppression at the time that you give the antigen. I think with a drug like Cyclophosphamide you could actually kill the clones that start to respond to the antigen. And, I think that that's a toxic drug so the question is: can you get some new generation drugs that might help you do the same thing? As I pointed out, rapamycin is a drug, for instance, that I think has some potential. The problem with a lot of people trying to use the overlay of immunosuppression with an antigen-specific therapy is often times the drug they're using will suppress the antigen receptor from signaling and, of course, without that happening you're not going to get specific tolerance. So, I think that those things need to be balanced. For instance, in mice, you can take small, resting B cells, and—it's been reported by a number of groups—infuse them into an allogeneic animal and get transplantation tolerance. Do that same thing in a pig and you get priming and hyperacute rejection. If you give the small, resting B cells but now under the cover of rapamycin, you can actually get some benefit later. So, I think this is not going to be easy and a fairly blunt instrument is going to be required until we understand better what's going on. The other thing is there is, a lot of interest in CTLA4-Ig *in vivo* use to interrupt the CD28 - B7 interaction. This is the perfect situation to try something like that, it's being tried in a lot of other situations. Again, getting enough of it into a person to get a biological effect will not be cheap.

Dr. Hilgartner: Do the other immunologists have something to add before we move on?

Dr. Rickles: I think one point that needs to be stressed one more time, is that nothing is going to do any good unless it can get to the cells. So, it's pretty clear that you have got to get the inhibitor level down in order to get anything to work. That's probably why the high zone tolerance is, or at least from my perspective, may have a better chance of working or may give a better efficacy on very high-level inhibitor patients. However, I would also tend to believe that any prophylactic measures that one takes along the way would also promote a more tolerant phenotype in the individual. Finally, if you are going to look at how much material you need to use, I know that the Bethesda units tell you more about the function but I can't stress strongly enough you need to know how much antigen to deliver. The antibodies don't care whether it's accurate or not, they'll just bind to what's there.

Unidentified speaker: What about the size of the antigen? Would it be more efficacious to use just the epitopes that are involved – small, recombinant epitopes?

Dr. Rickles: It may be more rapid and it may be more cost-effective, but overall the major point is that you've got antibodies to the entire structure, not just to the particular epitope.

Dr. Hilgartner: Dr. Kazatchkine, do you have something to add?

Dr. Kazatchkine: A few brief points. One is that I would be skeptical about the use of Cyclophosphamide. In order to induce tolerance in experimental animals, one should use Cyclophosphamide at specific amounts and at very specific times of administration versus that of the antigen. With regard to the point David [Scott] made about Cyclophosphamide preventing rebound phenomenon after plasmapheresis, . . . there is an ongoing trial in Kiel, Germany, in patients with severe lupus who are plasmaphersed and then, at the time of rebound, either receive Cyclophosphamide or receive IVIg. IVIg may be, in their experience, as effective as Cyclophosphamide in preventing rebound phenomenon after plasmapheresis. So maybe that's one of the ways by which IVIg is effective in the Malmö protocol. The third point relates to the general discussion. I hope the people in the audience are not too confused about the different pathways of tolerance that we are talking about. The discussion should at one point focus on ways of inducing T-cell or B-cell unresponsiveness, e.g., by blocking CD28-B7 interactions. However, one should also keep in mind, as discussed in the previous panel and as stated by Jean Guy Gilles that we do have clinical experience with induction of unresponsiveness. You have rendered patients unresponsive and those patients who were tolerized are patients who have not deleted their T or B cells specific for F.VIII. They're making anti-F.VIII antibodies. However, they're also making anti anti-F.VIII antibodies or regulatory/suppressive T cells. These are two different ways of looking at unresponsiveness and tolerance. I hope the discussion is not too confusing in this respect.

Dr. Gill: I would like to make some comments about Cyclophosphamide. I think there is some old literature that suggests that giving a dose of F.VIII to a patient who already has an established inhibitor and then treating with Cyclophosphamide has not been successful. Cyclophosphamide has not been given in the context of continued infusions of F.VIII, and that may be the difference in the Malmö protocol. Therefore, some of those experiments have already been done.

I also wanted to put in a plug for a protocol that Donna DiMichele is currently developing to be dispersed through the Hemophilia Research Society; we would welcome any other investigators who are interested in participating. We've taken the Malmö protocol and tried to dissect some of the components. We had to eliminate the removal of antibody because we don't have the antibody columns in this country. However, the protocol will include high dose F.VIII infusions plus intravenous gammaglobulin and randomization to receive Cyclophosphamide or not. This will address the question of whether or not Cytoxan is needed.

Dr. Hilgartner: The protocol is listed in the back of your booklet under the abstracts for those of you who may want to look at it a little bit more closely. Thank you, Joan.

Dr. Montgomery: We've been focusing on dealing with infusion of F.VIII but yet we're about to embark on an era of gene therapy. I'd like to hear some comments from the immunologists about the issue of going from recognition of an injected material versus dealing with a cell that is now making that same protein and also to make a case that since the RFA for gene therapy in hemophilia is going to be released in December [1993] to make sure that as we develop animal models to look at gene therapy that we also not lose the opportunity of studying the immune response to that gene therapy as well.

Unidentified speaker: What I am worried about is that you are only talking about the high-dose immune protocols regarding the highest inhibitor levels because I think there is certainly a group of patients that can also be treated at a much lower dose of immune-tolerance protocols as we have shown in our results in our group of Dutch patients. I think you should try to spread out the dose to groups of patients to go the cheapest way.

Dr. Mariani: I am a little bit concerned because everyone is speaking of tolerance as if we were able to give these patients just F.VIII. Whereas the clinical experience that dates from the mid-1970s demonstrated that we have been giving to these patients concentrates in which F.VIII is contained. It's a rare protein. It is contained in 1 to 10,000 of the total protein content. So, are we sure that fibronectin, fibrinogen, immunoglobulins that have been transfused into these patients did not play a role in inducing tolerance? Are we sure that immune tolerance, the way we have got it, is a specific stage?

Dr. Lusher: On the other hand, the more recent experience in children in the United States, and probably in other countries as well, has been to attempt to induce immune tolerance with a high-purity F.VIII, either recombinant F.VIII or monoclonal antibody purified F.VIII alone. Even though these products do contain albumin, they do not contain fibronectin or other plasma proteins, and they seem to be quite effective.

Dr. Hilgartner: That's perfectly true. But we have to remember that most of the patients in Sweden received an intermediate type product, intermediate- or low-purity product and that may have, in some way, made a difference. I don't think we have an answer to the question.

Dr. Lusher: However, it seems that both types are effective.

Dr. Hilgartner: Yes.

Dr. White: Just a partial answer to that. The several patients that we've done, when they have finished, we've looked at hepatitis antibodies or blood group antibodies, HIV antibodies and tetanus antibodies. That is certainly not a complete list. But the levels of those antibodies do not appear to change during the production of tolerance. I think that says there's some specificity to it. It' s certainly not a blunting of the entire immune response.

Dr. Hilgartner: We have just a few more minutes left before we have to wind up. Is there another subject we'd like to move on to?

Dr. Kasper: Could I question the immunologists again? We are beginning to understand which group of patients with inhibitors are most likely to respond to induction of immune tolerance. There are several criteria that have evolved with all of our experience, with Dr. Mariani's registry and so on. We have a few patients whose inhibitors do not respond – do not come down with protocols that are successful in the majority of patients. We have some patients whose inhibitors come down and then rebound after many months. They may be down to one unit or so–I've heard about some eight cases in the United States during this meeting–and then they zoom up again in spite of no change in the protocol. These are failures either from rebound or from lack of decent response in the first place. If we have these failure patients, what kind of approach might one use with them?

Dr. Brackmann: I think this may be due to the dose itself. We have never seen those patients in our population, and we have finished now [with] 44 patients, and we have never seen a rebound in those patients. So, I think it may be dependent on the dosage itself.

Dr. Hilgartner: I'm not so sure since I have one of those patients who has received enormous doses at the moment. But, for the sake of discussion, let's leave your question there.

Dr. Kasper: But I think Dr. Nilsson has seen failures. Is that true?

Dr. Nilsson: In one of our patients with hemophilia B, the inhibitor reappeared after 6 months. In all the other patients the tolerant state appears stable. Our first treated patient has now been tolerant for 11 years.

Dr. Kasper: But were there some, Dr. Nilsson, who did not respond to the tolerance protocol in the first place?

Dr. Nilsson: Altogether 19 patients have been treated using the Malmö protocol, and of these patients 3 did not respond.

Dr. Hilgartner: So there is a group of patients, obviously, that do not respond or do have a rebound that is not responsive after their protocol is finished. We still have not answered the question as to how you can help all of these patients who do have an inhibitor.

Dr. Lusher: But, thank goodness, we do have some other therapeutic options for treating or preventing bleeding episodes in those patients with "high responder" inhibitors. We have FEIBA and recombinant VIIa and porcine F.VIII. I think we're fortunate to at least have several arrows in the quiver of our therapeutic armamentarium for those who don't respond to immune tolerance.

Speaker unidentified: I guess what concerns me is that, and I have said this before, every new inhibitor is a disaster. How to prevent these things and how to do it in a way with the materials that we have off the shelf, that we have in our pharmacopoeia, and in things that won't be too difficult to get through human subjects so that we can think about strategies for trying to prevent more from coming. I guess some of the random thoughts that come to mind, in fact even in the discussion, about somehow interfering during ontogeny and enhancing the development of tolerance. Maybe the only practical thing that would get through a human subject committee would be to give mothers F.VIII concentrate during pregnancy or to do that as a study. The other thing that comes to mind is that it seems as though what we've learned here has a lot to do that the factors that are involved and has to do with the scheduling of factor, the age in which they are first given and perhaps also the purity of the factor. It may be that we ought to look at things, such as if we had to treat a baby in the newborn period, maybe we ought to say, now this is the time to just continue the treatment and not wait three months and that should be the time to begin the prophylaxis when we first had to treat. Would there be any value in, at the time of first treatment, giving intravenous gammaglobulin and keeping everybody from making a right turn as you pointed out? I don't know. But these are things perhaps that we can do off the shelf that we could think about.

Dr. Montgomery: One of the things that several of us were discussing last evening was the question of making a recombinant F.VIII without the von Willebrand factor binding sites that you didn't solid phase out and then to give proteolysed F.VIII in some way to a mother that in fact cannot bind to vWF. What we don't know is what it will bind to and obviously giving something like that to a mother will be a little tough to get through a human research review committee but it is something that might be tried as animal models.

Dr. Kasper: I'm full of questions for the immunologists. If I were going to look at those patients who have not responded well or have not responded to induction of immune tolerance, such as the notorious Chapel Hill nephew with the big gene deletion whose attempt at tolerization was done by our center, by Dr. Ewing and our group. His inhibitor didn't budge, but others who responded like a shot–fast. What would you look at? Would you look at the genotype, the fact that he has a big gene deletion? Would you look at something else? What would you look at between the successes and the not-so-good successes?

Dr. Preston: Just looking at the overall situation of immune tolerance, it seems to me that perhaps we've all got lucky, and we have no real idea who is going to respond and who is not. Some groups try to improve this by using agents such as Cytoxan, Prednisone and so on, but we're doing this against a background of ignorance. Rather than continuing with phenomenology, we should really try to understand exactly what is going on. In our patient groups, why is one patient a high responder, another a low responder? Why does the response to tolerization vary from patient to patient? I think that these are key issues and concerted, careful research, whether it is clinical or basic, needs to be carried out. We certainly don't want to harm our patients and the issue of tolerization needs to be addressed and watched with a lot more care.

Dr. Hilgartner: A final comment?

Dr. Gilles: To see if the treatment has succeeded or to have an idea of the future of the application of your treatment, you can follow the production of the quantity of the anti-idiotype antibody. It's very clear that if you have a sample of your patient before, during and after the treatment you can see a modification not only in the epitope specificity – because there is sometimes some switch which has been described effectively – but you will also see the modification in the production of this anti-idiotype antibody. In the patient who does not succeed, probably you will see a very poor increasing of this anti-idiotype antibody.

It's possible to have a good idea and to apply this method to detect the efficacy of the treatment on such patient or not. To decide if treatment has to be pursued or not you can have the response with such a method.

Dr. White: I will make a short comment so that you can wrap up. Carol [Kasper], it seems to me that if you look at the data on the tolerization of patients that two things are important. One is probably the affinity of the antibody for the antigen and that's going to be important in your ability to get rid of it. And, the second is the antibody load – the number of cells that are making antibody. My prediction would be with that patient that we sent to you is that he has a very large number of antibody-producing cells and I would speculate that you gave him a lower dose of F.VIII as a toleragen. And it may be that everybody ultimately will respond to F.VIII but if they have a large number of antibody-producing cells you have to give them an even larger amount of factor VIII; if they have a small number of antibody-producing cells you may be able to get away with a small amount of F.VIII. I think that's what Mariani's registry is telling us. That those are crucial factors. There may be other things as well but those are the things that distinguish those who respond and those who don't respond.

Dr. Lusher: So, theoretically, one would think that an infant who's developed an inhibitor should have fewer antibody-producing cells than say, someone who has had their inhibitor for 20 years. And yet, Professor Mariani told us yesterday that age was not a factor according to the data in his registry. I find that difficult to understand.

Dr. Hilgartner: There are many of us who would like to continue the discussion because it has obviously been a very fruitful one and I think you all have enjoyed this discussion as much as I have. Unfortunately time is short. From this panel I think we have learned that we must look more closely at the definition of the patients, at the variable responses of the patient to F.VIII, at the type of antibody that is produced before we can perhaps add as much as we would like to the current protocols already in place. We certainly have to develop additional protocols and I trust that many of you will think closely and confer with your colleagues as to how we can continue to look at this question of how to classify, stratify the patient material and how to improve on the protocols that already exist.

Dr. Reisner: With that, the meeting draws to a close. I'd like to thank all of the participants, the speakers, the members of the audience, the people who took part in the discussion for making this a successful meeting. I would like to close with one thought. I was trained, at least in part, as an immunologist so I suppose I am allowed to say this. I think if we had asked the immunologist today: Should we go ahead and try tolerance with individuals with high levels of inhibitor having hemophilia–without the evidence that we have? They would have said no, it will never work, you can't be successful 75 percent of the time in inducing lasting, permanent tolerance in many, many patients. You have to remember it was the work of Dr. Rizza, Dr. Brackmann, Dr. Nilsson who did the experiments, did them in humans and were highly, highly successful that led us to the point we are at today. We do have successful protocols. Undoubtedly we can make them better. Undoubtedly knowing the basic mechanisms will make them better but let's not always take the dark view. I think unfortunately the thing that has impeded much of this work has been the economic problem. It is expensive and that has inhibited many countries, many individuals from being able to carry this out successfully. I think we have to recognize that. I know it won't take 10 years until we have accumulated enough data that we'll want to get together. I hope when we do it that you'll consider coming back here. Thank you and have a pleasant journey home.

Poster Session Abstracts

1. Acquired Circulating Anticoagulants in Patients with B-Cell Chronic Lymphocytic Leukemia. Takeshi Wajima. Olin E. Teague Veterans' Center and Texas A&M University, Temple, Texas.

2. Acquired von Willebrand Disease Caused by an Antibody, Selectively Inhibiting the Binding of von Willebrand Factor to Collagen. J.J. Michiels, P.J.J. Van Genderen, A. Vink, M.B. Van 't Veer, H.H.D.M. van Vliet, J.J. Sixma. Department of Hematology, University Hospitals Rotterdam and Utrecht, The Netherlands.

3. Factor VIII Inhibitor Postpartum. J.J. Michiels, K. Hamulyak, H.K. Nieuwenhuis, I. Novakova and H.H.D.M. van Vliet. Departments of Hematology, University Hospitals Rotterdam, Maastricht, Utrecht, and Nijmegen, The Netherlands.

4. Blocking of Tissue Factor Pathway Inhibitor (TFPI) Shortens the Bleeding Time in Rabbits with Antibody Induced Hemophilia A. Elisabeth Erhardtsen, Mads T. Madsen, Mirella Ezban, Ulla Hedner, Viggo Diness, Steven Glazer, Ole Nordfang. Biopharmaceuticals Research, Novo Nordisk A/S, Gentofte, Denmark.

5. Reversible Acylation of Factor Xa: Characterization of a Novel Proenzyme Form of Factor Xa. Uma Sinha, Pei-Hua Lin, and David Wolf. COR Therapeutics Inc., South San Francisco, California 94080.

6. Hemostatic Efficacy of Recombinant Factor VIIa in Surgical Interventions in Hemophilia A Patients with Inhibitors. J. Ingerslev, O. Sneppen, L. Knudsen, S. Sindet-Pedersen, Hemophilia Centre, Department of Clinical Immunology, and Departments of Orthopedic Surgery and Oral and Maxillofacial Surgery, University Hospital Aarhus, Denmark.

7. Clinical Aspects of a Recent Population of Dutch Inhibitors Compared to Classic Inhibitors. E.P. Mauser-Bunschoten, G. Roosendall, K. Nieuwenhuis, F. Rosendall, E. Briët, H.M. van den Berg. Van Creveld Clinic, University Hospital, Utrecht; Department of Hematology, University Hospital, Utrecht; Departments of Clinical Epidemiology and Hematology, University Hospital, Lieden, The Netherlands.

8. A Prospective Study of the Treatment of Patients with Acquired Factor VIII (FVIII) Inhibitors Using High Dose Intravenous Immunoglobulin (IVIg)R.S. Schwartz, D.A. Gabriel, L.M. Aledort, D. Green, and C.M. Kessler. Miles Inc., Berkeley, California; The University of North Carolina at Chapel Hill, North Carolina; Mt. Sinai Medical School, New York, New York; Northwestern University, Chicago, Illinois; George Washington University, Washington, D.C.

9. Antibodies to Bovine Thrombin and Coagulation Factor V Associated with Surgical Use of Topical Bovine Thrombin or Fibrin "Glue": A Frequent Finding.W.L. Nichols, T.M. Daniels, P.K. Fisher, W. G. Owen, A.A. Pineda, and K.G. Mann. Hematology Research and Transfusion Medicine, Mayo Clinic, Rochester, Minnesota, and Department of Biochemistry, University of Vermont, Burlington, Vermont.

10. Presentation of a New Factor VIII Inhibitor with an Intraspinal Bleed. Brian Wicklund, M.D.; Joy Johnson, M.D.; Maxine Hetherington, M.D.; Wendy Wright, R.N.; M.S.N.; Mollie Spoor, R.N., B.S.N.; James McEntire, D.O. Children's Mercy Hospital, Kansas City, Missouri.

11. Successful Use of Purified Monoclonal Factor IX for Induction of Immune Tolerance in a Patient with Hemophilia B and a High Titer Inhibitor. Jacob Katz, M.D.;Victoria L. Castaneda, M.D.; Awilda Fagin, M.T. Department of Pediatrics, Division of Hematology/Oncology, University of California, Irvine, California.

12. Induction of Immune Tolerance with Kogenate® Recombinant Factor VIII (rFVIII) in Previously Untreated Patients (PUPs) with High Titer Inhibitors; Preliminary Results Using a Defined Protocol. Inwood M.J., Lusher, J.M., Rousell R.H., Abildgaard C.F., Aledort L.M., Hilgartner M.W., Zimmermann R.

13. Epitope Mapping Antibodies that Bind to the C-Type Domains of Coagulation Factor V and Factor VIII Using Recombinant Chimeras. Thomas L. Ortel, Mary Ann Quinn-Allen, Karen D. Moore, Frank G. Keller, and William H. Kane. Duke University Medical Center, Durham, North Carolina.

14. Evaluation of the Epitope Specificity of Human Anti-Factor VIII Antibodies by Using a Panel of Murine Monoclonal Antibodies. Jean Guy G. Gilles and Jean-Marie R. Saint-Remy. Allergy and Clinical Immunology Unit, International Institute of Cellular and Molecular Pathology, Universite Catholique de Louvain, Brussels, Belgium.

15. Cellular Immune Responses to Coagulation Factor VIII in FVIII Inhibitor Patients. Debra K. Newton-Nash, Ph.D.; Joan Cox Gill, M.D.; and Paul A. Foster, M.D. The Blood Research Institute, Milwaukee, Wisconsin.

16. The Use of Factor VIII Inhibitor By-Passing Activity (FEIBA, Immuno) and Porcine Factor VIII (HYATE: C, Porton) for Treatment of Patients with Inhibitors: The Canadian Experience, 1988-1993. G.E. Rivard,* S. Vick, M. David. Hopital Ste-Justine, Montreal, Quebec and The Canadian Red Cross, Ottawa, Ontario in Canada.

17. A Proposed Prospective Randomized Trial of Immune Tolerance Therapy in a Pediatric Hemophilia A Population with High Responding

Inhibitors. D.M. DiMichele, Northwestern University, Chicago, Illinois; B. Bell, Emory University, Atlanta, Georgia; K. Hoots, University of Texas Medical School, Houston, Texas.

18. A Feasibility Study for Continuous Ambulatory Factor VIII (CAFE) Therapy for Patients with Hemophilia A and Factor VIII Inhibitors. E. M. Gordon, A. Gilbert, R. Salazar and J.C. Goldsmith. Childrens Hospital Los Angeles and Quantum Health Resources, Los Angeles, California.

19. Incidence of Factor VIII Inhibitors in Patients with Hemophilia A. K. Peerlinck, F. Rosendaal*, and J. Vermylen. Center for Molecular and Vascular Biology, University of Leuven, Belgium and *Department of Clinical Epidemiology, University Hospital Leiden, The Netherlands.

20. Factor VIII Inhibitor Assays Using Plasma-Derived FVIII Versus Recombinant FVIII - A Comparative Study. Hillman-Wiseman C., Vitale C., Lusher J., Children's Hospital of Michigan, Detroit, Michigan, USA.

21. Immune Tolerance Induction (ITI) in Children with Hemophilia A (HA) and FVIII Inhibitors. Warrier I., Pfaffmann L., and Lusher J.M. Children's Hospital of Michigan and Wayne State University School of Medicine, Detroit, Michigan.

22. Determining Inhibitor-Development Risk in Previously Untreated Hemophilia A Patients (PUPs)Treated with Recombinant Factor VIII (Recombinate™). M. L. Lee, S. Courter, and E. Gomperts. Baxter/Hyland Division, Glendale, California.

23. Multicentric Retrospective Study on the Utilization of FEIBA in France in Patients with Factor VIII and Factor IX Inhibitors. C. Negrier (Hemophilia Care Center, Edouard Herriot Hospital, Lyon, France), and the French FEIBA study group (A. Ballocchi, D. Bastit, M.A. Bertrand, C. Bosser, D. Boveldieu, J.Y. Borg, G. Dirat, A. Durin, M. Gaillard, C. Gazengel, J. Goudemand, P. N. Stielties, Y. Sultan, M.F. Torchet and P. Tron).

24. FEIBA® VH Immuno: Partitioning and Inactivation of HIV-1. Turecek, P.L., Schwarz H.P., Barrett, N., Pölsler, G., Dorner, F., and Eibl, J. Immuno AG, Vienna, Austria.

Acquired circulating anticoagulants in patients with B-cell chronic lymphocytic leukemia

Takeshi Wajima
Olin E. Teague Veterans' Center and Texas A&M University, Temple, Texas

A circulating anticoagulant should be suspected when unexplained bleeding develops in patients with lymphoproliferative disorders in elderly.

Case 1. A 76-year-old white male with B-cell chronic lymphocytic leukemia (CLL), stage O, had a hip fracture, developed a large hematoma over the pelvis, and required 8 units of packed red cells and 10 units of fresh frozen plasma. Initial hemostasis studies were normal except for a prolonged activated partial thromboplastin time (APTT) 43.3 sec (control 29 sec). Prothrombin time (PT), thrombin time, bleeding time (BT), platelet aggregation by ADP, epinephrine, collagen, and ristocetin were normal. Coombs test, and lupus anticoagulant were negative. The APTT of the patient's plasma was not completely corrected by the addition of an equal part of normal fresh plasma. Incubation of mixture showed more prolongation of APTT. The factor VIII:C was 5%. The levels of factors IX, XI, XII, FSP and fibrinogen were normal. ANA, RF and SPE were negative. The antibody to VIII:C was IgG_{K}, and the titer of inhibitor was 4.7 Bethesda units (BU). After recovery from the bleeding, the patient was followed over a 3-month period without replacement therapy or immunosuppressive therapy or chemotherapy for CLL. The APTT remained prolonged ranging from 39 to 41 sec.

Case 2. A 65-year-old white male with B-cell CLL, stage II presented with spontaneous ecchymosis and nose bleeding. He had no prior personal or family history of bleeding. There were multiple ecchymoses of all extremities, lymphadenopathy, and splenomegaly, WBC 33.7 x 10^9/l with 90% lymphocytes. Hct 43%, platelet 360 x10^9/l, BT 15 min. PT 12 sec, aPTT 32 sec (control 32 sec), platelet aggregation with ADP, epinephrine, and collagen was normal, but aggregation by ristocetin was absent. Factor VIII:C was 48%. Ristocetin cofactor was 29%. When the patient's plasma was incubated with normal plasma, ristocetin cofactor activity in normal plasma was decreased. A circulating inhibitor to von Willebrand factor was suspected. He was treated with cryoprecipitate, chlorambucil, and prednisone. He responded clinically, but laboratory tests were not improved with the treatment. Incidence of acquired circulating anticoagulants in non-hemophilic patients is not unknown.

A higher incidence is reported in the elderly with immune disorders or lymphoproliferative disorders. A unique property of CLL B-cell is the presence of CD5 which may be involved in the generation of autoimmune phenomena in CLL. The clonal leukemic B lymphocytes might cause an altered F.VIII antigen that could have elicited an antibody that cross-reacted with the patient's own F.VIII. Inhibitors arising from autoimmune disease or lymphoproliferative disorders may respond to treatment of the underlying disorder, but the inhibitor to von Willebrand factor of the presented patient did not respond to chemotherapy for CLL.

Acquired von Willebrand disease caused by an antibody, selectively inhibiting the binding of von Willebrand factor to collagen

J.J. Michiels, P.J.J. Van Genderen, A.Vink, M.B. Van 't Veer, H.H.D.M. van Vliet, J.J. Sixma
Departments of Hematology, University Hospitals Rotterdam and Utrecht, The Netherlands

An 82-year-old man with a low-grade Non-Hodgkin's lymphoma and an IgG λ monoclonal gammopathy presented with a recently acquired bleeding tendency, consisting of recurrent epistaxis, bruises, and episodes of melena, requiring blood transfusions. Coagulation studies revealed a prolonged Ivy bleeding time, normal platelet count, prolonged APTT, very low levels of F.VIIIC, von Willebrand factor (vWF) antigen, vWF-ristocetin cofactor activity (vWF.RCF) and vWF-collagen binding activity (vWF.CBA), absence of RIPA and absence of high and intermediate size molecular weight vWF multimers. DDAVP administration induced transient reappearence and subsequent rapid disappearence of high and intermediate vWF multimers. These data were consistent with acquired von Willebrand's disease. The patient's plasma contained a circulating inhibitor against the vWF:CBA activity without inhibiting the F.VIIIC, vWF:Ag and vWF:RCF activity. Using recombinant vWF fragments it was demonstrated that the inhibitor reacted with both the GPlb (aa 422-717) and the A3 (aa 910-1112) domains, but not with the A2 (aa 716-908) and D4 (aa 1183-1535) domains of vWF. It is concluded that the inhibitor inhibits the vWF:CBA activity of vWF by reacting with an epitope on both the GPlb and A3 domains of vWF, which have both been implicated in the binding of von Willebrand factor to collagens type I and III.

The administration of vWF concentrates elicited only a short-lived rise of FVIII/vWF related activities without clinical benefit, whereas the administration of high-dose intravenous gammaglobulin (1 g/kg bodyweight) for 3 consecutive days resulted in a normalization of the acquired von Willebrand defect for 2 weeks and recovery of gastrointestinal bleeding without further transfusion requirement.

Factor VIII inhibitor postpartum

J.J.Michiels, K. Hamulyak, H.K. Nieuwenhuis, I. Novakova and H.H.D.M. van Vliet
Departments of Hematology, University Hospitals Rotterdam, Maastricht, Utrecht and Nijmegen, The Netherlands

The bleeding symptoms and time lapses between delivery and onset of bleedings in 34 reported cases of F.VIII inhibitor postpartum are the following: (N=number of patients)

Bleeding Symptoms	N	Time lapse delivery/bleeding	N
Postpartum bleeding	8	During pregnancy	1
Bruises, ecchymoses	27	Postabortum	1
Soft tissue bleeding	24	Postpartum <4 weeks	8
Secondary hemorrhages	18	Postpartum 2-4 months	16
Hemarthros	15	Postpartum 4-12 months	3
Hematuria	12	Postpartum >12 months	5

Spontaneous disappearence of F.VIII inhibitor occurred in 17 after 3 to 96 months. No remission was noted in 8 after a follow up of 1 to 108 months. Four women with F.VIII inhibitor postpartum died of uncontrolled bleeding.

Prednisone treatment was ineffective in 11 of 12 evaluable patients. Experience with chemotherapeutic immunosuppressive agents has been very scant and for the most part uninterpretable. Therefore, we performed a study on the natural history and treatment in 4 cases of F.VIII inhibitors postpartum. The hemorrhages became apparent 2 to 4 months after delivery. *In vivo* F.VIII levels and *in vitro* F.VIII inhibitor potency expressed in Bethesda Units (BU) were: 2% and 8 BU, 2% and 12 BU, <1% and 26 BU, and <1% and >40 BU, respectively. F.VIII inhibitors were type II in all 4 cases. High dose F.VIII concentrate substitution in 2 cases and high dose gammaglobulin intravenously in 1 case were effective in controlling bleeding episodes. Treatment with prednisone/ cyclophosphamide in 3 cases did not affect the inhibitor, which spontaneously disappeared 6 to 21 months after the onset of bleeding symptoms. Despite different treatment modalities, including cyclophosphamide the most potent F.VIII inhibitor, in case 4, persisted for several years. These results indicate that chemotherapy with prednisone/ cyclophosphamide does not influence the natural history of spontaneous disappearence of F.VIII inhibitors in healthy women postpartum.

Blocking of tissue factor pathway inhibitor (TFPI) shortens the bleeding time in rabbits with antibody induced hemophilia A

Elisabeth Erhardtsen, Mads T. Madsen, Mirella Ezban, Ulla Hedner, Viggo Diness, Steven Glazer, Ole Nordfang
Biopharmaceuticals Research, Novo Nordisk A/S, Gentofte, Denmark

Tissue factor (TF)/F.VIIa initiates coagulation by activating F.IX and F.X. Complexes between Tissue Factor Pathway Inhibitor (TFPI) and F.Xa are formed and will inhibit TF/F.VIIa. However, a normal intrinsic coagulation pathway including F.IX/F.VIII is necessary for a normal hemostatic control, which is indicated by the severe bleeding symptoms in hemophilia A (lack of F.VIII) and hemophilia B (lack of F.IX). Blocking the inhibition of the F.VIIa/TF dependent coagulation pathway, by blocking TFPI, may facilitate the hemostasis initiated by this pathway, thereby compensating for an impaired F.IX/F.VIII dependent coagulation. This hypothesis was tested in a study in rabbits made temporarily hemophilic by the injection of antibodies against F.VIII.

Polyclonal antibodies against human TFPI were obtained by immunizing goats with $rTFPI_{1-161}$. The antiserum showed cross reactivity with rabbit TFPI and monospecific IgG was prepared using precipitation, affinity chromatography on $TFPI_{1-161}$-Sepharose and ion exchange. The IgG preparation contained 1700 Inhibitory Units (IU)/mg IgG where one unit inhibits 50% of the TFPI activity in normal human plasma. Rabbits with temporary hemophilia A were given i.v. injections of 12000 IU/kg of anti-TFPI antibodies. Forty minutes later, bleeding was initiated by cutting a nail including the apex of the cuticle. The group receiving anti-TFPI (n=4) had a mean bleeding time of 11 minutes (range 6-15) and the group receiving anti-F.VIII alone (n=14) a bleeding time of 24 minutes (range 15-30). The normal mean bleeding time, measured in non-hemophilia rabbits (n=13), was 6 minutes (range 4-11). The effect of anti-TFPI was reflected in blood values. The APTT coagulation time was prolonged after dosing anti-FVIII, and shortened after dosing anti-TFPI to a value slightly higher than in the non-hemophilia group. In a dilute Tissue Factor clotting assay (dTF), the clotting time was prolonged after dosing anti-F.VIII. The value was shortened after dosing anti-TFPI to a value somewhat below the dTF measured in the non-hemophilia group. Neither injections of anti-F.VIII antibody nor anti-TFPI antibody had any effect on blood pressure. In conclusion, injections of anti-TFPI antibodies shortened bleeding time significantly in hemophiliac rabbits thus providing *in vivo* evidence of the importance of TFPI for bleeding in hemophilia.

Reversible acylation of factor Xa: Characterization of a novel proenzyme form of factor Xa

Uma Sinha, Pei-Hua Lin and David Wolf
COR Therapeutics Inc., South San Francisco, California 94080

We have investigated the improvement of conventional inhibitor bypass therapies for hemophilia with slow activating forms of factor Xa. These novel proenzyme forms of F.Xa occur upon chemical interaction with amidinophenyl derivatives resulting in reversible acylation of the catalytic serine residue. Nonenzymatic reactivation occurs by hydrolysis of the labile acyl group. We have detailed the *in vitro* activities of the p-amidinophenyl-p'-anisate modified form of human F.Xa (anisoyl-Xa). Factor Xa is rapidly inactivated (<0.2% amidolytic activity), within 2 minutes at room temperature. Anisoyl-Xa recovery rates for F.Xa catalytic activity are dependent upon time, temperature, and pH. Amidolytic, prothrombinase, and plasma clotting activities recover at similar rates. Prothrombinase associated anisoyl-Xa is also reactivated in a time-dependent manner. The data suggest that F.Xa, like other serine proteases, can be transiently inactivated and could lead to new therapies for the treatment of acute and congenital bleeding disorders.

Hemostatic efficacy of recombinant factor VIIa in surgical interventions in hemophilia A patients with inhibitors

J. Ingerslev, O. Sneppen, L. Knudsen, S. Sindet-Pedersen
Hemophilia Centre, Department of Clinical Immunology, and Departments of Orthopedic Surgery and Oral & Maxillofacial Surgery, University Hospital Aarhus, Denmark

In hemophilia A, the development of inhibitors against F.VIII contributes significantly to patient's risk of death from acute bleeding. Likewise, surgery may provoke uncontrolled bleeding. In spite of anecdotal reports of successful outcomes of surgery using unactivated and activated F.IX concentrates, rehabilitation surgery is generally avoided in these patients. We report our recent experiences using recombinant factor VIIa (rVIIa) to control hemostasis in five hemophilia A patients with high-responder F.VIII inhibitors. In total, 13 procedures are reported: 6 extractions of wisdom teeth, 3 central line (Port-A-Cath) insertions, 1 acute heriontomy, 1 complete condylar knee-joint replacement, and 1 case of bilateral total condylar arthroplasty. Due to the short *in vivo* half-life of rVIIa, frequent dosing was demanded. A general dosage scheme was selected in which the perioperative and immediate postoperative dose was in the range of 77-125 μg/kg that was administered at 2- to 3-hour intervals for the first 24 to 30 hours, followed by a reduced dose or an increase in the dosage interval on an individual basis. In dental extractions, rVIIa infusions were halted before the elapse of 34 hours, and in the central line implants, substitution by rVIIa was discontinued no later than 41 hours from surgery. In some of these cases, late re-bleeds were easily controlled by renewed short-term treatment with rVIIa. In cases of major surgery, prolonged use of rVIIa was preferred, even if hemostasis was maintained, and no rebleeds were encountered here. Tranexamic acid 25 mg/kg q 4h was used in all patients up till day 10 postoperatively. From our results a firm conclusion can be established that rVIIa has proven highly efficacious in control of hemostasis during and after surgery in these 13 instances of minor and major operative procedures in hemophilia A inhibitor patients. In our opinion, this opens a new perspective for the inhibitor patient.

Clinical aspects of a recent population of Dutch inhibitors compared to classic inhibitors

E.P. Mauser-Bunschoten,[1] G. Rosendaal,[1] K. Nieuwenhuis,[2] F. Rosendaal,[3] E. Briët,[4] Hm. van den Berg[1]

[1] *Van Creveld Clinic, University Hospital Utrecht, postbox 85500, 3508AG Utrecht Netherlands*
[2] *Department of Hematology, University Hospital, Utrecht, Netherlands*
[3] *Department of Clinical Epidemiology, University Hospital, Leiden, Netherlands*
[4] *Department of Hematology, University Hospital, Leiden, Netherlands*

Since June 1990 an increase in the incidence of inhibitors to F.VIII was seen in Dutch hemophilia patients. This increase seemed to be related to a new F.VIII concentrate (F.VIII CPS-P) which was introduced at that time. Twelve newly developed inhibitors were detected since then. The clinical aspects of these 12 inhibitor patients were compared with a group of 23 patients who had their inhibitors before June 1990.

Differences were seen not only in age and number of exposure days prior to inhibitor development, but also in the response to immune tolerance therapy.

In the group of newly developed inhibitors the response seemed to depend on the clotting product used, rather than on the immune tolerance therapy.

A prospective study of the treatment of patients with acquired factor VIII (F.VIII) inhibitors using high dose intravenous immunoglobulin (IVIg)

R.S. Schwartz, D.A. Gabriel, L.M. Aledort, D. Green, C.M. Kessler

Miles Inc. Pharmaceutical Division, Berkeley, California; University of North Carolina, Chapel Hill, North Carolina; Mt. Sinai Medical School, New York, New York; Northwestern University, Chicago, Illinois; George Washington University, Washington, D.C.

Nineteen patients were enrolled in a multicenter prospective study evaluating the safety and efficacy of high dose intravenous immunoglobulin (IVIg) for the treatment of acquired factor VIII (F.VIII) inhibitors. These patients were administered IVIg infusion using a dosing schedule of either 1000 mg / kg x 2 days or 400 mg / kg x 5 days. Serial measurements of F.VIII inhibitor levels were taken for all patients, and in selected cases F.VIII levels were also monitored. Of the 19 patients enrolled in the study, 16 were evaluable for efficacy: 2 patients died within 1 day of treatment and 1 patient did not receive the entire treatment regimen. Of the 16 evaluable patients, inhibitor titers prior to IVIg infusion ranged from 0.9 BU to 1228 BU with a median value of 25.5 BU. A significant reduction in inhibitor was observed in 6/16 (37.5%) patients: in 3 patients with initially low level inhibitor titers of 1.0, 1.0, and 0.9 BU, the inhibitor subsequent to treatment was undetectable, and in 3 patients with initial titers of 12, 102, and 280 BU, the inhibitor titer was reduced to < 6.0 BU. The relationship between the response rate (CR + PR) and the initial inhibitor titer was as follows: < 10 BU, 3/5 patients responded (60%); 10 - 50 BU, 1/5 patients responded (20%); and > 50 BU, 2/6 patients responded (33%). In the responder population there was significant variability in the time of response with nadir inhibitor levels achieved from 2 weeks to 1 year after the commencement of IVIg treatment. The median initial titer for responders was 6.5 BU and for non-responders was 44 BU. Of 19 patients evaluable for safety, 3/19 (16%) reported adverse reactions representing 10/88 (11.4%) infusions, the majority resulting from a single patient reporting 8 adverse reactions with headache being the most common complaint. These data confirm the safety of high dose IVIg and indicate that it may be of benefit to a subset of patients with acquired inhibitors to F.VIII.

Antibodies to bovine thrombin and coagulation factor V associated with surgical use of topical bovine thrombin or fibrin "glue": A frequent finding

W.L. Nichols, T.M. Daniels, P.K. Fisher, W.G. Owen, A.A. Pineda, and K.G. Mann

Hematology Research and Transfusion Medicine, Mayo Clinic, Rochester, Minnesota and Department of Biochemistry, University of Vermont, Burlington, Vermont

During 37 months (1990-1993), we identified 16 Mayo Clinic patients who postoperatively developed circulating inhibitors of bovine thrombin and/or Factor V (F.V). Inhibitors were initially detected by clot-based coagulation assays. Assays for coagulation factors II or XI, using commercial substrate plasmas containing either bovine or human F.V, were especially useful in detecting bovine F.V inhibitors, whereas thrombin time assays using either bovine or human thrombin were useful for identifying bovine thrombin inhibitors. Patient plasmas inhibited bovine F.V clotting activity in titers up to 1:21. Inhibition of human F.V by patient antibodies was generally minimal, although 10 patients had <30% plasma F.V activity (7 had ≤8% F.V) with concordantly decreased F.V antigen. Immunoassays employing solid-phase antigens (thrombin or F.V) were used to confirm and titer human plasma antibodies (IgG or IgM). All tested patients (14/14) had bovine F.V antibodies, in titers ranging from $1{:}10^2$ to $1{:}10^7$, although only 10 had definitely abnormal clot-based inhibitor assays. Bovine thrombin inhibitors were present by clot-based assays in 11 of 16 patients, but were not detectable or were equivocally present in 5 patients, although demonstrable by immunoassay in 3 of the latter. Immunoassays specific for bovine F.V/F.Va showed commercial preparations of topical bovine thrombin contained F.V which ranged from <1 ng to ~100 µg FV/mL. Cardiovascular surgery had been performed in 14 patients and orthopedic surgery in 2. In all evaluable cases, F.V containing topical bovine thrombin or fibrin "glue" (prepared using bovine thrombin) had been applied during surgery. Three patients died of hemostatic complications associated with inhibitors, and several patients had morbidity which could be attributed to inhibitors. Most patients were apparently hemostatically asymptomatic. These data suggest that antibodies to bovine F.V or thrombin may commonly develop following surgical use of topical thrombin, producing morbidity or mortality in some cases.

Presentation of a new factor VIII inhibitor with an intraspinal bleed

Brian Wicklund, M.D., Joy Johnson, M.D., Maxine Hetherington, M.D.; Wendy Wright, R.N., M.S.N.; Mollie Spoor, R.N., B.S.N.; James McEntire, D.O.
Children's Mercy Hospital, Kansas City, Missouri

A 3-year, 10-month-old moderate hemophilia A patient, with no previous family history or evidence of inhibitors, presented to the emergency room with the sudden onset of severe back pain and refusal to sit or stand. Five days before presentation, the patient fell off a tricycle and landed on his buttocks. Two days before presenting, he fell while playing on carpeting and landed flat on his back. During the previous month, he was playing every other day on a spring-suspended riding horse, bouncing strongly enough to lift the stabilizing platform for the toy off the ground. The evening of presentation, he complained of low back pain and refused to walk. The home nursing service gave 50 U/kg of monoclonal factor VIII (F.VIII) with seeming relief. Four hours later, he became hysterical with lower back pain and was brought to the ER. Fifty U/kg of F.VIII and IM meperidine were given to control pain. Physical exam showed a child who was irritable when aroused and in moderate distress. He had bruises over the coccyx and tenderness on palpation over L5 and the sacrum. Reflexes were normal, distal motor exam was grossly intact, but with extreme pain on passive and active motion. The sensory and rectal exams were normal.

CT scan of the spine and pelvis showed an intraspinal hematoma with subarachnoid extension. Bone films and head CT were normal. The patient was started on dexamethasone and continuous infusion F.VIII at 6 units/kg/hr. It was impossible to raise the F.VIII level beyond 25%, and a recovery study showed only a 0.36% increase in F.VIII activity for every unit/kg given. A F.VIII inhibitor was diagnosed, and measured at 1.58 Bethesda units. The F.VIII infusion was increased incrementally to 400 U/kg/day, to obtain a F.VIII level of >90%. An MRI done on the second hospital day showed an extradural hematoma extending from the mid-thoracic to sacral level with compression and displacement of the lower cord and thecal sac. Factor VIII was given as a continuous infusion until full symptomatic resolution occurred, and the patient was discharged from hospital after 10 days. He continued on an immune tolerance protocol using F.VIII concentrate, and had half-life studies done every 2 weeks. At present, he is fully recovered, is on 50 units/kg/day of F.VIII concentrate, and has no evidence of the inhibitor on Bethesda assays. He is being weaned further as half-life studies continue to stay in the normal range.

Successful use of purified monoclonal factor IX for induction of immune tolerance in a patient with hemophilia B and a high titer inhibitor

Jacob Katz, M.D.,Victoria L. Castaneda, M.D., Awilda Fagin, M.T.

Department of Pediatrics, Division of Hematology/Oncology, University of California, Irvine, California

The prevalence of inhibitors in hemophilia A is 15 to 20%; however, the occurrence of inhibitors in hemophilia B is rare. A telephone inquiry revealed that most large hemophilia treatment centers in the U.S. do have patients with hemophilia B and low titer inhibitors who respond to high dose F.IX or Autoplex/Feiba. Dr. Nilsson, Malmö, Sweden, has reported successful immune tolerance induction of F.IX inhibitors using the Malmö regimen. We report here the use of purified monoclonal F.IX in a 5-year-old white severe hemophilia B patient who presented with poor response to conventional F.IX dose administration and an initial inhibitor titer of 10 BU. The patient was enrolled on a study after IRB and FDA approval and parental consent was obtained that would use 50-100 units F.IX/kg/day. The protocol was activated 3 months after the detection of the inhibitor and during that time no F.IX was administered and a conservative therapy of bedrest, immobilization and pain medication was used for the 2 episodes of hemarthrosis that occurred. At commencement of the study, the inhibitor level was 6.6 BU and successful implantation of a central venous line (Broviac catheter) was accomplished with high dose mononine. High dose mononine (Armour) induction of immune tolerance was started 7 days post surgery using approximately 80 units/kg (the vials contained 1370 units F.IX) daily. After 1 month the inhibitor titer rose to 12 BU, at 2 months 9.8 BU, after 3 months 5.8 BU and at 4 months <1 BU. During the 4 months the patient required Autoplex treatments on 2 occasions for bleeding problems. At 6 weeks the Broviac catheter fell out and was replaced under cover with high dose mononine. Laboratory studies during the second Broviac surgery showed a dilution phenomenon, F.IX levels at 1/5 dilution were much lower than the 1/10 and subsequent dilutions and the APTT remained abnormal (61secs) when the F.IX level was 1/5 dilution (9%) and 1/10 dilution (43%), the patient developed a hematoma at the incision site. When the inhibitor titer was <1 BU, mononine administration showed a 50% recovery at 20 mins (1 unit/kg = 0.5% rise) and a T/2 of 10 hours. At present the patient is receiving mononine 30 units/kg/day. We conclude immune tolerance induction therapy can be accomplished with F.IX concentrates (mononine in this patient) and should be attempted in the young patient at the earliest detection of a significant inhibitor level.

Induction of immune tolerance with Kogenate® recombinant factor VIII in previously untreated patients (PUPs) with high titer inhibitors: Preliminary results using a defined protocol

Inwood M.J., Lusher J.M., Rousell R.H., Abildgaard C.F., Aledort L.M., Hilgartner M.W., Zimmerman R.

St. Joseph's Health Center, London, Ontario, Canada; Children's Hospital of Michigan, Detroit, MI; Miles Inc., Berkeley, CA; University of California at Davis, Sacramento, CA; The Mount Sinai Medical Center, New York, NY; New York Hospital-Cornell Medical Center, New York, NY; Rehabilitationsklinik, Heidelberg, Germany

High titer inhibitors, >10 BU, are a serious complication, interfering with continued treatment with F.VIII. We previously reported success in inducing immune tolerance in certain patients who developed inhibitors to rF.VIII using a variety of Kogenate® dosages. Encouraged by the successful response in two of the three PUPs treated, a standard protocol was developed defining the dosage of rF.VIII for induction of tolerance according to the highest Bethesda titer found prior to starting the regimen. Essentially the protocol recommends a dosage of 50 IU/kg/day for BU<10; 100 IU/kg/day for BU≥10-<50; 150IU/kg/day for BU≥50-<100; and 200 IU/kg/day for BU≥100. The Bethesda titer disappeared in 3 of 5 patients thus treated with partial response in the other 2. One continues on his initial dosage but variation in response in the other has, according to the protocol, required the dosage be doubled.

	Baseline	Treatment		Response		Current rF.VIII
Subject	BU titer	Dose (U/kg)	Start	Week	BU	Dose
70	19	50/day	1/23/91	2	0	40 U/kg every 8 days
110	19.5	100/day	7/8/91	52	0	50 U/kg every other day
126*	34	100/day**	3/18/92	4	0	PRN since 12/92
129*	154	50-100/day	3/22/91	No response at 65 wks; 1229 BU peak; FEIBA PRN		
142	22	100/day**	11/2/92	2	0	50 U/kg every 3 days
166	70	150/day**	6/10/93	6	144(819)	150 U/kg daily
171	8	200/day**	2/8/93	25	61.6(117)	400 U/kg daily since wk 26
201	11	100/day**	7/1/93	2	0	50 U/kg daily since wk 6

* Family History Of Inhibitor ** 9101 Protocol () = peak BU 1 to 2 wks post start

This is the first study where dosage of FVIII for induction of tolerance is administered in accordance with the Bethesda titer. Results of the International Registry survey (G. Mariani) suggest that successful tolerance is more likely with higher doses (>200 FVIII IU/kg). It seems logical that an attempt be made to relate the dosage to the inhibitor status in the individual subject.

Epitope mapping antibodies that bind to the C-type domains of coagulation factor V and factor VIII using recombinant chimeras

Thomas L. Ortel, Mary Ann Quinn-Allen, Karen D. Moore, Frank G. Keller, and William H. Kane
Duke University Medical Center, Durham, North Carolina, 27710, USA

Both F.V and F.VIII possess two C-type domains that are ~35-40% identical in amino acid sequence. These domains are ~150 amino acids in length and are present in tandem at the carboxyl-terminal end of the molecule. Many F.VIII inhibitors have been epitope mapped to the second C-type domain of F.VIII. In addition, we have epitope mapped a F.V inhibitor (H1) to the second C-type domain of F.V. Inhibitors that bind to the C2 domain of F.VIII inhibit the binding of F.VIII to phosphatidylserine, and we have shown that the H1 inhibitor blocks the binding of F.V to phosphatidylserine. However, not all antibodies that bind to the second C-type domains of these two cofactors interfere with binding to phosphatidylserine, although they may interfere with procoagulant activity. Because of our interest in defining structure-function relationships in F.V, we have used F.V inhibitors and monoclonal antibodies that bind to the C2 domain as molecular probes. To precisely map C2-domain directed antibodies, we have constructed a series of recombinant F.V-F.VIII C2 domain chimeras that contain exon-size substitutions of F.VIII for the corresponding F.V sequences in the C2 domain. Although the C-type domains of F.V and F.VIII possess ~40% amino acid sequence identity, the inhibitors and monoclonal antibodies are completely fastidious in their epitope recognition and do not cross-react with the homologous cofactor sequences. These constructs were prepared using the cDNA's for F.V and F.VIII with the polymerase chain reaction, and were expressed in COS cells. This approach offers several potential advantages over epitope mapping by immunoblotting with protein fragments from a prokaryotic source. First, this approach preserves the higher order structure of the domain while substituting different segments of primary sequence. The primary epitopes are therefore presented in a comparable structural environment as when they are *in situ*. Second, appropriate post-translational modifications are more likely to occur in a mammalian expression system than in a prokaryotic system. Third,the recombinant constructs are soluble, and some possess measurable F.Va procoagulant activity (although reduced) and/or the ability to bind to anionic phospholipids. This facilitates functional characterization of antibodies that bind to defined epitopes.

At present, we have used this approach to map eight factor inhibitors and four monoclonal antibodies that bind to the C2 domains of either F.V or F.VIII. Preliminary evidence suggests that the region in the F.V C2 domain that is recognized by most F.V antibodies is different from the region in the F.VIII C2 domain that is recognized by most factor VIII antibodies. These results confirm the utility of this approach to rapidly epitope map and characterize antibodies that bind to the C-type domains of F.V and F.VIII.

Evaluation of the epitope specificity of human anti-factor VIII antibodies by using a panel of murine monoclonal antibodies

Jean Guy G. Gilles and Jean-Marie R. Saint-Remy

Allergy and Clinical Immunology Unit, International Institute of Cellular and Molecular Pathology, Université Catholique de Louvain, Brussels, Belgium

The epitope specificity of human anti-Factor VIII (F.VIII) antibodies is usually evaluated by immunoblotting methods and, more recently, by immunoprecipitation with F.VIII fragments. These assays provide a useful but rough estimate of the precise specificity.

In an attempt to further define it, we used affinity-purified antibodies obtained from the plasma of three donor groups: hemophiliac patients with stable inhibitor levels, hemophiliac patients with recent inhibitors, and healthy blood donors. The F.VIII epitope specificity was determined by measuring the capacity of human antibodies to compete with a panel of murine monoclonal antibodies toward both the heavy and light F.VIII chains, recognizing a total of 10 non-overlapping epitopes on the F.VIII molecule.

We show that: (1) each of the affinity-purified human antibody sample binds to F.VIII, as measured in a direct ELISA, including samples prepared from healthy blood donors; (2) each sample presents an unique profile of reactivity; (3) antibodies from hemophiliac patients are frequently directed towards F.VIII determinants that do not participate in the functional activity of the molecule; (4) some regions on the F.VIII molecule seem to be more immunogenic than others; (5) the preparation procedure for F.VIII may sufficiently alter the molecule as to give an antibody response of restricted epitope specificity.

Monoclonal antibodies can therefore be used for the characterization of epitopes recognized by human antibodies and follow-up of desensitization treatments, as well as to evaluate F.VIII preparation procedures.

Cellular immune responses to coagulation Factor VIII in F.VIII inhibitor patients

Debra K. Newton-Nash, Ph.D., Joan Cox Gill, M.D. and Paul A. Foster, M.D.
The Blood Research Institute, Milwaukee, WI

As a consequence of treatment with F.VIII concentrates some individuals with hemophilia A develop F.VIII-specific alloantibodies (inhibitors) that block F.VIII procoagulant activity. To assess cellular immune responses that may regulate the production of F.VIII inhibitors, peripheral blood mononuclear cells (PBMC) isolated from normal individuals and from hemophilia A patients with F.VIII inhibitors were screened for proliferative responses to F.VIII. F.VIII in clotting factor concentrates and purified F.VIII elicited dose-dependent proliferative responses from PBMC isolated from one inhibitor patient (H1102) but not from PBMC isolated from normal control individuals. These results suggest that F.VIII-specific proliferative responses may correlate with the development of F.VIII inhibitors. Human CD4 T cells recognize fragments of foreign antigen complexed with products of the MHC class II locus. To identify potential T-cell epitopes within F.VIII, synthetic peptides corresponding to amino acid sequences within F.VIII that are predicted to form an amphipathic α helix or to contain a class II MHC binding motif were prepared. Peptides were coupled to keyhole limpet hemocyanin (KLH) so that their ability to function as serological determinants upon injection into rabbits could be coincidentally assessed. Of 19 peptides tested, dose-dependent proliferative responses of PBMC derived from patient H1102 but not from normal PBMC were observed in response to only one KLH-coupled peptide. The region of the F.VIII heavy chain corresponding to this peptide has previously, in another individual, been demonstrated to contain an inhibitor epitope as evidenced by peptide-dependent neutralization of F.VIII inhibitory activity present in plasma. Identification of T-cell epitopes and response pathways that correlate with development of pathologic F.VIII-specific antibodies can be used to develop therapeutic strategies that depend upon manipulation of the production of F.VIII inhibitors at the T-cell level.

The use of Factor VIII inhibitor by-passing activity (FEIBA, IMMUNO) and porcine factor VIII (HYATE:C, PORTON) for treatment of patients with inhibitors: The Canadian experience, 1988-1993

G.E. Rivard,*S. Vick, M. David

*Hôpital Ste-Justine, Montréal, Québec and *The Canadian Red Cross, Ottawa, Ontario in Canada*

A nationwide survey of the use of FEIBA and HYATE:C was conducted for the years 1988 to 1993. An excellent collaboration was obtained from medical directors of hemophilia treatment centers and the Canadian Red Cross Transfusion Centers. FEIBA was used as the first line product to treat 33 hemophilia A patients with inhibitors. Over 17,000,000 units were administered. The efficacy was judged as "Fair" to "Good" in most instances. No clinically significant adverse reactions were observed. HYATE:C was by far the most commonly used product to treat nonhemophiliacs with factor VIII inhibitors and to support hemostasis for surgical procedures in hemophiliacs with inhibitors. In both situations the efficacy of the product was judged as "Good" to "Excellent". No clinically significant adverse reactions were observed. Both products appear useful and safe. More data are needed to determine their specific indications in the management of these difficult clinical problems.

A proposed prospective randomized trial of immune tolerance therapy in a pediatric hemophilia A population with high responding inhibitors

D. M. DiMichele, B. Bell, and K. Hoots
Northwestern University, Chicago, IL; Emory University, Atlanta, GA; and University of Texas Medical School, Houston, TX

We propose to conduct a prospective randomized trial of immune tolerance therapy in a pediatric hemophilia A population with high responding inhibitors. The specific aim of this study is to compare the efficacy as well as short-term cost and effectiveness of immune tolerance induction using either daily high dose factor VIII replacement alone, or in conjunction with immune modulation therapy.

Our rationale for this study is based on the following: 1) Although all protocols used to date demonstrate similar efficacy in the rate of tolerance induction (62-81%), they differ appreciably with respect to the time, effort, and cost required for successful inhibitor eradication (1.3-19.6 months); 2) There have been no prospective controlled trials to date comparing immune tolerance regimens with respect to parameters of overall efficacy, required duration of therapy, and total cost of treatment; 3) For the pediatric patient, in whom peripheral venous access is frequently an impediment to the initiation of immune tolerance, and in whom the alternative of central venous access is commonly associated with life-threatening complications, the expedient and permanent eradication of an inhibitor is imperative; 4) With the growing need for cost-containment in health care, immune tolerance therapy must be proven cost-effective both in the short and in the long-term analysis of the cost of hemophilia care.

We plan to randomize inhibitor patients to one of 2 tolerance regimens. In the first regimen, subjects will receive what we believe to be the minimally-effective dose of factor VIII replacement therapy, 100 U/kg/day, without immune modulation. In the second regimen, adapted from the Malmö protocol, subjects will receive 100 U/kg/day of factor VIII, as well as short initial courses of low dose oral cyclophosphamide (2mg/kg/day po for 10 days) and high dose intravenous gammaglobulin (1gm/kg, day 4 and 5). To maximize the overall success rate of each regimen, we plan to enroll each subject for a period of 18 months. We will compare both regimens with respect to a) the overall success rate in inducing immune tolerance; b) the time required for successful immune tolerance induction; and; c) the cost of therapy. Study subjects will be recruited primarily from pediatric hemophilia treatment centers throughout the USA, Canada and other interested international centers. Forty-five subjects will be enrolled in each arm of the study. Study eligibility will be based on a set of 12 criteria.

We postulate that the addition of immune modulation to high dose daily factor VIII therapy will statistically decrease the median time to induction of immune tolerance from 4 to 2 months. We postulate, therefore, that there will be a reduction in cost of therapy without an increase in treatment-associated morbidity.

A feasibility study for continuous ambulatory factor VIII (CAFE) therapy for patients with hemophilia A and factor VIII inhibitors

E.M. Gordon, A. Gilbert, R. Salazar and J.C. Goldsmith
Childrens Hospital Los Angeles and Quantum Health Resources, Los Angeles, CA

Patients with hemophilia A and F.VIII inhibitors have increased morbidity. Immune tolerance may be induced by continuous exposure to F.VIII. The goal of this study is to determine the stability of ultra-pure factor VIII concentrates during continuous delivery via ambulatory infusion pumps. Four Walk-Med Model 350 ambulatory infusion pumps were used for simultaneous continuous delivery of 4 ultra-pure F.VIII concentrates (Monoclate P, Recombinate, Hemofil M and Kogenate). Factor VIII was continuously infused through the pump and collected into sterile glass test tubes for 4 days. The loaded pumps were placed in cloth pouches and carried on a belt around the waist of 2 volunteers for 4 days. The collection tubes were replaced every 24 hours, and the infused factor VIII was stored at -80°C until tested. The temperature of each F.VIII-containing plastic bag was measured daily. F.VIII coagulant activity (F.VIII:C) was measured in a one-stage assay, using F.VIII-deficient plasma. An aliquot of each product was sent for bacterial, viral and fungal cultures on the 1st and 4th days of the study.

The figure shows percent of initial F.VIII:C measured in aliquots of infused product over 4 days. The factor VIII:C measured immediately after reconstitution of each product (initial F.VIII:C) was standardized to

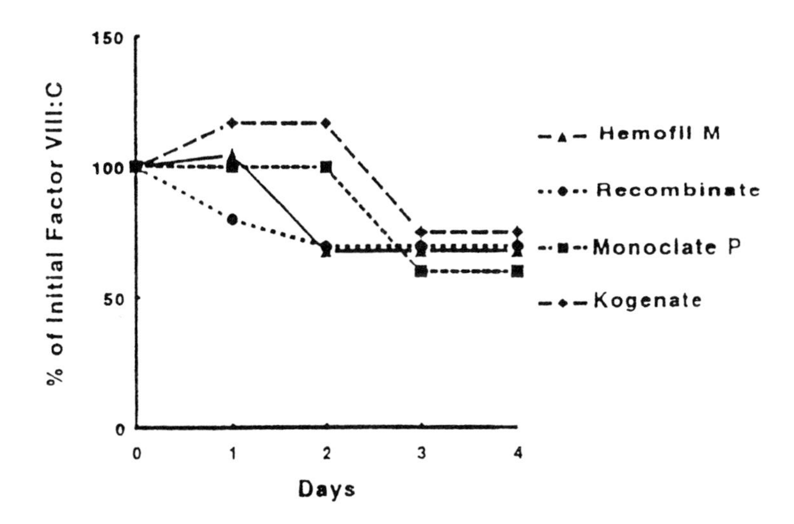

100%. F.VIII:C remained stable at 100% for 48 hours with Monoclate P and Kogenate. In contrast, F.VIII:C was stable for only 24 hours with Hemofil M, while Recombinate showed 20% less F.VIII:C 24 hours after reconstitution. A decline from normalized factor VIII:C was noted with all products by 72 hours. Cultures showed no growth of pathogenic organisms. The mean temperature of the bags containing factor VIII was 30°C throughout the study. These data indicate that Monoclate P and Kogenate may be more stable than Hemofil M or Recombinate. This information would be useful in the development of protocols for *in vivo* continuous ambulatory F.VIII (CAFE) infusion therapy in patients with hemophilia A with F.VIII inhibitors.

Incidence of factor VIII inhibitors in patients with hemophilia A

K. Peerlinck, F. Rosendaal,* and J. Vermylen
Center for Molecular and Vascular Biology, University of Leuven, Belgium and Department of Clinical Epidemiology, University Hospital Leiden, The Netherlands

An important issue at the introduction of new F.VIII concentrates is the frequency with which they induce neutralizing antibodies (inhibitors) to infused F.VIII. Previously untreated patients treated with recombinant F.VIII concentrates seem to have a higher incidence of inhibitor formation than patients treated with less pure products. However, informative data on incidence of inhibitors with older plasma-derived concentrates are scarce.

We performed an analysis of the incidence of inhibitor development in a cohort of 67 hemophilia A patients born between 1971 and 1990 and exclusively treated with lyophilized cryoprecipitate (Belgian Red Cross). The age-dependent cumulative incidence of developing an inhibitor was 4.6 % at age 4 and 6.7 % at age 8. Since these patients have been treated with a single F.VIII preparation, this is a suitable reference population.

In May 1990, two concentrates (F.VIII-P, Central Laboratory of the Netherlands Red Cross Blood Transfusion Service, Amsterdam, The Netherlands and F.VIII-S.D., Biotransfusion, Lille, France) were introduced in Belgium to replace cryoprecipitate. For both products cryoprecipitate from Belgian donors was used as starting material. Two-hundred-eighteen patients with hemophilia A regularly attending the Leuven Hemophilia Center were randomly assigned to a group receiving either of the two products. About 15 months after the introduction, five of the 109 patients receiving F.VIII-P developed a clinically significant inhibitor, whereas none of the 109 patients receiving F.VIII-S.D. developed an inhibitor. All patients who developed an inhibitor had previously been exposed to F.VIII infusions for more than 200 days. The prospective study thus allowed for the early detection of an increased immunogenicity of the F.VIII-P. A national study performed in the Netherlands, where the same product was used (F.VIII CPS-P) also showed an association of this product with the occurrence of inhibitors. To evaluate new concentrates, prospective cohort studies, in which hemophiliacs are treated with a single product and evaluated using a uniform protocol that takes into account whether patients have been previously treated or not, are necessary. Comparison with historical controls is only valid if these controls used only a single F.VIII product.

Factor VIII inhibitor assays using plasma-derived F.VIII versus recombinant F.VIII: A comparative study

C. Hillman-Wiseman, C. Vitale, J. Lusher
Children's Hospital of Michigan, Detroit, Michigan

While Abildgaard and Harrison found no difference between F.VIII inhibitor assays run using pooled normal plasma as a source of F.VIII and inhibitor assays using Miles' recombinant (r) F.VIII, others have found a discrepancy between assay results when Baxter's rF.VIII was used as compared to pooled normal plasma. We performed inhibitor assays on 24 samples from 11 patients with F.VIII inhibitors using 3 different sources of F.VIII: 1) pooled normal plasma, 2) Miles' rF.VIII, and 3) Baxter's rF.VIII. Samples with inhibitor levels as high as 218 Bethesda Units (BU) were assayed. All plasma samples were snap frozen and stored at -70°C until assays were performed in duplicate by the Bethesda methods. When rFVIII was used in the Bethesda assay system, rFVIII was reconstituted in distilled water as per reconstitution instructions on the label. Aliquots were then further diluted in hemophilic (F.VIII deficient) plasma such that the F.VIII level was 1 unit/ml. This was then substituted for pooled normal human plasma in the assay. All results were subjected to statistical analysis. One-way analysis of variance indicated no significant difference between source of F.VIII in the Bethesda assay ($F = 0.007$, $p = \leq 0.993$). Linear regression analysis also showed excellent correlation between assays utilizing the 3 sources of F.VIII:pooled human plasma (PNP) vs. Miles' rFVIII, $r^2 = 0.975$; PNP vs. Baxter's rF.VIII, $r^2 = 0.984$; and Miles' rF.VIII vs Baxter's rF.VIII, $r^2 = 0.997$. We conclude that Bethesda assay values show excellent comparison when patient samples are assayed using the usual source of F.VIII for such assays (PNP) or either of the two commercially available rF.VIII preparations.

Immune tolerance induction (ITI) in children with hemophilia A and F.VIII inhibitor

I. Warrier, L. Pfaffmann, and J.M. Lusher
Children's Hospital of Michigan and Wayne State University School of Michigan, Detroit, MI

High titer inhibitors (INH) (>10 Bethesda Units [B.U.]) develop in 15-25% of persons with severe HA. INH have been eradicated in some HA by ITI regimens. Several groups have reported variable success with varying ITI regimens and results of a recent international survey (conducted by Prof. G. Mariani) suggested that very high dosage regimens (200 U/Kg/day) were most effective. The ITI regimen used by us was a modification of that developed by Ewing et al., with daily doses of F.VIII being given until the INH is no longer detectable; then dosage and dosage intervals are extended as long as the inhibitor remains negative and F.VIII recovery is normal. Initial dosage of F.VIII was 50-100 U/Kg depending on the child's highest INH assay prior to ITI. Seven children (ages 3 months to 16 years) whose maximum INH concentrations were 24-207 B.U. were put on ITI. ITI was started within 1 to 12 months of inhibitor detection in 4 children; in 3 ITI was started 4 to13 years later. Six of 7 children have had an excellent response to ITI with disappearance of INH and normalization of F.VIII recovery within 1 to 7 months of starting ITI. The 7th child, now aged 8 months, has been on ITI for 5 months with declining but still detectable INH. One child has now been on demand treatment with F.VIII for 8 months without reappearance of INH. The only problems encountered were those of venous access in young children. All 5 children who were < 5 yrs of age required central lines (CL) and 1 required >1 CL to complete ITI. CL sepsis was the most common complication of ITI (7 episodes in 2 children). In summary, our experience with ITI in infants and children suggests that excellent responses to ITI in this age group can be achieved quite rapidly and with lower dosages of F.VIII than reported in the Mariani survey.

Determining inhibitor-development risk in previously untreated hemophilia A patients (PUPs) treated with recombinant Factor VIII (Recombinate™)

M.L. Lee, S. Courter, and E. Gomperts
Baxter/Hyland Division, Glendale, CA.

The usual method of determining the risk of developing a Factor VIII inhibitor after receiving recombinant Factor VIII (rF.VIII) or similar preparations is to compare the number of previously untreated hemophilia A patients (PUPs) who develop an inhibitor over some predetermined interval of time divided by the number of PUPs who are at risk during this time period. This estimate is biased, since the denominator of this estimate contains all PUPs who may not all be at the same risk because of the differential amount of product received, the amount of time on the study, the number of infusions of product received, the number of exposure days and the number of bleeds. In order to determine the probability of developing an inhibitor, routine survival analysis with progressive right censoring is employed. Before performing this analysis, a choice of the most important risk factor (time on study, number of bleeds, number of infusions, number of exposure days, or cumulative units received) is determined by stepwise logistic regression analysis. Based on the analysis of 68 PUPs, it was determined that the risk factors that best correlated with inhibitor development were number of infusions and exposure days. By using parametric (log-normal model) or non-parametric (Kaplan-Meier) survival analysis and allowing for right censoring in the development of an inhibitor, the probability of remaining inhibitor-free after N exposure days (or infusions) can be determined. In order to provide more precise estimates of these probabilities, the majority of patients need to receive a reasonable number of infusions or exposure days.

Multicentric retrospective study on the utilization of FEIBA in France in patients with Factor VIII and Factor IX inhibitors

C. Negrier (Hemophilia Care Center, Edouard Herriot Hospital, Lyon, France), and the French FEIBA study group (A. Ballocchi, D. Bastit, M.A. Bertrand, C. Bosser, D. Boyeldieu, J.Y. Borg, G. Dirat, A. Durin, M. Gaillard, C. Gazengel, J. Goudemand, P. Lauroua , L. de Lumley, A. Mahe , R. Navarro, P. Nguyen, A. Parquet, C. Potron, J. Reynaud, C. Rothschild, N. Stieltjes, Y. Sultan, M.F. Torchet and P. Tron).

We performed a multicenter retrospective study to evaluate the use of FEIBA (Factor Eight By-passing Activity) in France, with particular interests in tolerance and efficacy, and in viral HIV and hepatitis C seroprevalences. Sixty patients from 14 hemophilia centers have been evaluated, 52 with an allo-antifactor VIII antibody, 6 with an auto-antifactor VIII antibody and 2 with an allo-antifactor IX antibody. Twenty eight (51.8%) developed the inhibitor before the age of 10 years and 42 (78%) before 25 years old. We were able to evaluate the data from 434 bleeding episodes, including hemarthroses, muscle hematomas, epistaxes, hematurias, intestinal hemorrhages, and surgical procedures. The efficacy was judged as good or very good in 353 episodes (81.5%), poor in 73 episodes (16.8%) and non-existent in 8 episodes (1.85%). There was no correlation between the efficacy and the inhibitor titer. Some surgical procedures have been successfully performed using FEIBA after human or porcine F.VIII; ankle synoviorthesis (1 case), retroperitoneal hematoma (2 cases), kidney hematoma (1 case), tibial osteotomy (1 case) and splenectomy (1 case). In other cases FEIBA was used as the only substitution product; endoscopic procedures (3 cases), synoviortheses (10 cases), dental extractions (5 cases), knee arthroscopy (1 case), knee synovectomy (1 case), knee arthroplasty (1 case), muscular plasty (1 case) and prostatic adenomectomy (1 case). The tolerance was assessed as good in 428 episodes (98.8% of cases), but in 5 cases adverse effects were reported. Two out of these 5 adverse reactions occurred early after the FEIBA injection: pruritis and cutaneous erythema in 1 case and myocardial infarction in the other case (hemophilia B patient). In the remaining 3 cases, biological signs of disseminated intravascular coagulation were noted. Concerning the HIV seroprevalence, only 3 patients out of 52 regularly evaluated (5.76%) were HIV positive, and for two of them the seroconversion occurred prior to the first use of FEIBA. The last patient used many blood products (cryoprecipitate, red blood cells, Autoplex) during the same period of time and it was impossible to attribute the HIV seroconversion to a particular one. These data evoke the possibility of a HIV inactivation step during the fabrication process of this product. In contrast, concerning the HCV seroprevalence, 80.4% of the patients were seropositive. Finally, an anamnestic response after the administration of FEIBA was noted in 31.4% of cases. In conclusion, this study points out the main features of the use of FEIBA in France and particularly the low HIV seroprevalence in the patients treated. The good efficacy and the excellent tolerance of this product should earn it consideration among the therapeutic options for the treatment of inhibitor-developing hemophiliacs, and even for some surgical procedures.

FEIBA® VH IMMUNO: Partitioning and inactivation of HIV-1

P.L. Turecek, H.P. Schwarz, N. Barrett, G. Polsler, F. Dorner, and J. Eibl
Immuno AG, Vienna, Austria

In the manufacture of plasma-derivatives from pooled human plasma, viruses are partitioned to varying degrees depending on the fractionation processes and fractionation steps used. Partitioning in the individual manufacturing steps can be determined using model viruses.

Partitioning alone may be sufficient to render certain plasma products safe. Thus, in the early 1980s, although HIV-1 was transmitted on a large scale through antihemophilic factor concentrates, millions of doses of immune globulin manufactured from the same plasma pools never transmitted HIV-1.[1] With this background, it appears desirable to also investigate the virus partitioning potential of the processes used to manufacture other plasma fractions, specifically their potential for partitioning HIV-1.

Since the 1970s, Anti-Inhibitor Coagulant Complex, FEIBA®, has been among the mainstays of treatment for bleeding episodes in patients with F.VIII or F.IX inhibitors. No HIV transmission related to FEIBA® has ever been reported, even before the introduction of specific virus inactivation steps.

The capacity of various production steps to remove and/or inactivate HIV-1 was investigated. Whereas a factor of 5 is believed to be sufficient to render a blood product safe as far as HIV transmission is concerned,[2] when samples from selected manufacturing steps of FEIBA® were spiked with HIV-1, the log virus reduction was determined to be > 6. The product distributed today is vapor heated. Since this proprietary virus inactivation process yields a log virus reduction of 12.8, the overall log virus reduction through partitioning and virus inactivation is 18.8. These results exclude the possibility of transmission of HIV-1 by FEIBA® even prior to vapor heating.

References

1. Wells, M.A., et al., Inactivation and partition of human T-cell lymphotrophic virus, type III, during ethanol fractionation, Transfusion, 26/2:210-213 (1986)
2. Petricciani, J.C., et al., Case for concluding that heat-treated, licensed anti-haemophilic factor is free from HTLV-III, The Lancet, II:890:891 (1985)

Index